Dental Implant Materials 2019

Dental Implant Materials 2019

Editor

In-Sung Yeo

MDPI • Basel • Beijing • Wuhan • Barcelona • Belgrade • Manchester • Tokyo • Cluj • Tianjin

Editor
In-Sung Yeo
Seoul National University
Korea

Editorial Office
MDPI
St. Alban-Anlage 66
4052 Basel, Switzerland

This is a reprint of articles from the Special Issue published online in the open access journal *Materials* (ISSN 1996-1944) (available at: https://www.mdpi.com/journal/materials/special_issues/dental_materials_2019).

For citation purposes, cite each article independently as indicated on the article page online and as indicated below:

LastName, A.A.; LastName, B.B.; LastName, C.C. Article Title. *Journal Name* **Year**, *Volume Number*, Page Range.

ISBN 978-3-0365-0416-2 (Hbk)
ISBN 978-3-0365-0417-9 (PDF)

© 2021 by the authors. Articles in this book are Open Access and distributed under the Creative Commons Attribution (CC BY) license, which allows users to download, copy and build upon published articles, as long as the author and publisher are properly credited, which ensures maximum dissemination and a wider impact of our publications.

The book as a whole is distributed by MDPI under the terms and conditions of the Creative Commons license CC BY-NC-ND.

Contents

About the Editor vii

In-Sung Luke Yeo
Special Issue: Dental Implant Materials 2019
Reprinted from: *Materials* **2020**, *13*, 5790, doi:10.3390/ma13245790 1

In-Sung Luke Yeo
Modifications of Dental Implant Surfaces at the Micro- and Nano-Level for Enhanced Osseointegration
Reprinted from: *Materials* **2020**, *13*, 89, doi:10.3390/ma13010089 5

Taek-Ka Kwon, Jung-Yoo Choi, Jae-Il Park and In-Sung Luke Yeo
A Clue to the Existence of Bonding between Bone and Implant Surface: An In Vivo Study
Reprinted from: *Materials* **2019**, *12*, 1187, doi:10.3390/ma12071187 21

Jun-Beom Lee, Ye-Hyeon Jo, Jung-Yoo Choi, Yang-Jo Seol, Yong-Moo Lee, Young Ku, In-Chul Rhyu and In-Sung Luke Yeo
The Effect of Ultraviolet Photofunctionalization on a Titanium Dental Implant with Machined Surface: An In Vitro and In Vivo Study
Reprinted from: *Materials* **2019**, *12*, 2078, doi:10.3390/ma12132078 29

Chang-Bin Cho, Sung Youn Jung, Cho Yeon Park, Hyun Ki Kang, In-Sung Luke Yeo and Byung-Moo Min
A Vitronectin-Derived Bioactive Peptide Improves Bone Healing Capacity of SLA Titanium Surfaces
Reprinted from: *Materials* **2019**, *12*, 3400, doi:10.3390/ma12203400 43

Annalena Bethke, Stefano Pieralli, Ralf-Joachim Kohal, Felix Burkhardt, Manja von Stein-Lausnitz, Kirstin Vach and Benedikt Christopher Spies
Fracture Resistance of Zirconia Oral Implants In Vitro: A Systematic Review and Meta-Analysis
Reprinted from: *Materials* **2020**, *13*, 562, doi:10.3390/ma13030562 55

Jung-Ju Kim, Jae-Hyun Lee, Jeong Chan Kim, Jun-Beom Lee and In-Sung Luke Yeo
Biological Responses to the Transitional Area of Dental Implants: Material- and Structure-Dependent Responses of Peri-Implant Tissue to Abutments
Reprinted from: *Materials* **2020**, *13*, 72, doi:10.3390/ma13010072 77

Nak-Hyun Choi, Hyung-In Yoon, Tae-Hyung Kim and Eun-Jin Park
Improvement in Fatigue Behavior of Dental Implant Fixtures by Changing Internal Connection Design: An In Vitro Pilot Study
Reprinted from: *Materials* **2019**, *12*, 3264, doi:10.3390/ma12193264 93

Ki-Seong Kim and Young-Jun Lim
Axial Displacements and Removal Torque Changes of Five Different Implant-Abutment Connections under Static Vertical Loading
Reprinted from: *Materials* **2020**, *13*, 699, doi:10.3390/ma13030699 105

Pietro Montemezzi, Francesco Ferrini, Giuseppe Pantaleo, Enrico Gherlone and Paolo Capparè
Dental Implants with Different Neck Design: A Prospective Clinical Comparative Study with 2-Year Follow-Up
Reprinted from: *Materials* **2020**, *13*, 1029, doi:10.3390/ma13051029 115

Pei-Ching Kung, Shih-Shun Chien and Nien-Ti Tsou
A Hybrid Model for Predicting Bone Healing around Dental Implants
Reprinted from: *Materials* **2020**, *13*, 2858, doi:10.3390/ma13122858 .... **125**

Hadas Heller, Adi Arieli, Ilan Beitlitum, Raphael Pilo and Shifra Levartovsky
Load-Bearing Capacity of Zirconia Crowns Screwed to Multi-Unit Abutments with and without a Titanium Base: An In Vitro Pilot Study
Reprinted from: *Materials* **2019**, *12*, 3056, doi:10.3390/ma12193056 .... **141**

Yong-Seok Jang, Sang-Hoon Oh, Won-Suck Oh, Min-Ho Lee, Jung-Jin Lee and Tae-Sung Bae
Effects of Liner-Bonding of Implant-Supported Glass–Ceramic Crown to Zirconia Abutment on Bond Strength and Fracture Resistance
Reprinted from: *Materials* **2019**, *12*, 2798, doi:10.3390/ma12172798 .... **153**

João P. M. Tribst, Amanda M. O. Dal Piva, Alexandre L. S. Borges, Lilian C. Anami, Cornelis J. Kleverlaan and Marco A. Bottino
Survival Probability, Weibull Characteristics,
Stress Distribution, and Fractographic Analysis of Polymer-Infiltrated Ceramic Network Restorations Cemented on a Chairside Titanium Base: An In Vitro and In Silico Study
Reprinted from: *Materials* **2020**, *13*, 1879, doi:10.3390/ma13081879 .... **167**

About the Editor

In-Sung Yeo served as a dental officer (Captain) at Special Forces Commands in the Korean Army for three years (2003–2006). He got his Ph.D. in biomaterials, in vivo studies of implant surfaces, at Seoul National University in the year, 2007, and received his prosthodontic specialty certificate in 2017 from the Korean government. He was an Assistant Professor at Seoul National University from 2010 to 2014 and an Associate for 5 years until 2019. Now, Dr. In-Sung Yeo is a Professor at the same university. His major research is about biologic responses to artificial biocompatible surfaces. Additionally, he is trying to find physical or statistical solutions for biologic phenomena in prosthodontics and implantology.

Editorial

Special Issue: Dental Implant Materials 2019

In-Sung Luke Yeo

Department of Prosthodontics, School of Dentistry and Dental Research Institute, Seoul National University, 101, Daehak-Ro, Jongro-Gu, Seoul 03080, Korea; pros53@snu.ac.kr; Tel.: +82-2-2072-2661

Received: 17 December 2020; Accepted: 18 December 2020; Published: 18 December 2020

The Special Issue, "Dental Implant Materials 2019", has tried to introduce recent developments in material science and implant dentistry with biologic and clinical aspects. Biocompatibility, design and surface characteristics of implant materials are very important in the long-term clinical service of dental implants. Ten original research articles and three review articles in this issue are considered to show well the significance of such factors from the clinical point of view.

Hard tissue response to implant surface is one of main fields many researchers are involved in. Surface modification technologies for implants have begun to be applied to titanium at the micro-level for about four decades. Currently, implant surfaces are being topographically and chemically modified at the micro- and nano-levels. The modified surfaces used globally in dental clinics are well described and comprehensively reviewed in a review article of this Special Issue [1]. This review article also explores some modified implant surfaces that are highly possible to be clinically used, which are very interesting to the readers investigating biologic interfaces.

In fact, the nature of bone-to-implant contact remains unknown. Whether or not a real bond exists between hard tissues and implants is still under investigation. Although some researchers suggest that the bone-to-implant contact would be a simple physical attachment at the bone–implant interface, Kwon et al. proposed that an actual bond might exist between a bone and an implant surface by showing different shear bond strength values of the grades 2 and 4 commercially pure titanium surfaces that have similar topographies [2].

Although the nature of bone response to an implant surface is still under investigation, various methodological approaches are being developed to enhance the bone healing around the surface. For example, ultraviolet photofunctionalization of the grade 4 commercially pure titanium surface eliminates contaminating hydrocarbon on the surface and highly increases surface hydrophilicity, resulting in the acceleration of osseointegration in vivo, which is shown in an article of this Special Issue [3]. A functional peptide that is involved in cell adhesion is very useful to speed up the bone healing process. This Special Issue contains the evaluation of early bone response to a vitronectin-derived functional peptide-treated sandblasted, large-grit, acid-etched titanium surface [4]. A systematic review of zirconia dental implants describes that the clinical use of implants which are more aesthetic than titanium metal ones will increase [5].

The stable peri-implant soft tissue is another key to the long-term success of dental implants, which is closely associated with the implant-abutment connection structure. Both the soft and hard tissue responses, depending on the structures and abutment material characteristics, are becoming another focused topic in clinical implant dentistry. A review of this Special Issue summarizes the relevant literature to establish guidelines regarding the effects of connection type between abutments and implants in soft and hard tissues [6]. Biomechanical behaviours of implant-abutment connection designs are shown in two articles, and clinical outcomes are presented in one article, depending on the connection designs [7–9]. It is necessary for researchers and clinicians to interpret the clinical data in implantology in the light of the old axioms that pocket formation is the initiator for peri-implant or periodontal inflammation and that bone responds to strain, not to stress itself.

Masticatory forces are transferred from superstructures or artificial teeth to bone via implants. A biomechanical model was introduced in the study of Kung et al. for the prediction of bone healing around a dental implant system composed of an artificial crown cemented to a one-body implant, where an abutment and an implant are fused together [10]. Various materials are being developed for superstructures that are usually cemented to abutments. Two major materials are zirconia and glass ceramics, which have been recently supported by digital technology. Interesting mechanical results are shown in an article of this Special Issue, when the zirconia superstructures are cemented or when the superstructures are screw-retained [11]. Intriguingly, Jang et al. evaluated a cemented interface between an artificial crown and an abutment, investigating the effects of cementation methods on the bond strength and fracture resistance between glass–ceramic superstructures and zirconia abutments [12]. In addition, Tribst et al. estimated implant-supported polymer-infiltrated ceramic crowns in vitro when the crowns were cemented to the titanium abutments [13]. These materials and skills were tested in laboratories to reduce the frequent clinical complications of implant-supported superstructures, which are material chipping, crown dislodgement and crown fracture. Long-term studies in clinics designed to evaluate the performances of these materials and skills are being waited for.

Conflicts of Interest: The author declares no conflict of interest.

References

1. Yeo, I.-S.L. Modifications of Dental Implant Surfaces at the Micro- and Nano-Level for Enhanced Osseointegration. *Materials* **2020**, *13*, 89. [CrossRef]
2. Kwon, T.-K.; Choi, J.-Y.; Park, J.-I.; Yeo, I.-S.L. A Clue to the Existence of Bonding between Bone and Implant Surface: An In Vivo Study. *Materials* **2019**, *12*, 1187. [CrossRef]
3. Lee, J.-B.; Jo, Y.-H.; Choi, J.-Y.; Seol, Y.-J.; Lee, Y.-M.; Ku, Y.; Rhyu, I.-C.; Yeo, I.-S.L. The Effect of Ultraviolet Photofunctionalization on a Titanium Dental Implant with Machined Surface: An In Vitro and In Vivo Study. *Materials* **2019**, *12*, 2078. [CrossRef] [PubMed]
4. Cho, C.-B.; Jung, S.Y.; Park, C.Y.; Kang, H.K.; Yeo, I.-S.L.; Min, B.-M. A Vitronectin-Derived Bioactive Peptide Improves Bone Healing Capacity of SLA Titanium Surfaces. *Materials* **2019**, *12*, 3400. [CrossRef] [PubMed]
5. Bethke, A.; Pieralli, S.; Kohal, R.-J.; Burkhardt, F.; von Stein-Lausnitz, M.; Vach, K.; Spies, B.C. Fracture Resistance of Zirconia Oral Implants In Vitro: A Systematic Review and Meta-Analysis. *Materials* **2020**, *13*, 562. [CrossRef] [PubMed]
6. Kim, J.-J.; Lee, J.-H.; Kim, J.C.; Lee, J.-B.; Yeo, I.-S.L. Biological Responses to the Transitional Area of Dental Implants: Material- and Structure-Dependent Responses of Peri-Implant Tissue to Abutments. *Materials* **2020**, *13*, 72. [CrossRef] [PubMed]
7. Choi, N.-H.; Yoon, H.-I.; Kim, T.-H.; Park, E.-J. Improvement in Fatigue Behavior of Dental Implant Fixtures by Changing Internal Connection Design: An In Vitro Pilot Study. *Materials* **2019**, *12*, 3264. [CrossRef] [PubMed]
8. Kim, K.-S.; Lim, Y.-J. Axial Displacements and Removal Torque Changes of Five Different Implant-Abutment Connections under Static Vertical Loading. *Materials* **2020**, *13*, 699. [CrossRef] [PubMed]
9. Montemezzi, P.; Ferrini, F.; Pantaleo, G.; Gherlone, E.; Capparè, P. Dental Implants with Different Neck Design: A Prospective Clinical Comparative Study with 2-Year Follow-Up. *Materials* **2020**, *13*, 1029. [CrossRef] [PubMed]
10. Kung, P.-C.; Chien, S.-S.; Tsou, N.-T. A Hybrid Model for Predicting Bone Healing around Dental Implants. *Materials* **2020**, *13*, 2858. [CrossRef] [PubMed]
11. Heller, H.; Arieli, A.; Beitlitum, I.; Pilo, R.; Levartovsky, S. Load-Bearing Capacity of Zirconia Crowns Screwed to Multi-Unit Abutments with and without a Titanium Base: An In Vitro Pilot Study. *Materials* **2019**, *12*, 3056. [CrossRef] [PubMed]
12. Jang, Y.-S.; Oh, S.-H.; Oh, W.-S.; Lee, M.-H.; Lee, J.-J.; Bae, T.-S. Effects of Liner-Bonding of Implant-Supported Glass–Ceramic Crown to Zirconia Abutment on Bond Strength and Fracture Resistance. *Materials* **2019**, *12*, 2798. [CrossRef] [PubMed]

13. Tribst, J.P.M.; Dal Piva, A.M.O.; Borges, A.L.S.; Anami, L.C.; Kleverlaan, C.J.; Bottino, M.A. Survival Probability, Weibull Characteristics, Stress Distribution, and Fractographic Analysis of Polymer-Infiltrated Ceramic Network Restorations Cemented on a Chairside Titanium Base: An In Vitro and In Silico Study. *Materials* **2020**, *13*, 1879. [CrossRef] [PubMed]

Publisher's Note: MDPI stays neutral with regard to jurisdictional claims in published maps and institutional affiliations.

© 2020 by the author. Licensee MDPI, Basel, Switzerland. This article is an open access article distributed under the terms and conditions of the Creative Commons Attribution (CC BY) license (http://creativecommons.org/licenses/by/4.0/).

Review

Modifications of Dental Implant Surfaces at the Micro- and Nano-Level for Enhanced Osseointegration

In-Sung Luke Yeo

Department of Prosthodontics, School of Dentistry and Dental Research Institute, Seoul National University, Seoul 03080, Korea; pros53@snu.ac.kr; Tel.: +82-2-2072-2662

Received: 30 October 2019; Accepted: 20 December 2019; Published: 23 December 2019

Abstract: This review paper describes several recent modification methods for biocompatible titanium dental implant surfaces. The micro-roughened surfaces reviewed in the literature are sandblasted, large-grit, acid-etched, and anodically oxidized. These globally-used surfaces have been clinically investigated, showing survival rates higher than 95%. In the past, dental clinicians believed that eukaryotic cells for osteogenesis did not recognize the changes of the nanostructures of dental implant surfaces. However, research findings have recently shown that osteogenic cells respond to chemical and morphological changes at a nanoscale on the surfaces, including titanium dioxide nanotube arrangements, functional peptide coatings, fluoride treatments, calcium–phosphorus applications, and ultraviolet photofunctionalization. Some of the nano-level modifications have not yet been clinically evaluated. However, these modified dental implant surfaces at the nanoscale have shown excellent in vitro and in vivo results, and thus promising potential future clinical use.

Keywords: surface modification; osseointegration; SLA; TiO_2 nanotube; fluoride; photofunctionalization

1. Introduction

The surface quality of titanium (Ti) dental implants, which replace missing teeth, is one of the keys to the long-term clinical success of implants in a patient's mouth [1]. The bone response to the Ti implant surface depends on its surface characteristics: Contact (bone formation on the implant surface towards the bone) and distance osteogenesis occur around micro-roughened Ti surfaces while only distance osteogenesis (bone formation from the old bone toward the implant surface) appear around turned Ti [2]. Although contact osteogenesis seems to require other factors to be triggered, modification of the implant surface is very important to accelerate osseointegration [3].

Ti is known to be stable in biologic responses and not to trigger a foreign body reaction when inserted into the human body [4,5]. Therefore, osseointegration was originally defined as the direct contact between a loaded implant surface and bone at the microscopic level of resolution [1]. Recently, this term has been interpreted from a new point of view: Osseointegration is essentially a demarcation response to a foreign body of Ti when the Ti implant is immobile in bone [6]. This demarcation is immune-driven and is classified as a type IV hypersensitivity [7]. Based on the original definition, the modification of a Ti implant surface implies that the surface would be more biocompatible, thereby increasing the bioaffinity of the hard tissue and accelerating the bone response to the surface. The new standpoint on osseointegration suggests that the modified Ti surface would be recognized more sensitively by the hard tissue, which would isolate this foreign body with a faster and stronger accumulation of bone substances. Thus, the nature of osseointegration is under investigation at present [8]. The detection of the actual bond between the bone and implant surfaces could support the bioaffinitive nature of bone response to the surfaces [9,10]. Only friction and physical contact would exist at the interface if the bony demarcation hypothesis is correct.

To date, implant surfaces have been modified in various ways under the bioaffinity concept for osseointegration. Conventionally, the topography of the surface has been changed at the micro-level

(1–10 μm). At present, some chemical features and nanotechnologies have been added to the surfaces. This review introduces several recent advancements of biocompatible implant surfaces with a few representative micro-roughened modified surfaces. Since most implant surfaces used in the global market have been made of commercially pure Ti (cp-Ti), especially grade 4 cp-Ti, this review is based on the modification of a grade 4 cp-Ti surface.

2. Micro-Roughened Modification

2.1. Sandblasted, Large-Grit, Acid-Etched (SLA) Surface

The computer numerical controlled milling of cp-Ti manufactures screw-shaped endosseous dental implants. The surface machined by this milling procedure, which is now called a turned Ti surface, shows many parallel grooves in scanning electron microscopy (SEM). The turned surface experiences no modification process, which has frequently served as a control to evaluate the biocompatibility of modified surfaces. When an implant is inserted into the bone and the implant surface becomes juxtaposed to the bone, bone healing (or osseointegration) on the surface is known to be fulfilled by two mechanisms: distance and contact osteogenesis [2,11]. In distance osteogenesis, new bone starts to be formed on the surfaces of bone. The direction of bone growth is from the bone towards the implant surface (Figure 1A). In contact osteogenesis, or de novo bone formation, new bone formation begins on the implant surface. The direction of bone growth is from the implant towards the bone, opposite to that for distance osteogenesis (Figure 1B). When an endosseous implant with a turned surface is placed into the jawbone, only distance osteogenesis occurs, which implies that more time is needed for sufficient osseointegration to withstand masticatory forces [2,12]. The necessity of reduction in the patient's edentulous period has led the modification of an implant surface to accelerate bone healing.

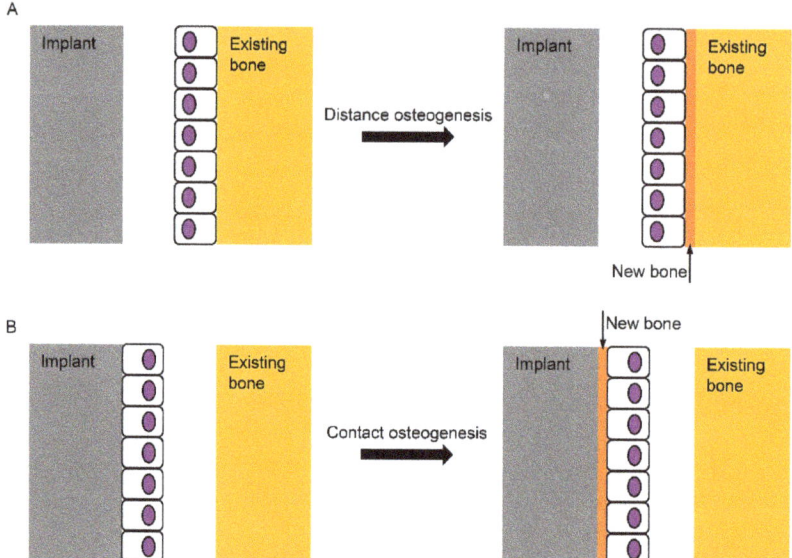

Figure 1. Schematic diagram for the healing mechanisms of the bone surrounding an implant. (**A**) In distance osteogenesis, the direction of bone formation is from the existing bone to the implant; (**B**) in contact osteogenesis, however, the direction is opposite, from the implant to the existing bone, which is known not to occur on the turned Ti (Titanium) surface without any modification.

The traditional approach to the surface modification of a Ti implant has been roughening at the micro-level. One of the most successful surfaces in clinical dentistry is the sandblasted, large-grit, and acid-etched (or SLA) surface. An SLA Ti surface is made by sandblasting the turned Ti surface with large-grit particles, the sizes of which range from 250 µm to 500 µm in general, and by acid-etching the blasted surface. The acids for etching are usually strong acids including hydrochloric, sulfuric, and nitric acids. SEM shows topographically changed irregularities on the SLA surface, with large dips, small micropits, sharp edges, and pointed tips. Sa, one of the surface parameters defined as the arithmetic mean height of the surface, is approximately 1.5 µm to 2 µm. Osteogenic cells migrate to the roughened Ti surface through the fibrin clot that is formed at the peri-implant site after bone drilling for implant insertion, and these cells appear to recognize the irregularities of the SLA surface as lacunae to be filled with bone materials [2,13]. Contact osteogenesis occurs as the osteogenic cells secrete a bone matrix. The occurrence of both contact and distance osteogenesis accelerates the osseointegration on the SLA surface compared to the turned surface.

The Ti surface of a dental implant is originally hydrophobic [14]. Water (H_2O) is considered to have initial contact with the implant surface when the implant is inserted into the bone [15]. Therefore, there have been attempts to add hydrophilicity to an SLA surface, since hydrophilicity is expected to help accelerate the bone healing process [14,16]. A dental implant with a hydrophilic SLA surface, commercially called SLActive (Institute Straumann AG, Basel, Switzerland), is made with a water rinse of the original SLA implant in a nitrogen chamber and a packaging technique of storing the implant in an isotonic sodium chloride solution with no atmospheric contact, and this hydrophilic implant is being clinically used in the global market [17].

Regardless of whether an SLA surface is hydrophobic or hydrophilic, this dental implant surface has shown excellent long-term clinical results [18–22]. A previous 10-year retrospective study investigating more than 500 SLA Ti implants concluded that both the survival and success rates were 97% or higher [18]. The 10-year survival rate of SLA Ti implants was reported to be higher than 95%, even in periodontally compromised patients, although strict periodontal interventions were applied to these patients [20]. Similar results were found in 10-year prospective studies investigating the survival rates of dental implants with SLA surfaces [19,21,22]. This modified surface, roughened at the micro-scale, is one of the dental implant surfaces that has been most frequently tested in clinics for the longest period.

2.2. Anodic Oxidation

The genuine biocompatible surface on the Ti dental implant is Ti oxide (TiO_2), not Ti itself, which is spontaneously formed when the Ti surface is exposed to the atmosphere. However, this Ti oxide layer is very thin (a few nm in thickness) and is imperfect with defects [23]. Also, chemically unstable Ti^{3+} and Ti^{2+} are known to exist in the oxide layer [24]. Therefore, there have been several techniques developed to thicken and stabilize the Ti oxide layer, which is considered to increase the biocompatibility of the surface [25–27]. When Ti becomes the anode under an electric potential in an electrochemical cell, Ti is oxidized to be Ti^{4+}, and the TiO_2 layer is able to be thickened and roughened [15]. Topographically, the oxidized Ti surface for a dental implant has many volcano-like micropores with various sizes, which are observed in SEM. The surface characteristics of the anodized Ti surface depend on the applied potential, surface treatment time, concentrations, and types of electrolytes [15,27]. The arithmetic mean height of this surface, or Sa, is evaluated to be approximately 1 to 1.5 µm for dental use [28–31].

Osteogenic cells appear to recognize the topography of a dental implant surface although we do not yet know which surface topography is more proper in bone healing, or if the irregularities of the SLA surface are more effective for the osteogenic cell response than the microporous structure of the anodized surface [32]. To date, no in vivo model has found any significant differences in bone responses to the microtopographies of Ti dental implant surfaces [33,34]. What is definitely known about implant surface topography is that the cp-Ti surfaces topographically modified at the microscale accelerate osseointegration more than the turned surface, and these modified surfaces show superior results to the turned surface during in vitro, in vivo, and clinical studies.

The anodically oxidized Ti surface has shown superior results to the turned surface in various in vitro tests and in vivo histomorphometry [31,34–36]. A previous meta-analytic study reported lower failure rates of the oxidized Ti implants than those of the turned implants from the included 38 clinical investigations [37]. A prior retrospective and a 10-year prospective study concluded that that success rates were higher than 95% for the TiUnite surface (Brånemark System, Nobel Biocare, Göteborg, Sweden), which is a trade name for the oxidized Ti surface [38,39]. However, a recent 20-year randomized controlled clinical trial notably reported a similar marginal bone loss between micro-roughened and turned Ti implants [40]. This clinical study used an identical implant design with an implant-abutment connection structure and internal friction connection [40]. Identifying which of the two factors (surface characteristics and implant design) is a major contributor to the long-term clinical success of dental implants needs to be thoroughly investigated, although higher success or survival rates have been steadily published for Ti dental implants with modified surfaces at the micro-scale, compared to the turned implant [19,41,42].

3. Molecular Modification

3.1. TiO$_2$ Nanotube

Anodic oxidation is extended to the modification of a Ti dental implant at the nanoscale (1–100 nm). The electric current of the electrochemical cell, temperature, the pH values of electrolyte solutions, the electrolytes, oxidation voltage, and oxidation time affect the nanotopographies of the Ti surface [43,44]. In an electrochemical cell composed of Ti at the anode and platinum (or Ti) at the cathode, the TiO$_2$ layer is normally formed on the Ti implant surface of the anode [43]. In an appropriate fluoride-based electrolyte, the nano-morphology of the TiO$_2$ layer is changed, and the aligned TiO$_2$ nanotube layer is developed (Figure 2) [43].

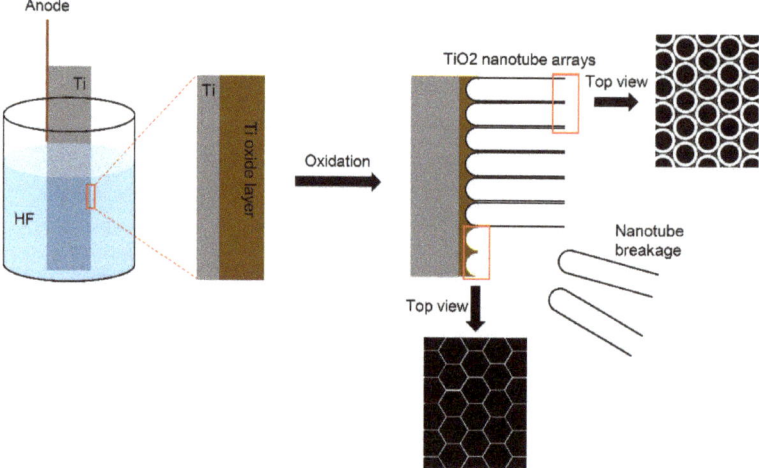

Figure 2. Schematic diagram showing the formation of TiO$_2$ nanotube arrays. In the electrolyte solution containing hydrogen fluoride (HF), regular tube structures are formed on the Ti surface of the anode at a nanoscale. When the structures are viewed on top, the circular forms of the tubules are found via scanning electron microscopy. The binding between the nanotube arrays and Ti surface is generally weak, and breakdown is frequent at the interface. The morphology underneath the tubes is hexagonal.

In the past, implant surface nanostructures were reported to have no effect on cell responses and bone responses to dental implant surfaces and were thought to depend on the microtopographies of the surfaces [45,46]. Optimal micro-roughness is known at present to be 1.5 μm in Sa and approximately

4 µm in diameter of the surface irregularities [30,47]. However, a previous review article noted that the microtopographies of the dental implant surfaces have a limited influence on the initial responses of the in vivo hard tissue environment [48]. Presently, the nanotopographical features of Ti implant surfaces have been known to be contributors to the initial biologic responses of the hard tissue, including osteoblast activities and osteoclast reactions [44,49].

This modified surface with TiO_2 nanotube arrays is highly biocompatible [44,50,51]. Both osteoblasts and osteoclasts showed maximal cellular responses to Ti surfaces with TiO_2 nanotubes that were 15 nm in diameter [52]. Interestingly, smaller TiO_2 nanotubes, which were approximately 30 nm in diameter, were more effective in the adhesion and growth of mesenchymal stem cells than larger TiO_2 nanotubes that ranged from 70 nm to 100 nm, while the latter TiO_2 nanotubes were more inductive in the differentiation into osteoblast-like cells, although there is contrary to previous studies [52,53]. The modified TiO_2 nanotubular surface showed excellent bone-to-implant contact in the osteoporotic bone in an in vivo study using ovariectomized rats [54].

Another characteristic of this nano-modified surface is a drug delivery effect [55–58]. Drug release from TiO_2 nanotubes is associated with the dimensions of TiO_2 nanotube arrays regardless of the direct release or indirect discharge by nanocarriers [59]. The diameter and length of TiO_2 nanotubes generally increase as the voltage and duration of the oxidation process increase, and the drug release has been found to be effective when the diameter is larger than approximately 100 nm [56,59,60]. A combination of this nano-modified TiO_2 surface and carrier molecules, including micelles, is being actively investigated for drug delivery at a constant rate, unrelated to the drug concentration and release period [57,58,60].

The nanotopography of the TiO_2 nanotubular surface has antibacterial properties alongside delivering antibiotic drugs [61]. Streptococcus mutans, which are associated with the initial formation of biofilm in the oral cavity, were reported to adhere to the TiO_2 nanotube arrays less than to a micro-roughened SLA surface [62]. The hydrophilic properties of TiO_2 nanotubes seems to hinder bacterial adhesion to the nanotubular surface [62]. However, it is notable that many studies have described the wettability of the TiO_2 nanotube arrays, showing conflicting results in cellular and bacterial responses to the nanotubular surface [61,63,64]. Although the hydrophilicity of the TiO_2 nanotube arrays is adjustable, some studies reported that the reduction of bacterial adhesion was due to the hydrophilic properties of the surface, whereas other studies described that such a result was due to the hydrophobic properties [61,63,64]. Further investigation is required to determine the mechanism of bacterial and cellular responses to the wettability of Ti surfaces.

Despite that the modified surface with TiO_2 nanotube arrays has very useful advantages (e.g., high biocompatibility, the capability of drug delivery, and antibacterial properties), this surface has been neither applied nor tested clinically. The mechanical strength between the TiO_2 nanotubes and the base Ti surface is too weak for this surface to be applied to a dental implant [43]. Recently, the hexagonal nano-structure of the base Ti surface was evaluated to be adequate for biologic application when the TiO_2 nanotube arrays are removed from the base surface in order to prevent the delamination of the TiO_2 nanotube coating in an in vivo environment (Figure 2) [44]. The aligned TiO_2 nanotube-layered surface has great potential in biologic and clinical applications [55–57,65]. However, it is necessary to overcome this delamination problem before this TiO_2 nanotubular surface is clinically used in the field of dental implantology.

3.2. Functional Peptides

Water and ions have first contact with the implant surface when the bone is drilled for implant insertion and a screw-shaped endosseous dental implant is placed into the bone. Then, the plasma proteins adhere to the surface through ionic bridges (like a calcium ion linkage), and the fibrin clot is formed. During hemostasis, extracellular matrix (ECM) proteins gradually replace the plasma proteins [15]. The adhesion proteins, including fibronectin and vitronectin (which are also ECM proteins), are recognized by the transmembrane proteins of osteogenic cells like integrins. Through binding of the transmembrane proteins to the osteogenic cells, the cells interact with ECM, which controls the cellular activities for bone healing [66]. Therefore, the bone healing process starts from the adhesion of the osteogenic cells to surfaces, and these adhesion proteins can play a role in accelerating osseointegration into dental implants when the proteins are applied to the implant surfaces. Core amino acid sequences, which are extracted from the original adhesion proteins and still have binding activities to the transmembrane receptors, are very useful in rapid bone healing when the core sequences are treated on the implant surfaces. These core functional peptides are considered to be more promising candidates for implant surface treatment than the original proteins because of the lower antigenicity and simpler adjustability of the peptides [67].

A functional peptide derived from the fibronectin, arginyl-glycyl-aspartic acid sequence, revealed improved histomorphometric results when this peptide was coated on a Ti dental implant surface and when this peptide-treated surface was compared to the uncoated surface [68]. Two functional amino acid sequences derived from another adhesion protein, laminin, showed excellent results as accelerating modifiers for Ti implant surfaces for osseointegration [35,67]. These functional peptides based on the adhesion of osteogenic cells seem to surpass the effects of the microtopographical features of the underlying Ti implant surfaces in bone healing, although further studies are definitely needed [35,69]. The mechanism behind the superior bone cell responses has been tried to be explained, based on the hypothesized tunable allosteric control of the receptor proteins [67,70]. A recent investigation evaluating a functional peptide from vitronectin found a Janus effect of this peptide for bone formation, activating osteoblasts and inhibiting osteoclasts, that is, controlling the osteoporotic environment locally to be favorable for osseointegration [71].

Cytokines, particularly growth factors, are another class of bioactive proteins. Bone morphogenetic proteins (BMPs) are available for bone healing in the field of dental implantology. Human recombinant BMP-2 (rhBMP-2) is used in the global market for bone regeneration. BMP-2 is known to have a direct effect on osteogenic cells to promote bone formation with various interactions between this protein and other bioactive molecules, including osteogenic genes [72,73]. However, these growth factors have many problems to be solved before clinical application to Ti dental implant surfaces. BMP-2 has complicated biologic effects depending on its concentrations and surroundings; osteogenesis, adipogenesis, and chodrogenesis, but osteolysis also occurs [72,74,75]. The rhBMP-2-treated Ti surface was reported to make bone healing around a dental implant faster in an in vivo model [76,77]. However, it is recognizable that growth factors are usually active in free forms, not in bound forms. Therefore, these molecules are ineffective or, if any, limitedly active when the factors are bound or attached to implant surfaces [78]. The cell transmembrane proteins that recognize these growth factors are disengaged in the attachment of the cells [78]. Because of the multiple enigmatic effects of these growth factors on living tissues and the growth factor receptors' lack of involvement in cell adhesion, growth factor-treated implant surfaces have not been used clinically until now.

Although these bioactive molecules, including the adhesion molecules and growth factors, have the potential to be applied to dental implants for accelerated osseointegration, the Ti dental implants on which these molecules are coated have not been clinically tested; there have been no published clinical trials to report the results of such implants. The functional peptides from the adhesion molecules are to be clinically tried and applied in dental implantology in the near future due to the simplicity in their biologic effects and their low probability of side effects. For growth factors, it seems to be necessary to find core amino acid sequences from growth factors to increase the clinical applicability of these factors. Before these derived peptides are clinically tried, further studies are required on release strategies for the molecules from the implant surfaces and on the biologic activities of the core peptides.

3.3. Fluoride Treatment (Cathodic Reduction)

When a Ti implant is a cathode in the hydrofluoric acid solution of an electrochemical cell, a fluoride ion gives an electron to the cathode, where the reduction of a Ti ion occurs. As a result, a trace amount of fluoride ions adheres to the Ti implant surface when the concentration of hydrogen fluoride is low in the solution. This trace amount of fluoride ions is known to primarily affect osteoprogenitor cells and undifferentiated osteoblasts to enhance bone formation, rather than highly differentiated osteoblasts [79,80]. Furthermore, fluoride is helpful for bone mineralization because of its properties that are attractive for calcium [78]. However, fluoride ions are thought to become cytotoxic as the number of ions increases on the Ti implant surface.

Clinically, a modified surface is used as a dental implant surface (Osseospeed, Astra Tech, Dentsply, Waltham, MA, USA), which is fluoride-treated after the grade 4 cp-Ti is sandblasted with TiO_2 particles. This fluoride-modified surface has a very low amount of fluoride, which is difficult to find by energy dispersive spectroscopy, while x-ray photoelectron spectroscopy is able to detect this trace amount [81,82]. The average mean height of this marketed surface has been investigated to be approximately 1.5 µm [30,82]. The fluoride-treated Ti surface has shown stronger binding between the bone and this surface than the control Ti surface without fluoride-treatment [83,84]. However, finding any significant differences in the histomorphometric results has been very rare when the fluoride-treated dental implants have been compared in vivo to other modified implants, including SLA implants, while some previous studies have been found to show more favorable results in bone responses to the fluoride-treated surface than those to its predecessor with no application of fluoride [78,82,85,86].

Dental implants with a fluoride-treated surface have exhibited high rates of success and survival rates in clinical trials. These fluoride-treated implants have supported prosthodontic restorations in edentulous mandibles with a 100% survival rate for ten years [87]. Regardless of the maxilla or the mandible, high survival rates of over 95% have been reported for the surface-modified dental implants in the prospective clinical studies, the observation periods of which are longer than 5 years [87–89]. It is notable and very interesting that these previous clinical studies have consistently reported the vertical loss of bone surrounding the implants of less than 0.5–1 mm, which is interpreted as almost no change of the bone level [40,87,89]. Importantly, these clinical studies used fluoride-treated dental implants with the same implant macro-design, including an identical thread shape and internal friction implant–abutment connection, so care must be taken when interpreting data in comparison studies of the biologic responses between the dental implant systems [82]. It remains uncertain which factor (surface chemistry in fluoride treatment, surface topography, implant–abutment connection tectonics) is a major contributor to the biologic responses in humans to this marketed fluoride-treated dental implant.

3.4. Hydroxyapatite and Other Calcium–Phosphorus Compounds

This idea of hydroxyapatite (HA) coating on a Ti dental implant surface is based on the fact that the main component of bone is HA. HA ($Ca_{10}(PO_4)_6(OH)_2$) is still the most commonly-utilized coating material for Ti dental implant surfaces [90]. HA and other calcium-phosphorus coating materials are basically osteoconductive to the surrounding bone. The biologic features of these materials, such as their biodegradation properties and foreign body reactions, seem to depend on calcium/phosphorus ratios, crystallinities, and coating thicknesses [31,90–93]. Plasma spraying (a conventional atmospheric plasma-spray method) is one of the most widely used methods to coat HA on a Ti implant surface [90]. HA particles that are contained and heated in a plasma flame whose temperature is approximately 15,000 to 20,000 Kelvin are sprayed on the Ti surface, resulting in a HA coating layer that is 50–100 µm in thickness [94]. The spray parameters, including the flame combination and spraying flow rate, affect the chemical and physical features of the HA coating [92].

The HA coating is biocompatible with the hard tissue, showing direct contact with bone and the attachment of osteoblasts on the coating surface [95,96]. Many studies have reported enhanced bone apposition and the prevention of metal-ion release into the bone from metal implants with an HA coated surface [97–101]. However, the HA coating layer has some critical issues to be addressed. Like the TiO_2 nanotube arrays, the delamination of the coating layer from the Ti dental implant surface is one of the problems (adhesive failure) [92]. Delaminated or worn HA particles hinder bone healing and provoke inflammation around the implant inserted into the bone [92,102]. The thick coating layer is able to make a breakage inside the layer, especially at the implant in a load-bearing area (cohesive failure) [92]. Recently, a thin calcium–phosphorus coating layer has been achieved and investigated using various coating techniques [76,93,101,103]. Compared to the plasma sprayed HA coating, however, the other calcium–phosphorus coatings are considered to lack long term clinical results [104,105].

The five-year clinical success rate of the HA coated implant has been evaluated to be approximately 95% [106]. However, this success rate has dropped markedly to below 80% after 10 years of implant placement [106–108]. Such a low success rate may result from the above-mentioned problems of the HA coating layer. It is notable, however, that these clinical evaluations resulted from the data of cylindrical implants [107,108]. A previous study using HA coated screw-shaped implants (MicroVent, Zimmer Dental, Carlsbad, CA, USA) reported that the long-term clinical success rate (> 10 years) was higher than 90% [109]. Nevertheless, clinical trials are certainly necessary to evaluate the calcium–phosphorus coating more precisely.

3.5. Photofunctionalization

In 1997, it was determined that the wettability of the TiO_2 surface is increased by ultraviolet (UV) radiation [110]. Originally, the UV-induced TiO_2 surface is amphiphilic—both hydrophilic and oleophilic [110]. However, the enhanced biologic effect of this surface is considered to be caused by the hydrophilic properties. Such hydrophilicity and elimination of hydrocarbon contamination on the TiO_2 surface are known to be the mechanisms behind further activated bone responses to a dental implant in UV-mediated photofunctionalization. The hard tissue affinity drops for an aged Ti surface that has been stored for longer than two weeks [111]. UV irradiation on the Ti implant surface appears to make the Ti surface reactivate, as the implant is freshly made.

UV radiation is subcategorized into three types according to its wavelengths and dermal biologic reactions to the electromagnetic waves: UVA, UVB, and UVC [112]. The wavelengths of UVA range from 320 to 400 nm, and those of UVC range from 200 to 280 nm [113]. Both UVA and UVC contribute to increasing the hydrophilicity of the Ti surface. However, considering the fact that some reports show the promoted osteogenic activities on hydrophobic surfaces, the removal of carbon from the Ti surface, which is caused by UVC, is likely a fundamental mechanism behind excellent osseointegration [114–117]. Strictly, neither UVA nor UVC appears to make a topographic change at the nano-scale on the Ti surface [115,117]. Friction force microscopy shows a nano-scale modification that UV irradiation may produce by converting Ti^{4+} to Ti^{3+} [110]. UV treatment on the Ti surface enhances the adsorption of proteins, such as albumin and fibronectin, which are plasma proteins in the human body [118]. UV-photofunctionalized implant surfaces show improved osteogenic cell attachment, spreading, and proliferation [117]. The antibacterial effects are described for the UV activation of the Ti surface [112]. Faster bone responses to UV-treated Ti surfaces are reported in various in vivo studies, some of which show almost 100% bone-to-implant contact [117–120].

A previous clinical study showed that the stability of implants inserted into the patients' jaw bones increased more rapidly when the implants were UV-photofunctionalized [121]. The retrospective clinical studies concluded that UV-mediated photofunctionalization reduced early implant failure, and the success rate of the photofunctionalized implants was 97.6% during the functional loading period of approximately 2.5 years [122,123]. No prospective long-term clinical study (published in English) evaluating UV-mediated photofunctionalization has yet been found in the field of implant dentistry. However, a prospective clinical evaluation of UV-treated implants over more than 5 years is expected to be published shortly.

3.6. Laser Ablation

For laser ablation, an implant whose collar, or neck area, was treated by laser micromachining to generate nano-channels is used (Laser-Lok, BioHorizons, Birmingham, AL, USA) [124,125].

Laser ablation is also able to produce micro-scale patterns by controlling laser processing parameters [126]. This approach was intended to promote not only fast osseointegration, but also connective tissue attachment [124,127]. The connective tissue fiber direction in the soft tissue attachment is known to be perpendicular to the laser-microtextured Ti implant surface, which is characteristically different from the general orientation of the fibers parallel to implant surfaces [127,128].

This marketed laser-modified surface (Laser-Lok) showed significantly improved bone-to-implant contact in a previous in vivo study, compared to a turned Ti surface [129]. The survival rate was evaluated to be 95.6% in a two-year retrospective multicenter study and to be 94% in another 5-year retrospective controlled study [127,130]. Recently, the prospective three-year results of a randomized clinical trial were reported for single implant-supported restorations with the laser-modified Ti implant surface, where the survival rate was estimated to be 96.1% [131]. Both the hard and soft tissue responses to the laser-modified Ti surface appear to be favorable [127,130–132]. However, long-term prospective clinical results of laser micromachining are still needed.

4. Concluding Remarks

When the bone is prepared for implant placement, surgical trauma provokes bleeding and hemostasis. Moreover, this surgical trauma activates the growth and differentiation factors released from the bone debris and matrix [15]. Surface modification of the Ti dental implant focuses on improving such initial biologic responses to the implant surface. Researchers and dental clinicians anticipate the best performance of the implant surfaces during these initial events and more readily establish these events by changing the physical and chemical properties of the surfaces, thereby boosting the speed and strengthening the quality of the healing process [133]. However, as the long-term clinical studies show, implant-supported prostheses have been used for a long time in patients' mouths. Therefore, the modified surfaces also need to harmonize with the bone remodeling process, which has not yet been investigated. This paper reviews several modified surfaces of dental implants that are widely used in the global market or are highly possible to be clinically used. All these reviewed surfaces are targeted to accelerate early bone responses. The late responses of the hard tissue to the surfaces, including bone remodeling, need to be investigated. Moreover, long-term clinical trials are still required for these implant surfaces.

Funding: This research was supported by a grant from the Korea Health Technology R&D Project through the Korea Health Industry Development Institute (KHIDI), which is funded by the Ministry of Health and Welfare of the Republic of Korea (grant number: HI15C1535).

Conflicts of Interest: The authors report no conflict of interest related to this study.

References

1. Albrektsson, T.; Branemark, P.I.; Hansson, H.A.; Lindstrom, J. Osseointegrated titanium implants. Requirements for ensuring a long-lasting, direct bone-to-implant anchorage in man. *Acta Orthop. Scand.* **1981**, *52*, 155–170. [CrossRef] [PubMed]
2. Davies, J.E. Mechanisms of endosseous integration. *Int. J. Prosthodont.* **1998**, *11*, 391–401. [PubMed]
3. Choi, J.Y.; Sim, J.H.; Yeo, I.L. Characteristics of contact and distance osteogenesis around modified implant surfaces in rabbit tibiae. *J. Periodontal Implant Sci.* **2017**, *47*, 182–192. [CrossRef] [PubMed]
4. Albrektsson, T.; Wennerberg, A. Oral implant surfaces: Part 1–review focusing on topographic and chemical properties of different surfaces and in vivo responses to them. *Int. J. Prosthodont.* **2004**, *17*, 536–543. [PubMed]
5. Kulkarni, M.; Mazare, A.; Gongadze, E.; Perutkova, S.; Kralj-Iglic, V.; Milosev, I.; Schmuki, P.; Iglic, A.; Mozetic, M. Titanium nanostructures for biomedical applications. *Nanotechnology* **2015**, *26*, 062002. [CrossRef]
6. Albrektsson, T.; Jemt, T.; Molne, J.; Tengvall, P.; Wennerberg, A. On inflammation-immunological balance theory-A critical apprehension of disease concepts around implants: Mucositis and marginal bone loss may represent normal conditions and not necessarily a state of disease. *Clin. Implant Dent. Relat. Res.* **2019**, *21*, 183–189. [CrossRef]

7. Albrektsson, T.; Chrcanovic, B.; Molne, J.; Wennerberg, A. Foreign body reactions, marginal bone loss and allergies in relation to titanium implants. *Eur. J. Oral Implantol.* **2018**, *11*, S37–S46.
8. Davies, J.E. INVITED COMMENTARY: Is Osseointegration a Foreign Body Reaction? *Int. J. Prosthodont.* **2019**, *32*, 133–136. [CrossRef]
9. Brunski, J. On Implant Prosthodontics: One Narrative, Twelve Voices - 2. *Int. J. Prosthodont.* **2018**, *31*, s15–s22.
10. Kwon, T.K.; Choi, J.Y.; Park, J.I.; Yeo, I.L. A Clue to the Existence of Bonding between Bone and Implant Surface: An In Vivo Study. *Mater.* **2019**, *12*, 1187. [CrossRef]
11. Osborn, J.; Newesely, H. Dynamic aspects of the implant-bone interface. *Dental implants* **1980**, *111*, 123.
12. Branemark, P.I.; Adell, R.; Breine, U.; Hansson, B.O.; Lindstrom, J.; Ohlsson, A. Intra-osseous anchorage of dental prostheses. I. Experimental studies. *Scand. J. Plast. Reconstr. Surg.* **1969**, *3*, 81–100. [CrossRef] [PubMed]
13. Sims, N.A.; Gooi, J.H. Bone remodeling: Multiple cellular interactions required for coupling of bone formation and resorption. *Semin. Cell Dev. Biol.* **2008**, *19*, 444–451. [CrossRef] [PubMed]
14. Rupp, F.; Scheideler, L.; Olshanska, N.; de Wild, M.; Wieland, M.; Geis-Gerstorfer, J. Enhancing surface free energy and hydrophilicity through chemical modification of microstructured titanium implant surfaces. *J. Biomed. Mater. Res. A.* **2006**, *76*, 323–334. [CrossRef] [PubMed]
15. Yeo, I.-S. Surface modification of dental biomaterials for controlling bone response. In *Bone Response to Dental Implant Materials*; Woodhead Publishing: Cambridge, UK, 2017; pp. 43–64.
16. Rupp, F.; Scheideler, L.; Rehbein, D.; Axmann, D.; Geis-Gerstorfer, J. Roughness induced dynamic changes of wettability of acid etched titanium implant modifications. *Biomaterials* **2004**, *25*, 1429–1438. [CrossRef] [PubMed]
17. Wall, I.; Donos, N.; Carlqvist, K.; Jones, F.; Brett, P. Modified titanium surfaces promote accelerated osteogenic differentiation of mesenchymal stromal cells in vitro. *Bone* **2009**, *45*, 17–26. [CrossRef]
18. Buser, D.; Janner, S.F.; Wittneben, J.G.; Bragger, U.; Ramseier, C.A.; Salvi, G.E. 10-year survival and success rates of 511 titanium implants with a sandblasted and acid-etched surface: A retrospective study in 303 partially edentulous patients. *Clin. Implant Dent. Relat. Res.* **2012**, *14*, 839–851. [CrossRef]
19. Nicolau, P.; Guerra, F.; Reis, R.; Krafft, T.; Benz, K.; Jackowski, J. 10-year outcomes with immediate and early loaded implants with a chemically modified SLA surface. *Quintessence Int.* **2018**, *50*, 2–12.
20. Roccuzzo, M.; Bonino, L.; Dalmasso, P.; Aglietta, M. Long-term results of a three arms prospective cohort study on implants in periodontally compromised patients: 10-year data around sandblasted and acid-etched (SLA) surface. *Clin. Oral Implants Res.* **2014**, *25*, 1105–1112. [CrossRef]
21. Rossi, F.; Lang, N.P.; Ricci, E.; Ferraioli, L.; Baldi, N.; Botticelli, D. Long-term follow-up of single crowns supported by short, moderately rough implants-A prospective 10-year cohort study. *Clin. Oral Implants Res.* **2018**, *29*, 1212–1219. [CrossRef]
22. Van Velzen, F.J.; Ofec, R.; Schulten, E.A.; Ten Bruggenkate, C.M. 10-year survival rate and the incidence of peri-implant disease of 374 titanium dental implants with a SLA surface: A prospective cohort study in 177 fully and partially edentulous patients. *Clin. Oral Implants Res.* **2015**, *26*, 1121–1128. [CrossRef] [PubMed]
23. Li, S.M.; Yao, W.H.; Liu, J.H.; Yu, M.; Wu, L.; Ma, K. Study on anodic oxidation process and property of composite film formed on Ti-10V-2Fe-3Al alloy in SiC nanoparticle suspension. *Surf. Coat Tech.* **2015**, *277*, 234–241. [CrossRef]
24. Hanawa, T. Metal ion release from metal implants. *Mat. Sci. Eng. C-Bio. S.* **2004**, *24*, 745–752. [CrossRef]
25. Manhabosco, T.M.; Tamborim, S.M.; dos Santos, C.B.; Muller, I.L. Tribological, electrochemical and tribo-electrochemical characterization of bare and nitrided Ti6Al4V in simulated body fluid solution. *Corros. Sci.* **2011**, *53*, 1786–1793. [CrossRef]
26. Wang, J.W.; Ma, Y.; Guan, J.; Zhang, D.W. Characterizations of anodic oxide films formed on Ti6A14V in the silicate electrolyte with sodium polyacrylate as an additive. *Surf. Coat Tech.* **2018**, *338*, 14–21. [CrossRef]
27. Zhang, L.; Duan, Y.; Gao, R.; Yang, J.; Wei, K.; Tang, D.; Fu, T. The Effect of Potential on Surface Characteristic and Corrosion Resistance of Anodic Oxide Film Formed on Commercial Pure Titanium at the Potentiodynamic-Aging Mode. *Materials* **2019**, *12*, 370. [CrossRef]
28. Kwon, T.K.; Lee, H.J.; Min, S.K.; Yeo, I.S. Evaluation of early bone response to fluoride-modified and anodically oxidized titanium implants through continuous removal torque analysis. *Implant Dent.* **2012**, *21*, 427–432. [CrossRef]

29. Lee, H.J.; Yang, I.H.; Kim, S.K.; Yeo, I.S.; Kwon, T.K. In vivo comparison between the effects of chemically modified hydrophilic and anodically oxidized titanium surfaces on initial bone healing. *J. Periodontal. Implant Sci.* **2015**, *45*, 94–100. [CrossRef]
30. Wennerberg, A.; Albrektsson, T. On implant surfaces: A review of current knowledge and opinions. *Int. J. Oral Maxillofac. Implants.* **2010**, *25*, 63–74.
31. Yeo, I.S.; Han, J.S.; Yang, J.H. Biomechanical and histomorphometric study of dental implants with different surface characteristics. *J. Biomed. Mater. Res. B. Appl. Biomater.* **2008**, *87*, 303–311. [CrossRef]
32. Cooper, L.F. A role for surface topography in creating and maintaining bone at titanium endosseous implants. *J. Prosthet. Dent.* **2000**, *84*, 522–534. [CrossRef] [PubMed]
33. Koh, J.W.; Kim, Y.S.; Yang, J.H.; Yeo, I.S. Effects of a calcium phosphate-coated and anodized titanium surface on early bone response. *Int. J. Oral Maxillofac. Implants.* **2013**, *28*, 790–797. [CrossRef] [PubMed]
34. Yeo, I.S. Reality of dental implant surface modification: A short literature review. *Open Biomed. Eng. J.* **2014**, *8*, 114–119. [CrossRef] [PubMed]
35. Kang, H.K.; Kim, O.B.; Min, S.K.; Jung, S.Y.; Jang, D.H.; Kwon, T.K.; Min, B.M.; Yeo, I.S. The effect of the DLTIDDSYWYRI motif of the human laminin alpha2 chain on implant osseointegration. *Biomaterials* **2013**, *34*, 4027–4037. [CrossRef]
36. Min, S.K.; Kang, H.K.; Jang, D.H.; Jung, S.Y.; Kim, O.B.; Min, B.M.; Yeo, I.S. Titanium surface coating with a laminin-derived functional peptide promotes bone cell adhesion. *Biomed Res Int* **2013**, *2013*, 638348. [CrossRef]
37. Chrcanovic, B.R.; Albrektsson, T.; Wennerberg, A. Turned versus anodised dental implants: A meta-analysis. *J. Oral Rehabil.* **2016**, *43*, 716–728. [CrossRef]
38. Degidi, M.; Nardi, D.; Piattelli, A. 10-year follow-up of immediately loaded implants with TiUnite porous anodized surface. *Clin. Implant. Dent. Relat. Res.* **2012**, *14*, 828–838. [CrossRef]
39. Shibuya, Y.; Kobayashi, M.; Takeuchi, J.; Asai, T.; Murata, M.; Umeda, M.; Komori, T. Analysis of 472 Branemark system TiUnite implants:a retrospective study. *Kobe J. Med. Sci.* **2010**, *55*, E73–E81.
40. Donati, M.; Ekestubbe, A.; Lindhe, J.; Wennstrom, J.L. Marginal bone loss at implants with different surface characteristics - A 20-year follow-up of a randomized controlled clinical trial. *Clin. Oral Implants Res.* **2018**, *29*, 480–487. [CrossRef]
41. Adell, R.; Lekholm, U.; Rockler, B.; Branemark, P.I. A 15-year study of osseointegrated implants in the treatment of the edentulous jaw. *Int. J. Oral Surg.* **1981**, *10*, 387–416. [CrossRef]
42. Rocci, A.; Rocci, M.; Rocci, C.; Scoccia, A.; Gargari, M.; Martignoni, M.; Gottlow, J.; Sennerby, L. Immediate loading of Branemark system TiUnite and machined-surface implants in the posterior mandible, part II: A randomized open-ended 9-year follow-up clinical trial. *Int. J. Oral Maxillofac. Implants.* **2013**, *28*, 891–895. [CrossRef] [PubMed]
43. Li, T.; Gulati, K.; Wang, N.; Zhang, Z.; Ivanovski, S. Understanding and augmenting the stability of therapeutic nanotubes on anodized titanium implants. *Mater. Sci. Eng. C. Mater. Biol. Appl.* **2018**, *88*, 182–195. [CrossRef] [PubMed]
44. Shin, Y.C.; Pang, K.M.; Han, D.W.; Lee, K.H.; Ha, Y.C.; Park, J.W.; Kim, B.; Kim, D.; Lee, J.H. Enhanced osteogenic differentiation of human mesenchymal stem cells on Ti surfaces with electrochemical nanopattern formation. *Mater. Sci. Eng. C. Mater. Biol. Appl.* **2019**, *99*, 1174–1181. [CrossRef] [PubMed]
45. Rice, J.M.; Hunt, J.A.; Gallagher, J.A.; Hanarp, P.; Sutherland, D.S.; Gold, J. Quantitative assessment of the response of primary derived human osteoblasts and macrophages to a range of nanotopography surfaces in a single culture model in vitro. *Biomaterials* **2003**, *24*, 4799–4818. [CrossRef]
46. Wennerberg, A.; Albrektsson, T. Suggested guidelines for the topographic evaluation of implant surfaces. *Int. J. Oral Maxillofac. Implants.* **2000**, *15*, 331–344.
47. Hansson, S.; Norton, M. The relation between surface roughness and interfacial shear strength for bone-anchored implants. A mathematical model. *J. Biomech.* **1999**, *32*, 829–836. [CrossRef]
48. Mendonca, G.; Mendonca, D.B.; Aragao, F.J.; Cooper, L.F. Advancing dental implant surface technology–from micron- to nanotopography. *Biomaterials* **2008**, *29*, 3822–3835. [CrossRef]
49. Liu, H.; Webster, T.J. Nanomedicine for implants: A review of studies and necessary experimental tools. *Biomaterials* **2007**, *28*, 354–369. [CrossRef]
50. Ahn, T.K.; Lee, D.H.; Kim, T.S.; Jang, G.C.; Choi, S.; Oh, J.B.; Ye, G.; Lee, S. Modification of Titanium Implant and Titanium Dioxide for Bone Tissue Engineering. *Adv. Exp. Med. Biol.* **2018**, *1077*, 355–368.

51. Awad, N.K.; Edwards, S.L.; Morsi, Y.S. A review of TiO$_2$ NTs on Ti metal: Electrochemical synthesis, functionalization and potential use as bone implants. *Mater. Sci. Eng. C. Mater. Biol. Appl.* **2017**, *76*, 1401–1412. [CrossRef]
52. Park, J.; Bauer, S.; Schlegel, K.A.; Neukam, F.W.; von der Mark, K.; Schmuki, P. TiO$_2$ nanotube surfaces: 15 nm–an optimal length scale of surface topography for cell adhesion and differentiation. *Small* **2009**, *5*, 666–671. [CrossRef] [PubMed]
53. Oh, S.; Brammer, K.S.; Li, Y.S.; Teng, D.; Engler, A.J.; Chien, S.; Jin, S. Stem cell fate dictated solely by altered nanotube dimension. *Proc. Natl. Acad. Sci. USA* **2009**, *106*, 2130–2135. [CrossRef]
54. Jiang, N.; Du, P.; Qu, W.; Li, L.; Liu, Z.; Zhu, S. The synergistic effect of TiO$_2$ nanoporous modification and platelet-rich plasma treatment on titanium-implant stability in ovariectomized rats. *Int. J. Nanomed.* **2016**, *11*, 4719–4733.
55. Gulati, K.; Ivanovski, S. Dental implants modified with drug releasing titania nanotubes: Therapeutic potential and developmental challenges. *Expert Opin. Drug Deliv.* **2017**, *14*, 1009–1024. [CrossRef] [PubMed]
56. Kwon, D.H.; Lee, S.J.; Wikesjo, U.M.E.; Johansson, P.H.; Johansson, C.B.; Sul, Y.T. Bone tissue response following local drug delivery of bisphosphonate through titanium oxide nanotube implants in a rabbit model. *J. Clin. Periodontol.* **2017**, *44*, 941–949. [CrossRef] [PubMed]
57. Wang, Q.; Huang, J.Y.; Li, H.Q.; Zhao, A.Z.; Wang, Y.; Zhang, K.Q.; Sun, H.T.; Lai, Y.K. Recent advances on smart TiO$_2$ nanotube platforms for sustainable drug delivery applications. *Int. J. Nanomed.* **2017**, *12*, 151–165. [CrossRef]
58. Yang, W.; Deng, C.; Liu, P.; Hu, Y.; Luo, Z.; Cai, K. Sustained release of aspirin and vitamin C from titanium nanotubes: An experimental and stimulation study. *Mater. Sci. Eng. C. Mater. Biol. Appl.* **2016**, *64*, 139–147. [CrossRef]
59. Hamlekhan, A.; Sinha-Ray, S.; Takoudis, C.; Mathew, M.T.; Sukotjo, C.; Yarin, A.L.; Shokuhfar, T. Fabrication of drug eluting implants: Study of drug release mechanism from titanium dioxide nanotubes. *J. Phys. D. Appl. Phys.* **2015**, *48*, 275401. [CrossRef]
60. Aw, M.S.; Gulati, K.; Losic, D. Controlling drug release from titania nanotube arrays using polymer nanocarriers and biopolymer coating. *J. Biomater. Nanobiotechnol.* **2011**, *2*, 477. [CrossRef]
61. Kunrath, M.F.; Leal, B.F.; Hubler, R.; de Oliveira, S.D.; Teixeira, E.R. Antibacterial potential associated with drug-delivery built TiO$_2$ nanotubes in biomedical implants. *AMB Express.* **2019**, *9*, 51. [CrossRef]
62. Miao, X.; Wang, D.; Xu, L.; Wang, J.; Zeng, D.; Lin, S.; Huang, C.; Liu, X.; Jiang, X. The response of human osteoblasts, epithelial cells, fibroblasts, macrophages and oral bacteria to nanostructured titanium surfaces: A systematic study. *Int. J. Nanomed.* **2017**, *12*, 1415–1430. [CrossRef] [PubMed]
63. Gittens, R.A.; Scheideler, L.; Rupp, F.; Hyzy, S.L.; Geis-Gerstorfer, J.; Schwartz, Z.; Boyan, B.D. A review on the wettability of dental implant surfaces II: Biological and clinical aspects. *Acta Biomater.* **2014**, *10*, 2907–2918. [CrossRef] [PubMed]
64. Kulkarni, M.; Patil-Sen, Y.; Junkar, I.; Kulkarni, C.V.; Lorenzetti, M.; Iglic, A. Wettability studies of topologically distinct titanium surfaces. *Colloids Surf. B Biointerfaces* **2015**, *129*, 47–53. [CrossRef] [PubMed]
65. Kaur, G.; Willsmore, T.; Gulati, K.; Zinonos, I.; Wang, Y.; Kurian, M.; Hay, S.; Losic, D.; Evdokiou, A. Titanium wire implants with nanotube arrays: A study model for localized cancer treatment. *Biomaterials* **2016**, *101*, 176–188. [CrossRef]
66. Stephansson, S.N.; Byers, B.A.; Garcia, A.J. Enhanced expression of the osteoblastic phenotype on substrates that modulate fibronectin conformation and integrin receptor binding. *Biomaterials* **2002**, *23*, 2527–2534. [CrossRef]
67. Yeo, I.S.; Min, S.K.; Kang, H.K.; Kwon, T.K.; Jung, S.Y.; Min, B.M. Identification of a bioactive core sequence from human laminin and its applicability to tissue engineering. *Biomaterials* **2015**, *73*, 96–109. [CrossRef]
68. Ryu, J.J.; Park, K.; Kim, H.S.; Jeong, C.M.; Huh, J.B. Effects of anodized titanium with Arg-Gly-Asp (RGD) peptide immobilized via chemical grafting or physical adsorption on bone cell adhesion and differentiation. *Int. J. Oral Maxillofac. Implants.* **2013**, *28*, 963–972. [CrossRef]
69. Kim, S.; Choi, J.Y.; Jung, S.Y.; Kang, H.K.; Min, B.M.; Yeo, I.L. A laminin-derived functional peptide, PPFEGCIWN, promotes bone formation on sandblasted, large-grit, acid-etched titanium implant surfaces. *Int. J. Oral Maxillofac. Implants.* **2019**, *34*, 836–844. [CrossRef]
70. Motlagh, H.N.; Wrabl, J.O.; Li, J.; Hilser, V.J. The ensemble nature of allostery. *Nature* **2014**, *508*, 331–339. [CrossRef]

71. Min, S.K.; Kang, H.K.; Jung, S.Y.; Jang, D.H.; Min, B.M. A vitronectin-derived peptide reverses ovariectomy-induced bone loss via regulation of osteoblast and osteoclast differentiation. *Cell Death Differ.* **2018**, *25*, 268–281. [CrossRef]
72. Rogers, M.B.; Shah, T.A.; Shaikh, N.N. Turning Bone Morphogenetic Protein 2 (BMP2) on and off in Mesenchymal Cells. *J. Cell Biochem.* **2015**, *116*, 2127–2138. [CrossRef] [PubMed]
73. Song, R.; Wang, D.; Zeng, R.; Wang, J. Synergistic effects of fibroblast growth factor-2 and bone morphogenetic protein-2 on bone induction. *Mol. Med. Rep.* **2017**, *16*, 4483–4492. [CrossRef] [PubMed]
74. Kang, J.D. Another complication associated with rhBMP-2? *Spine J.* **2011**, *11*, 517–519. [CrossRef] [PubMed]
75. Kawaguchi, H.; Jingushi, S.; Izumi, T.; Fukunaga, M.; Matsushita, T.; Nakamura, T.; Mizuno, K.; Nakamura, T.; Nakamura, K. Local application of recombinant human fibroblast growth factor-2 on bone repair: A dose-escalation prospective trial on patients with osteotomy. *J. Orthop. Res.* **2007**, *25*, 480–487. [CrossRef]
76. Choi, J.Y.; Jung, U.W.; Kim, C.S.; Jung, S.M.; Lee, I.S.; Choi, S.H. Influence of nanocoated calcium phosphate on two different types of implant surfaces in different bone environment: An animal study. *Clin. Oral Implants Res.* **2013**, *24*, 1018–1022. [CrossRef]
77. Kim, J.E.; Kang, S.S.; Choi, K.H.; Shim, J.S.; Jeong, C.M.; Shin, S.W.; Huh, J.B. The effect of anodized implants coated with combined rhBMP-2 and recombinant human vascular endothelial growth factors on vertical bone regeneration in the marginal portion of the peri-implant. *Oral Surg. Oral Med. Oral Pathol. Oral Radiol.* **2013**, *115*, e24–e31. [CrossRef]
78. Ellingsen, J.E.; Thomsen, P.; Lyngstadaas, S.P. Advances in dental implant materials and tissue regeneration. *Periodontol. 2000.* **2006**, *41*, 136–156. [CrossRef]
79. Bellows, C.G.; Heersche, J.N.; Aubin, J.E. The effects of fluoride on osteoblast progenitors in vitro. *J. Bone Miner. Res.* **1990**, *5*, S101–S105. [CrossRef]
80. Kassem, M.; Mosekilde, L.; Eriksen, E.F. Effects of fluoride on human bone cells in vitro: Differences in responsiveness between stromal osteoblast precursors and mature osteoblasts. *Eur. J. Endocrinol.* **1994**, *130*, 381–386. [CrossRef]
81. Choi, J.Y.; Lee, H.J.; Jang, J.U.; Yeo, I.S. Comparison between bioactive fluoride modified and bioinert anodically oxidized implant surfaces in early bone response using rabbit tibia model. *Implant Dent.* **2012**, *21*, 124–128. [CrossRef]
82. Choi, J.Y.; Kang, S.H.; Kim, H.Y.; Yeo, I.L. Control variable implants improve interpretation of surface modification and implant design effects on early bone responses: An in vivo study. *Int. J. Oral Maxillofac. Implants.* **2018**, *33*, 1033–1040. [CrossRef] [PubMed]
83. Ellingsen, J.E. Pre-treatment of titanium implants with fluoride improves their retention in bone. *J. Mater. Sci-Mater. M.* **1995**, *6*, 749–753. [CrossRef]
84. Ellingsen, J.E.; Johansson, C.B.; Wennerberg, A.; Holmen, A. Improved retention and bone-to-implant contact with fluoride-modified titanium implants. *Int. J. Oral Max. Impl.* **2004**, *19*, 659–666.
85. Hong, Y.S.; Kim, M.J.; Han, J.S.; Yeo, I.S. Effects of hydrophilicity and fluoride surface modifications to titanium dental implants on early osseointegration: An in vivo study. *Implant Dent.* **2014**, *2*, 529–533. [CrossRef] [PubMed]
86. Taxt-Lamolle, S.F.; Rubert, M.; Haugen, H.J.; Lyngstadaas, S.P.; Ellingsen, J.E.; Monjo, M. Controlled electro-implementation of fluoride in titanium implant surfaces enhances cortical bone formation and mineralization. *Acta Biomater.* **2010**, *6*, 1025–1032. [CrossRef]
87. Windael, S.; Vervaeke, S.; Wijnen, L.; Jacquet, W.; De Bruyn, H.; Collaert, B. Ten-year follow-up of dental implants used for immediate loading in the edentulous mandible: A prospective clinical study. *Clin. Implant Dent. Relat. Res.* **2018**, *20*, 515–521. [CrossRef]
88. Mertens, C.; Steveling, H.G. Early and immediate loading of titanium implants with fluoride-modified surfaces: Results of 5-year prospective study. *Clin. Oral Implants Res.* **2011**, *22*, 1354–1360. [CrossRef]
89. Oxby, G.; Oxby, F.; Oxby, J.; Saltvik, T.; Nilsson, P. Early loading of fluoridated implants placed in fresh extraction sockets and healed bone: A 3- to 5-year clinical and radiographic follow-up study of 39 consecutive patients. *Clin. Implant Dent. Relat. Res.* **2015**, *17*, 898–907. [CrossRef]
90. Xuereb, M.; Camilleri, J.; Attard, N.J. Systematic review of current dental implant coating materials and novel coating techniques. *Int. J. Prosthodont.* **2015**, *28*, 51–59. [CrossRef]
91. Alizadeh-Osgouei, M.; Li, Y.; Wen, C. A comprehensive review of biodegradable synthetic polymer-ceramic composites and their manufacture for biomedical applications. *Bioact. Mater.* **2019**, *4*, 22–36. [CrossRef]

92. Sun, L.; Berndt, C.C.; Gross, K.A.; Kucuk, A. Material fundamentals and clinical performance of plasma-sprayed hydroxyapatite coatings: A review. *J. Biomed. Mater. Res.* **2001**, *58*, 570–592. [CrossRef]
93. You, C.; Yeo, I.S.; Kim, M.D.; Eom, T.K.; Lee, J.Y.; Kim, S. Characterization and in vivo evaluation of calcium phosphate coated cp-titanium by dip-spin method. *Curr. Appl. Phys.* **2005**, *5*, 501–506. [CrossRef]
94. Gupta, A.; Dhanraj, M.; Sivagami, G. Status of surface treatment in endosseous implant: A literary overview. *Indian J. Dent. Res.* **2010**, *21*, 433–438. [CrossRef] [PubMed]
95. Geesink, R.G.; de Groot, K.; Klein, C.P. Bonding of bone to apatite-coated implants. *J. Bone Joint Surg. Br.* **1988**, *70*, 17–22. [CrossRef] [PubMed]
96. Manero, J.M.; Salsench, J.; Nogueras, J.; Aparicio, C.; Padros, A.; Balcells, M.; Gil, F.J.; Planell, J.A. Growth of bioactive surfaces on dental implants. *Implant Dent.* **2002**, *11*, 170–175. [CrossRef] [PubMed]
97. Dalton, J.E.; Cook, S.D. In vivo mechanical and histological characteristics of HA-coated implants vary with coating vendor. *J. Biomed. Mater. Res.* **1995**, *29*, 239–245. [CrossRef]
98. Ducheyne, P.; Healy, K.E. The effect of plasma-sprayed calcium phosphate ceramic coatings on the metal ion release from porous titanium and cobalt-chromium alloys. *J Biomed Mater Res* **1988**, *22*, 1137–1163. [CrossRef]
99. Ducheyne, P.; Hench, L.L.; Kagan, A., II; Martens, M.; Bursens, A.; Mulier, J.C. Effect of hydroxyapatite impregnation on skeletal bonding of porous coated implants. *J. Biomed. Mater. Res.* **1980**, *14*, 225–237. [CrossRef]
100. Oonishi, H.; Yamamoto, M.; Ishimaru, H.; Tsuji, E.; Kushitani, S.; Aono, M.; Ukon, Y. The effect of hydroxyapatite coating on bone growth into porous titanium alloy implants. *J. Bone Joint Surg. Br.* **1989**, *71*, 213–216. [CrossRef]
101. Yeo, I.S.; Min, S.K.; An, Y. Influence of bioactive material coating of Ti dental implant surfaces on early healing and osseointegration of bone. *J. Korean Phys. Soc.* **2010**, *57*, 1717–1720. [CrossRef]
102. Yeung, W.K.; Reilly, G.C.; Matthews, A.; Yerokhin, A. In vitro biological response of plasma electrolytically oxidized and plasma-sprayed hydroxyapatite coatings on Ti-6Al-4V alloy. *J. Biomed. Mater. Res. B Appl. Biomater.* **2013**, *101*, 939–949. [CrossRef]
103. Wennerberg, A.; Jimbo, R.; Allard, S.; Skarnemark, G.; Andersson, M. In vivo stability of hydroxyapatite nanoparticles coated on titanium implant surfaces. *Int. J. Oral Maxillofac. Implants.* **2011**, *26*, 1161–1166.
104. Ostman, P.O.; Wennerberg, A.; Ekestubbe, A.; Albrektsson, T. Immediate occlusal loading of NanoTite tapered implants: A prospective 1-year clinical and radiographic study. *Clin. Implant Dent. Relat. Res.* **2013**, *15*, 809–818. [CrossRef]
105. Oztel, M.; Bilski, W.M.; Bilski, A. Risk factors associated with dental implant failure: A study of 302 implants placed in a regional center. *J. Contemp. Dent. Pract.* **2017**, *18*, 705–709. [CrossRef]
106. Van Oirschot, B.A.; Bronkhorst, E.M.; van den Beucken, J.J.; Meijer, G.J.; Jansen, J.A.; Junker, R. A systematic review on the long-term success of calcium phosphate plasma-spray-coated dental implants. *Odontology* **2016**, *104*, 347–356. [CrossRef]
107. Artzi, Z.; Carmeli, G.; Kozlovsky, A. A distinguishable observation between survival and success rate outcome of hydroxyapatite-coated implants in 5–10 years in function. *Clin. Oral Implants Res.* **2006**, *17*, 85–93. [CrossRef]
108. Binahmed, A.; Stoykewych, A.; Hussain, A.; Love, B.; Pruthi, V. Long-term follow-up of hydroxyapatite-coated dental implants–a clinical trial. *Int. J. Oral Maxillofac. Implants.* **2007**, *22*, 963–968.
109. Schwartz-Arad, D.; Mardinger, O.; Levin, L.; Kozlovsky, A.; Hirshberg, A. Marginal bone loss pattern around hydroxyapatite-coated versus commercially pure titanium implants after up to 12 years of follow-up. *Int. J. Oral Maxillofac. Implants.* **2005**, *20*, 238–244.
110. Wang, R.; Hashimoto, K.; Fujishima, A.; Chikuni, M.; Kojima, E.; Kitamura, A.; Shimohigoshi, M.; Watanabe, T. Light-induced amphiphilic surfaces. *Nature* **1997**, *388*, 431–432. [CrossRef]
111. Att, W.; Hori, N.; Takeuchi, M.; Ouyang, J.; Yang, Y.; Anpo, M.; Ogawa, T. Time-dependent degradation of titanium osteoconductivity: An implication of biological aging of implant materials. *Biomaterials* **2009**, *30*, 5352–5363. [CrossRef]
112. Flanagan, D. Photofunctionalization of dental implants. *J. Oral Implantol.* **2016**, *42*, 445–450. [CrossRef]
113. Clydesdale, G.J.; Dandie, G.W.; Muller, H.K. Ultraviolet light induced injury: Immunological and inflammatory effects. *Immunol. Cell Biol.* **2001**, *79*, 547–568. [CrossRef]

114. Aita, H.; Att, W.; Ueno, T.; Yamada, M.; Hori, N.; Iwasa, F.; Tsukimura, N.; Ogawa, T. Ultraviolet light-mediated photofunctionalization of titanium to promote human mesenchymal stem cell migration, attachment, proliferation and differentiation. *Acta Biomater.* **2009**, *5*, 3247–3257. [CrossRef]
115. Jain, S.; Williamson, R.S.; Marquart, M.; Janorkar, A.V.; Griggs, J.A.; Roach, M.D. Photofunctionalization of anodized titanium surfaces using UVA or UVC light and its effects against Streptococcus sanguinis. *J. Biomed. Mater. Res. B Appl. Biomater.* **2018**, *106*, 2284–2294. [CrossRef]
116. Jansen, E.J.; Sladek, R.E.; Bahar, H.; Yaffe, A.; Gijbels, M.J.; Kuijer, R.; Bulstra, S.K.; Guldemond, N.A.; Binderman, I.; Koole, L.H. Hydrophobicity as a design criterion for polymer scaffolds in bone tissue engineering. *Biomaterials* **2005**, *26*, 4423–4431. [CrossRef]
117. Ogawa, T. Ultraviolet photofunctionalization of titanium implants. *Int. J. Oral Maxillofac. Implants.* **2014**, *29*, e95–e102. [CrossRef]
118. Aita, H.; Hori, N.; Takeuchi, M.; Suzuki, T.; Yamada, M.; Anpo, M.; Ogawa, T. The effect of ultraviolet functionalization of titanium on integration with bone. *Biomaterials* **2009**, *30*, 1015–1025. [CrossRef]
119. Park, K.H.; Koak, J.Y.; Kim, S.K.; Han, C.H.; Heo, S.J. The effect of ultraviolet-C irradiation via a bactericidal ultraviolet sterilizer on an anodized titanium implant: A study in rabbits. *Int. J. Oral Maxillofac. Implants.* **2013**, *28*, 57–66. [CrossRef]
120. Lee, J.B.; Jo, Y.H.; Choi, J.Y.; Seol, Y.J.; Lee, Y.M.; Ku, Y.; Rhyu, I.C.; Yeo, I.L. The effect of ultraviolet photofunctionalization on a titanium dental implant with machined surface: An in vitro and in vivo study. *Materials* **2019**, *12*, 2078. [CrossRef]
121. Hirota, M.; Ozawa, T.; Iwai, T.; Ogawa, T.; Tohnai, I. Implant stability development of photofunctionalized implants placed in regular and complex cases: A case-control study. *Int. J. Oral Maxillofac. Implants.* **2016**, *31*, 676–686. [CrossRef]
122. Funato, A.; Yamada, M.; Ogawa, T. Success rate, healing time, and implant stability of photofunctionalized dental implants. *Int. J. Oral Maxillofac. Implants.* **2013**, *28*, 1261–1271. [CrossRef]
123. Hirota, M.; Ozawa, T.; Iwai, T.; Ogawa, T.; Tohnai, I. Effect of photofunctionalization on early implant failure. *Int. J. Oral Maxillofac. Implants.* **2018**, *33*, 1098–1102. [CrossRef]
124. Asensio, G.; Vazquez-Lasa, B.; Rojo, L. Achievements in the Topographic Design of Commercial Titanium Dental Implants: Towards Anti-Peri-Implantitis Surfaces. *J. Clin. Med.* **2019**, *8*, 1982. [CrossRef]
125. Smeets, R.; Stadlinger, B.; Schwarz, F.; Beck-Broichsitter, B.; Jung, O.; Precht, C.; Kloss, F.; Grobe, A.; Heiland, M.; Ebker, T. Impact of Dental Implant Surface Modifications on Osseointegration. *Biomed. Res. Int.* **2016**, *2016*, 6285620. [CrossRef]
126. Souza, J.C.M.; Sordi, M.B.; Kanazawa, M.; Ravindran, S.; Henriques, B.; Silva, F.S.; Aparicio, C.; Cooper, L.F. Nano-scale modification of titanium implant surfaces to enhance osseointegration. *Acta Biomater.* **2019**, *94*, 112–131. [CrossRef]
127. Iorio-Siciliano, V.; Matarasso, R.; Guarnieri, R.; Nicolo, M.; Farronato, D.; Matarasso, S. Soft tissue conditions and marginal bone levels of implants with a laser-microtextured collar: A 5-year, retrospective, controlled study. *Clin. Oral Implants Res.* **2015**, *26*, 257–262. [CrossRef]
128. Degidi, M.; Piattelli, A.; Scarano, A.; Shibli, J.A.; Iezzi, G. Peri-implant collagen fibers around human cone Morse connection implants under polarized light: A report of three cases. *Int. J. Periodontics Restorative Dent.* **2012**, *32*, 323–328.
129. Nevins, M.; Kim, D.M.; Jun, S.H.; Guze, K.; Schupbach, P.; Nevins, M.L. Histologic evidence of a connective tissue attachment to laser microgrooved abutments: A canine study. *Int. J. Periodontics Restorative Dent.* **2010**, *30*, 245–255.
130. Guarnieri, R.; Placella, R.; Testarelli, L.; Iorio-Siciliano, V.; Grande, M. Clinical, radiographic, and esthetic evaluation of immediately loaded laser microtextured implants placed into fresh extraction sockets in the anterior maxilla: A 2-year retrospective multicentric study. *Implant Dent.* **2014**, *23*, 144–154. [CrossRef]
131. Guarnieri, R.; Grande, M.; Ippoliti, S.; Iorio-Siciliano, V.; Ricciotello, F.; Farronato, D. Influence of a Laser-Lok Surface on Immediate Functional Loading of Implants in Single-Tooth Replacement: Three-Year Results of a Prospective Randomized Clinical Study on Soft Tissue Response and Esthetics. *Int. J. Periodontics Restorative Dent.* **2015**, *35*, 865–875. [CrossRef]
132. Farronato, D.; Mangano, F.; Briguglio, F.; Iorio-Siciliano, V.; Ricciotello, F.; Guarnieri, R. Influence of Laser-Lok surface on immediate functional loading of implants in single-tooth replacement: A 2-year prospective clinical study. *Int. J. Periodontics Restorative Dent.* **2014**, *34*, 79–89. [CrossRef]

133. Kunrath, M.F.; Hubler, R. A bone preservation protocol that enables evaluation of osseointegration of implants with micro- and nanotextured surfaces. *Biotech. Histochem.* **2019**, *94*, 261–270. [CrossRef]

 © 2019 by the author. Licensee MDPI, Basel, Switzerland. This article is an open access article distributed under the terms and conditions of the Creative Commons Attribution (CC BY) license (http://creativecommons.org/licenses/by/4.0/).

Article

A Clue to the Existence of Bonding between Bone and Implant Surface: An In Vivo Study

Taek-Ka Kwon [1,†], Jung-Yoo Choi [2,†], Jae-Il Park [3] and In-Sung Luke Yeo [4,*]

1. Division of Prosthodontics, Department of Dentistry, St. Catholic Hospital, Catholic University of Korea, Suwon 16247, Korea; tega95@naver.com
2. Dental Research Institute, Seoul National University, Seoul 03080, Korea; jychoi55@snu.ac.kr
3. Animal Facility of Aging Science, Korea Basic Science Institute, Gwangju 61186, Korea; jaeil74@kbsi.re.kr
4. Department of Prosthodontics, School of Dentistry and Dental Research Institute, Seoul National University, 101 Daehak-ro, Jongro-gu, Seoul 03080, Korea
* Correspondence: pros53@snu.ac.kr; Tel.: +82-2-2072-2661; Fax: +82-2-2072-3860
† These authors contributed equally to this work.

Received: 1 April 2019; Accepted: 9 April 2019; Published: 11 April 2019

Abstract: We evaluated the shear bond strength of bone–implant contact, or osseointegration, in the rabbit tibia model, and compared the strength between grades 2 and 4 of commercially pure titanium (cp-Ti). A total of 13 grades 2 and 4 cp-Ti implants were used, which had an identical cylinder shape and surface topography. Field emission scanning electron microscopy, X-ray photoelectron spectroscopy, and confocal laser microscopy were used for surface analysis. Four grades 2 and 4 cp-Ti implants were inserted into the rabbit tibiae with complete randomization. After six weeks of healing, the experimental animals were sacrificed and the implants were removed en bloc with the surrounding bone. The bone–implant interfaces were three-dimensionally imaged with micro-computed tomography. Using these images, the bone–implant contact area was measured. Counterclockwise rotation force was applied to the implants for the measurement of removal torque values. Shear bond strength was calculated from the measured bone–implant contact and removal torque data. The t-tests were used to compare the outcome measures between the groups, and statistical significance was evaluated at the 0.05 level. Surface analysis showed that grades 2 and 4 cp-Ti implants have similar topographic features. We found no significant difference in the three-dimensional bone–implant contact area between these two implants. However, grade 2 cp-Ti implants had a higher shear bond strength than grade 4 cp-Ti implants ($p = 0.032$). The surfaces of the grade 2 cp-Ti implants were similar to those of the grade 4 implants in terms of physical characteristics and the quantitative amount of attachment to the bone, whereas the grade 2 surfaces were stronger than the grade 4 surfaces in the bone–surface interaction, indicating osseointegration quality.

Keywords: osseointegration; titanium; bone–implant interface; shear strength; torque

1. Introduction

Modern endosseous dental implants have been used widely in dental practice since the Toronto conference in 1982 [1,2]. Modern implants have demonstrated reliable longevity and success, and have become a routine dental therapeutic protocol in edentulous patients. However, the bonding mechanism between the bone and the dental implant is still unclear.

Bone-to-implant contact (BIC) in dental implantology is defined as the direct attachment of bone to an implant observed on an undecalcified histologic slide under a light microscopic view without intervening soft tissue. The nature of BIC, whether a real bond exists between the bone and implant or only simple contact, remains unknown [3]. BIC has been suggested to be hard tissue encapsulation, or the bony isolation of the osseointegrated dental implant [4–6]. If such a foreign body reaction is the case, it is highly possible that BIC should be a simple physical attachment between the bone and implant. Therefore, if the implants have identical physical features, such as surface topography and implant design, the shear bond strength or removal torque per unit area of implant would be similar, regardless of the material compositions on implant surfaces. Conversely, different bond strengths imply osseointegration quality at the interface, which suggests the bone has biologic affinity depending on materials, rather than the bony isolation. To the best of our knowledge, no calculation of the interfacial binding per unit area has been published to test whether BIC is only frictional or has its own quality.

This study was designed to test the hypothesis that no significant difference would be found in the removal torque per unit of bone contact area, calculated by micro computed tomography (CT), in a comparison of interfacial bindings composed of grades 2 and 4 commercially pure titanium (cp-Ti), if BIC is a simple contact between the bone and implant. The grades 2 and 4 cp-Ti implants used had an identical geometry and microstructure.

2. Materials and Methods

2.1. Implant Preparation

Twenty-six experimental implants (Deep Implant System, Seongnam, Korea) were prepared, composed of a different grade of cp-Ti: Grade 2 and 4 (n = 13 for each grade). All the implants were conventional external hex designed, 3.0 mm in diameter, and 4.0 mm in length. The implants were formed without thread and surface modification: Solid rod or cylindrical design and turned surface (Figure 1A).

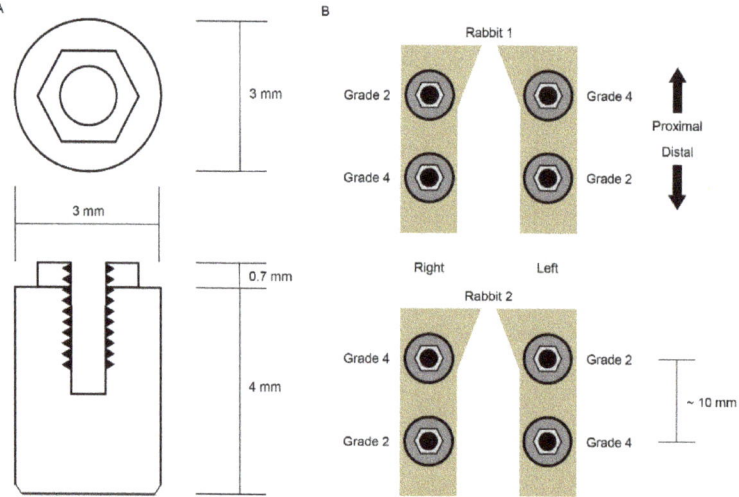

Figure 1. *Cont.*

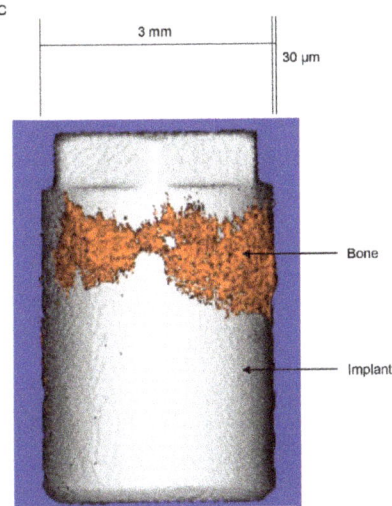

Figure 1. Schematic diagram of the specimens used in this study and the arrangement of the specimens. (**A**) The implant specimens had a simple geometric form: a cylinder that facilitates biomechanical calculations. (**B**) The grades 2 and 4 commercially pure titanium implants were installed into rabbit tibiae according to the Latin Square randomization technique. The distance between the centers of the proximal and distal implants was approximately 10 mm. (**C**) The three-dimensional (3D) bone–implant interfaces were reconstructed via digital image processing. A bone–implant contact surface is shown here. Note that the image of the bone is cut and processed in 30-μm thicknesses.

2.2. Surface Characteristics Analysis

Nine cp-Ti grades 2 and 4 implants were used in the surface character analysis test. Three surface analysis tests were performed for implants of each cp-Ti grade: Field emission scanning electron microscopy (FE-SEM), X-ray photoelectron spectroscopy (XPS), and confocal laser microscopy (CLSM). The FE-SEM (model S-4700, Hitachi, Tokyo, Japan) was used to capture several scaled images of each implant surface ($n = 3$). XPS (Sigma Probe, Thermo Fisher Scientific, Waltham, MA, USA) was used to identify the elemental content and quantify the atomic concentration of the tested surfaces; measurements were repeated three times for each specimen ($n = 3$). CLSM (LSM 800, Carl Zeiss AG, Oberkochen, Germany) was used to measure the surface topographical features of implant sides on three different areas (measurement area: 150 μm × 150 μm on a 200 × optically- and 3 × digitally-magnified image) for each specimen ($n = 3$). The images were filtered using a Gaussian low-pass filter with a cut-off wavelength of 80 μm. The average surface deviation (Sa) and developed surface area ratio (Sdr) were measured.

2.3. In Vivo Implant Surgery and Euthanasia

The animal study was approved by the Ethics Committee of the Animal Experimentation of the Institutional Animal Care and Use Committee (CRONEX-IACUC 201702001; Cronex, Hwasung, Korea). All the animal study procedures, including animal selection, management, preparation, and surgical protocols, were performed according to the guidelines of Animal Research: Reporting In Vivo Experiments (ARRIVE) [7]. We installed grades 2 and 4 cp-Ti implants in two male New Zealand white rabbits, each weighing 2.5 to 3.0 kg and aged about six months. They showed no sign of disease or illness before the experiment. Two implants were installed in each rabbit tibia using a standard Latin Square design (Figure 1B). Prior to surgery, the rabbits were anesthetized with an intramuscular injection of tiletamine/zolazepam (15 mg/kg; Zoletil 50, Virbac Korea, Seoul, Korea) and xylazine

(33 mg/kg; Rompun, Bayer Korea Ltd., Seoul, Korea). The animals received an intramuscular injection of 33 mg/kg Cefazolin (Yuhan, Seoul, Korea), a preoperative prophylactic antibiotic. The skin of each rabbit's proximal tibia area was shaved with an electric shaving machine and sterilized with povidone iodine solution, and local anesthetic, lidocaine (1:100,000 epinephrine; Yuhan, Seoul, Korea), was injected into each surgical site. The skin was incised with a surgical blade, and full-thickness periosteal flap reflection was performed to expose each tibia. The implant preparation drilling was conducted on the flat surface of the tibia using a 3-mm-diameter final dental implant drill under simultaneous sterile saline irrigation. The implant was inserted into the drill hole so that the top of the implant was 0.5 mm above the upper cortex of the rabbit tibia without contacting the lower cortex using a silicone ring. Four implants were installed into each rabbit—two grade 2 cp-Ti implants and two grade 4 cp-Ti implants—in a Latin Square design (2 × 2 Latin Square, n = 4). After the insertion of the implants, the surgical sites were sutured layer by layer. We allowed a relatively long healing period to present similar bone–implant contacts for the two different types of implants. Rabbits were sacrificed after six weeks of bone healing by intravenous administration of potassium chloride following anesthetization. The implants were exposed by full thickness periosteal elevation and retrieved en bloc with the adjacent bone collar.

2.4. Measurement of Three-Dimensional Bone–Implant Contact Area

CT imaging of the harvested implants and bone was performed using a Quantum GX μCT imaging system (PerkinElmer, Hopkinton, MA, USA), located at the Korea Basic Science Institute (Gwangju, Korea). The X-ray source was set to 90 kV and 88 μA with a 10 mm field of view (voxel size = 20 μm; scanning time = 57 min). The CT data were visualized using the Quantum GX's three-dimensional (3D) viewing software. 3D images of the implant specimens and bone growth were constructed. Image processing for the calculation of the bone–implant contact area was described in a previous study [8]. Briefly, following scanning, the images were segmented using Analyze software version 12.0 (AnalyzeDirect, Overland Park, KS, USA) and filtered to reduce imaging noise. Then, the dataset was manually reoriented using Analyze software to visualize standard coronal, sagittal, and horizontal planes through the implants. The images were reformatted to cubic volume (3D) with a resliced 30-μm image thickness of the surface area on the implant outer surface for the cross-sectional and longitudinal axes. The segmentations of implant and bone-growth images were also performed on Analyze software. As such, 3D rendering of the implants and bone growth was completed. To determine the bone volume on the implant surface in 30-μm thicknesses, the original scanned images were rotated by 10 degrees and the segmentations and 3D rendering were repeated. These 18 repetitions produced the overall 3D bone growth image on the implant surface in 30-μm thicknesses (Figure 1C). The 3D BIC area was the bone formation area on the cylindrical surface of the implant. Assuming that the volume in 30-μm thicknesses was homogeneously filled with bone, the 3D BIC area was determined by dividing the bone volume by the thickness (30 μm).

2.5. Measurement of Shear Bond Strength of Bone and Implant

After CT scanning, implant removal torque values were measured within 24 h. The experimental implants had no thread and no fixation via screw action. The constant-speed counter-clockwise rotation of the implant or collar bone results in the disintegration of the BIC or shear bond failure. This continuous removal torque test produces accurate and uniform results [9,10]. Implant removal torque was measured with a motorized torque test stand (TSTM, Mark-10 Co., Long Island, NY, USA). The rotation speed was 0.3 rpm for the lowest angular speed to ensure peak implant removal torque was not skipped between sampling intervals (65 samples/s). The peak implant removal torque was selected from the time–torque curve data. The shear bond strength (MPa) between the bone and osseointegrated dental implant, or binding force per unit area in this study, was calculated by dividing the peak implant removal torque (Ncm) by the determined 3D BIC area (mm^2) and the implant radius (1.5 mm). The units were adjusted before calculation.

2.6. Statistics

The independent t-test was used to determine the statistical significance of the surface roughness parameters, Sa and Sdr, between the grade 2 and 4 cp-Ti implants. The paired t-test was used to compare the 3D BIC area and the shear bond strength between the groups. All data were evaluated at the significance level of 0.05.

3. Results

3.1. Surface Characteristics Analysis

The FE-SEM image of each surface area is shown in Figure 2. The grades 2 and 4 cp-Ti surfaces show similar surface characteristics, which resulted from the computer numerical control machining manufacturing process. The results of the CLSM analysis showed that the means of Sa were 0.49 ± 0.082 μm for grade 2 cp-Ti implants and 0.45 ± 0.088 μm for grade 4 cp-Ti implants. No significant difference was found for Sa between the Ti grades (p = 0.63). The means of Sdr were 16.9% ± 1.5% for the grade 2 implant group and 14.4% ± 2.1% for the grade 4 implant group, and no significant difference was found between the groups (p = 0.18). These results indicate that the grades 2 and 4 cp-Ti implants used in this study had similar surface topographical features. However, surface chemistry depended on the Ti grades. The XPS showed that the atomic composition of the grade 2 cp-Ti surface was significantly different from that of the grade 4 surface (Table 1).

Figure 2. Field emission scanning electron microscopy (FE-SEM) images of commercially pure titanium of (**A**) grade 2 and (**B**) grade 4. Many machining grooves were observed. Similar surface characteristics were found for both grades, which implies that the influence of topographical features on the biological response should not be different between the commercially pure titanium grades used in this study.

Table 1. The atomic composition of surfaces of grades 2 and 4 commercially pure titanium (cp-Ti) obtained by X-ray photoelectron spectroscopy (XPS).

Element	Cp-Ti Grade 2	Cp-Ti Grade 4	p-Value
Carbon (C)	3.66% ± 0.21%	4.03% ± 0.39%	2.3×10^{-1}
Oxygen (O)	1.15% ± 0.10%	3.21% ± 0.29%	3.0×10^{-4} *
Iron (Fe)	0.26% ± 0.02%	0.39% ± 0.05%	2.0×10^{-2} *
Titanium (Ti)	94.93% ± 0.32%	92.38% ± 0.13%	2.1×10^{-4} *

* Statistically significant.

3.2. Three-Dimensional Bone–Implant Contact Area and Shear Bond Strength

The bone–implant contact surface area and three-dimensional bone–implant contact ratio are summarized in Table 2. The means and standard deviations of the contacted surface area for grades 2 and 4 cp-Ti implants were 12.7 ± 1.2 mm^2 and 11.5 ± 1.6 mm^2, respectively. The 3D bone–implant contact percentages of grades 2 and 4 cp-Ti implants were 33.6% ± 3.2% and 30.5% ± 4.3%, respectively. Neither the contacted surface area nor the 3D bone–implant contact showed any statistical difference between the two grades. The peak implant removal torques of grades 2 and 4 cp-Ti implants were 2.9 ± 0.4 Ncm and 1.9 ± 0.3 Ncm, respectively. The shear bond strength of grade 2 cp-Ti implants was 1.5 ± 0.2 MPa, which was statistically significantly higher than that of the grade 4 cp-Ti implants, which was 1.1 ± 0.1 MPa (p = 0.032) (Table 2).

Table 2. Three-dimensional bone-to-implant contact (BIC) area and shear bond strength.

	Grade 2 cp-Ti	Grade 4 cp-Ti	p-Value
3D BIC area (mm^2)	12.7 ± 1.2	11.5 ± 1.6	0.35
3D BIC ratio (%)	33.6 ± 3.2	30.5 ± 4.3	0.35
Implant removal torque (Ncm)	2.9 ± 0.4	1.9 ± 0.3	0.052
Torque per unit (Ncm/cm^2)	23.2 ± 3.5	16.5 ± 1.0	0.032 *
Shear bond strength (MPa)	1.5 ± 0.2	1.1 ± 0.1	0.032 *

* Statistically significant.

4. Discussion

The grades 2 and 4 cp-Ti implants used in this study showed similar surface topographic features in SEM images and CLSM analysis. The grade 2 cp-Ti implants showed higher shear bond strength than the grade 4 implants despite the similar 3D bone–implant contact ratios between these two grades. Considering the significant differences in the compositions between the grades 2 and 4 cp-Ti surfaces, the bone response was evaluated to be stronger to grade 2 than to grade 4. The results of this study suggest that the bone would have its own affinity, depending on the materials, and that an actual bond would exist in addition to the physical contact and friction between bone and an implant surface [3]. Although further verification studies are required, the osseointegration phenomenon of a Ti dental implant appears to be a bioaffinitive response to the Ti surface beyond the bony isolation.

The pull-out test of integrated implants has been widely adopted to test osseointegration strength [11–15]. However, we used rotating force instead of pull-out force because the removal direction for the pull-out test could be nonparallel to the long axis of the implant, imposing unintentional lateral force on the implant, which can introduce measurement error. This adverse effect was minimized in this study by using rotational force for implant removal and a motorized torque test stand to simplify adjustment to the long axis [10].

This study compared the grades 2 and 4 implants using the rabbit tibia model. Sample size determination and randomization are important to reduce the number of sacrificed animals [7]. Following the guidelines of the 3Rs (replacement, reduction, and refinement) of ARRIVE, this study used the 2 × 2 Latin Square design, which minimized the sample size and accomplished complete randomization in the arrangement of the implant groups [7]. Considering the standard deviation of the 3D BIC area and ratio measures, the number of sacrificed animals (two) was estimated to be adequate although sample size calculation was not performed in this study, which would require more animals than two.

Many studies reported significant differences in the amount of bone–implant contact when implant surfaces were topographically changed [16]. Further studies are needed to evaluate the effects of surface topography on the quality of osseointegration. A roughened surface increases the bone–implant area, increasing the physical interlocking between bone and the surface, which would result in the same shear bond strength or removal torque per unit area when the implant surfaces, topographically changed or not, have the same chemical composition. Conversely, the topographical change could have a qualitative influence on the bone–implant contact, which remains to be verified. In addition, further studies are required to evaluate osseointegration qualities by comparing the bone–implant contact and shear bond strength at the different phases of bone healing after implant placement.

5. Conclusions

Grades 2 and 4 cp-Ti surfaces with similar topographical features showed different bond strengths in the bone–implant contact. Considering the results of this study, an actual bond may occur between bone and a Ti dental implant surface beyond the physical attachment of bone to the surface.

Author Contributions: T.-K.K. and I.-S.L.Y. designed the experiments; T.-K.K., J.-Y.C., and I.-S.L.Y. performed the experiments; T.-K.K. and J.-I.P. obtained the data; J.-Y.C. and I.-S.L.Y. analyzed the data; T.-K.K., J.-Y.C., J.-I.P., and I.-S.L.Y. wrote the manuscript. T.-K.K. and J.-Y.C. equally contributed to this study and I.-S.L.Y. had correspondence.

Funding: This work was supported by a National Research Foundation of Korea (NRF) grant funded by the Korean Government (MSIP) [NRF-2016R1A2B4014330].

Conflicts of Interest: There are no conflicts of interest to declare related to this study.

References

1. Block, M.S. Dental Implants: The Last 100 Years. *J. Oral Maxillofac. Surg.* **2018**, *76*, 11–26. [CrossRef] [PubMed]
2. Brånemark, P.I.; Zarb, G.A.; Albrektsson, T. *Tissue-Integrated Prostheses: Osseointegration in Clinical Dentistry*; Quintessence Publishing: Chicago, IL, USA, 1985.
3. Brunski, J. On Implant Prosthodontics: One Narrative, Twelve Voices-2. *Int. J. Prosthodont.* **2018**, *31*, s15–s22. [PubMed]
4. Albrektsson, T. On Implant Prosthodontics: One Narrative, Twelve Voices-1. *Int. J. Prosthodont.* **2018**, *31*, s11–s14. [PubMed]
5. Jemt, T. On Implant Prosthodontics: One Narrative, Twelve Voices-4. *Int. J. Prosthodont.* **2018**, *31*, s31–s34. [PubMed]
6. Trindade, R.; Albrektsson, T.; Galli, S.; Prgomet, Z.; Tengvall, P.; Wennerberg, A. Osseointegration and foreign body reaction: Titanium implants activate the immune system and suppress bone resorption during the first 4 weeks after implantation. *Clin. Implant. Dent. Relat. Res.* **2018**, *20*, 82–91. [CrossRef] [PubMed]
7. Kilkenny, C.; Browne, W.J.; Cuthill, I.C.; Emerson, M.; Altman, D.G. Improving bioscience research reporting: The ARRIVE guidelines for reporting animal research. *Osteoarthr. Cartil.* **2012**, *20*, 256–260. [CrossRef] [PubMed]
8. Choi, J.Y.; Park, J.I.; Chae, J.S.; Yeo, I.L. Comparison of micro-computed tomography and histomorphometry in the measurement of bone–implant contact ratios. *Oral Surg. Oral Med. Oral Pathol. Oral Radiol.* **2019**, in press. [CrossRef] [PubMed]
9. Kwon, T.K.; Lee, H.J.; Min, S.K.; Yeo, I.S. Evaluation of early bone response to fluoride-modified and anodically oxidized titanium implants through continuous removal torque analysis. *Implant. Dent.* **2012**, *21*, 427–432. [CrossRef] [PubMed]
10. Lee, H.J.; Yeo, I.S.; Kwon, T.K. Removal Torque Analysis of Chemically Modified Hydrophilic and Anodically Oxidized Titanium Implants with Constant Angular Velocity for Early Bone Response in Rabbit Tibia. *Tissue Eng. Regen. Med.* **2013**, *10*, 252–259. [CrossRef]
11. Gracco, A.; Giagnorio, C.; Incerti Parenti, S.; Alessandri Bonetti, G.; Siciliani, G. Effects of thread shape on the pullout strength of miniscrews. *Am. J. Orthod. Dentofac. Orthop.* **2012**, *142*, 186–190. [CrossRef] [PubMed]
12. Nonhoff, J.; Moest, T.; Schmitt, C.M.; Weisel, T.; Bauer, S.; Schlegel, K.A. Establishment of a new pull-out strength testing method to quantify early osseointegration-An experimental pilot study. *J. Cranio-Maxillofac. Surg.* **2015**, *43*, 1966–1973. [CrossRef] [PubMed]
13. Velasco, E.; Monsalve-Guil, L.; Jimenez, A.; Ortiz, I.; Moreno-Munoz, J.; Nunez-Marquez, E.; Pegueroles, M.; Perez, R.A.; Gil, F.J. Importance of the Roughness and Residual Stresses of Dental Implants on Fatigue

and Osseointegration Behavior. In Vivo Study in Rabbits. *J. Oral Implantol.* **2016**, *42*, 469–476. [CrossRef] [PubMed]
14. Watanabe, T.; Nakada, H.; Takahashi, T.; Fujita, K.; Tanimoto, Y.; Sakae, T.; Kimoto, S.; Kawai, Y. Potential for acceleration of bone formation after implant surgery by using a dietary supplement: An animal study. *J. Oral Rehabil.* **2015**, *42*, 447–453. [CrossRef] [PubMed]
15. Yashwant, A.V.; Dilip, S.; Krishnaraj, R.; Ravi, K. Does Change in Thread Shape Influence the Pull Out Strength of Mini Implants? An In vitro Study. *J. Clin. Diagn. Res.* **2017**, *11*, ZC17–ZC20. [CrossRef] [PubMed]
16. Yeo, I.S. Reality of dental implant surface modification: A short literature review. *Open Biomed. Eng. J.* **2014**, *8*, 114–119. [CrossRef] [PubMed]

© 2019 by the authors. Licensee MDPI, Basel, Switzerland. This article is an open access article distributed under the terms and conditions of the Creative Commons Attribution (CC BY) license (http://creativecommons.org/licenses/by/4.0/).

Article

The Effect of Ultraviolet Photofunctionalization on a Titanium Dental Implant with Machined Surface: An In Vitro and In Vivo Study

Jun-Beom Lee [1], Ye-Hyeon Jo [2], Jung-Yoo Choi [3], Yang-Jo Seol [1], Yong-Moo Lee [1], Young Ku [1], In-Chul Rhyu [1,*] and In-Sung Luke Yeo [2,*]

1 Department of Periodontology, Seoul National University School of Dentistry, Seoul 03080, Korea
2 Department of Prosthodontics, School of Dentistry and Dental Research Institute, Seoul National University, Seoul 03080, Korea
3 Dental Research Institute, Seoul National University, Seoul 03080, Korea
* Correspondence: icrhyu@snu.ac.kr (I.-C.R.); pros53@snu.ac.kr (I.-S.L.Y.); Tel.: +82-2-2072-2641 (I.-C.R.); +82-2-2072-2662 (I.-S.L.Y.)

Received: 7 June 2019; Accepted: 26 June 2019; Published: 28 June 2019

Abstract: Ultraviolet (UV) photofunctionalization has been suggested as an effective method to enhance the osseointegration of titanium surface. In this study, machined surface treated with UV light (M + UV) was compared to sandblasted, large-grit, acid-etched (SLA) surface through in vitro and in vivo studies. Groups of titanium specimens were defined as machined (M), SLA, and M + UV for the disc type, and M + UV and SLA for the implant. The discs and implants were assessed using scanning electron microscopy, confocal laser scanning microscopy, electron spectroscopy for chemical analysis, and the contact angle. Additionally, we evaluated the cell attachment, proliferation assay, and real-time polymerase chain reaction for the MC3T3-E1 cells. In a rabbit tibia model, the implants were examined to evaluate the bone-to-implant contact ratio and the bone area. In the M + UV group, we observed the lower amount of carbon, a 0°-degree contact angle, and enhanced osteogenic cell activities ($p < 0.05$). The histomorphometric analysis showed that a higher bone-to-implant contact ratio was found in the M + UV implant at 10 days ($p < 0.05$). In conclusion, the UV photofunctionalization of a Ti dental implant with M surface attained earlier osseointegration than SLA.

Keywords: dental implants; titanium; osseointegration; photofunctionalization; ultraviolet light; surface treatment

1. Introduction

Dental implant restorations to replace missing teeth have become a routine practice in dental clinics. Using the implants as a prosthesis helps patients feel more comfortable and these implants are more functional compared to the traditional removable dentures [1–3]. For successful implant restorations, osseointegration must be achieved between the bone and the implant. Osseointegration is the direct structural and functional connection between living bone and the surface of a load-carrying implant [4], and it is an essential factor in achieving a successful implant. Generally, it is necessary to wait for several months after implant placement for osseointegration to be achieved [5]. Unsuccessful osseointegration leads to the early failure of implants, meaning that the implants cannot endure masticatory forces, resulting in implant mobility or pain [6,7]. This includes other time-consuming situations, such as an edentulous area with limited bone quantity, or problems in patients with osteoporosis, diabetes, cancer, irradiation, old age, and heavy smokers [8–10].

The original implant surface was a smooth machined surface, with an approximate Sa value of 0.5 μm [11,12]. This machined surface has certain advantages, including a simple manufacturing process (turning and polishing) and the ability to maintain a good hygienic state, resulting in a low incidence of peri-implant disease [12,13]. However, implants with a machined surface have shown a low bone-to-implant contact ratio (BIC) and a frequent failure of osseointegration before loading [14]. To enhance the osseointegration process, various surface modifying techniques have since been developed, where a roughened surface has demonstrated the best clinical long-term results. There are various roughening techniques, although the sandblasted, large-grit, acid-etched (SLA) surface is the most widely used and reported technique. The SLA surface sufficiently differentiates the pre-osteoblastic cells, enhances the osseointegration process, and leads to a higher BIC compared to the machined surface [15,16]. However, the roughened surface has been reported to accelerate plaque accumulation, wherein it is more difficult to remove plaque on the roughened surface than on the machined surface [12]. In this regard, reports show a greater incidence of peri-implant disease stemming from the use of the roughened surface compared to the machined surface [17].

Meanwhile, the photofunctionalization of implants using ultraviolet (UV) light has been highlighted as a simple and effective method to stimulate osseointegration in machined surfaces [18–20]. UV photofunctionalization is a phenomenon of changes occurring in titanium (Ti) surfaces after UV treatment. The process was discovered in 1977, where UV treatment transforms the natural hydrophobic properties of Ti surfaces into superhydrophilic properties by altering the physicochemical properties of the Ti. The process has been applied in environmental engineering and microbiology [21,22]. UV treatment creates the hydrophilic phase on the surface structure, thereby transforming the surface into a hydrophilic surface [23]. Photofunctionalization is also reported to enhance biological capabilities [18,19]. Consequently, the purpose of this study was to evaluate the effect of UV photofunctionalization on implants with a machined surface compared to the SLA surface, using an in vitro and in vivo study.

2. Materials and Methods

2.1. Ti Samples, Surface Analysis, and UV Treatment

2.1.1. Preparation of the Ti Disc and Implant

In this experiment, commercially pure grade 4 Ti was tested in the shape of a disc (10 mm in diameter and 1 mm in thickness) and a screw-shaped implant (3.3 mm in diameter and 7 mm in length). The surface of the specimen was treated using the following methods: (a) M: The machined surface was turned and polished using sandpaper (600–1000 times); (b) SLA: The surface was sandblasted with alumina (Al_2O_3) particles, which were 50 μm in size and acid-etched, using hydrochloric acid and sulfuric acids (SLA surface; Point Implant Co., Seoul, Korea); and (c) M + UV: Machined surface treated with ultraviolet (UV) light. For the disc type, all the three surface treatments were examined for a negative control (M), a positive control (SLA), and an experimental group (M + UV). For the implant type sample, two surface treatments, i.e., the control (SLA) and the experimental group (M + UV), were investigated.

2.1.2. Surface Analysis

Three samples were used in each group for each examination. We performed field emission scanning electron microscopy (FE–SEM; Hitachi S-4700, Hitachi, Tokyo, Japan) for a qualitative evaluation of the overall surface image. This was followed by a confocal laser scanning microscope (CLSM; LSM 800, Carl Zeiss AG, Oberkochen, Germany), where the surface roughness was quantitatively measured. The surface roughness parameters—arithmetical mean height (Sa), root mean square height (Sq), and developed interfacial area ratio (Sdr)—were measured at three randomly selected points in each sample. In addition, the chemical composition was analyzed using electron spectroscopy for chemical analysis (Sigma Probe, Thermo VG, East Grinstead, UK). Furthermore, the surface wettability of the Ti discs was examined using the contact angle from the sessile drop method, as measured by a contact angle meter (Pheonix 150, SEO, Kyunggido, Korea). All the procedures were performed under controlled conditions of 20 °C temperature and 46% humidity.

2.1.3. UV Light Treatment

UV light treatment was achieved by irradiating the Ti discs in a specially manufactured generator using four 15 W bactericidal lamps (G15T8, Sankyo Denki, Tokyo, Japan), for at least 48 h. The intensity was approximately 5 mW/cm^2 (λ = 254 ± 20 nm).

2.2. In Vitro Experiment

2.2.1. Cell Culture

Murine pre-osteoblast MC3T3-E1 cells were purchased from ATCC (American Type Culture Collection; Manassas, VA, USA). The cells were seeded onto the discs (1×10^4 cells/well) in a 12-well culture plate (Nunc, Roskilde, Denmark), and then cultured in α-minimum essential medium (α-MEM; Thermo Fisher Scientific, Waltham, MA, USA) supplemented with a 10% fetal bovine serum (FBS) and 1% penicillin/streptomycin. The cells were incubated at 37 °C under a humidified atmosphere of 95% air and 5% CO_2. The culture medium was replaced every three days, and the osteogenic medium contained 10 mM β-glycerophosphate and 50 µg/mL ascorbic acid in the α-MEM.

2.2.2. Cell Attachment

At 24 h after being seeded, the cell attachment was dual-stained using fluorescent dyes: 4′,6-diamidino-2-phenylindole (DAPI; Invitrogen, Carlsbad, CA, USA) and Alexa Fluor 568 phalloidin (Invitrogen, Carlsbad, CA, USA) to detect the nucleus and actin filaments, respectively. Fluorescence was visualized by a CLSM (LSM 800, Carl Zeiss AG, Oberkochen, Germany), and analyzed with the ZEN2010 software (Carl Zeiss, Oberkochen, Germany).

2.2.3. Cell Proliferation

The proliferative activity of cells was measured using a methyl thiazolyl tetrazolium (MTT) assay (Sigma-Aldrich, St. Louis, MO, USA) at 1, 3, and 7 days after being seeded. The culture media was replaced with an MTT solution and incubated for 3 h at 37 °C. After removing the MTT solution, 0.5 mL of 10% dimethyl sulfoxide in isopropanol (iDMSO) was added for 30 min at 37 °C. Then, the proliferation rate was assessed by its optical density (OD) at 570 nm. The value of the OD was measured using a microplate reader (BioTek, Winooski, VT, USA).

2.2.4. Cell Differentiation

Total RNA in the cell cultures was extracted using the TRIzol method described by Chomczynski at 1, 4, 7, 10, and 14 days after osteoblast differentiation [24]. A reverse transcriptase–polymerase chain reaction (RT–PCR) was performed with primer sets for type I collagen (Col), alkaline phosphatase (Alp), and osteocalcin (Ocn), as described in Table 1. Quantitative real-time PCR was performed using a Takara SYBR premix Ex Taq (Takara Bio, Kusatsu, Japan) on a 7500 real-time PCR system (Applied Biosystems, Foster City, CA, USA). Each primer contained a final concentration of 200 nM, and a quantity of cDNA corresponding to 50 ng of total RNA. The PCR primers were synthesized using Integrated DNA Technology (Coralville, IA, USA). According to the manufacturer's instructions, the PCR cycling conditions comprised 40 cycles at 95 °C for 5 s, and 60 °C for 34 s after denaturation at 95 °C for 30 s. The cycle threshold (Ct) values were acquired using the automated threshold analysis in the Sequence Detection software version 1.4 (Applied Biosystems, Foster, CA USA). Each target mRNA expression was calculated using the comparative cycle threshold method according to the manufacturer's instructions. The relative mRNA expression levels were normalized to glyceraldehyde-3-phosphate dehydrogenase (GAPDH). The GAPDH mRNA expression levels remained steady during the osteoblast differentiation, showing similar Ct values.

Table 1. Primer sequences for the reverse transcriptase–polymerase chain reaction (RT–PCR).

Gene	Forward Primer (5′-3′)	Reverse Primer (5′-3′)
Col [1]	GCTCCTCTTAGGGGCCACT	CCACGTCTCACCATTGGGG
Alp [2]	GGCTACATTGGTCTTGAGCTTTT	CCAACTCTTTTGTGCCAGAGA
Ocn [3]	CTGACAAAGCCTTCATGTCCAA	GCGCCGGAGTCTGTTCACTA

[1] Type I collagen; [2] Alkaline phosphatase; [3] Osteocalcin.

2.3. In Vivo Experiment

2.3.1. Animals

The rabbit tibia model was used. All the procedures were conducted with the approval of the Ethics Committee of Animal Experimentation of the Institutional Animal Care and Use Committee (CRONEX-IACUC 201803003; Cronex, Hwasung, Korea), according to the guidelines of Animal Research: Reporting In Vivo Experiments (ARRIVE).

Thereafter, four female New Zealand white rabbits (3–4 months old and 2.5–3.0 kg in weight) were anesthetized via a 1 mL intramuscular injection with a dose of 15 mg/kg of tiletamine/zolazepam (Zoletil 50, Vibrac Korea, Seoul, Korea) and 5 mg/kg of xylazine (Rompun, Bayer Korea, Seoul, Korea). Then the tibiae of the rabbits were shaved and disinfected with povidone iodine solution. Local anesthesia was administered in the surgical area with 2% lidocaine containing 1:100,000 epinephrine (Yuhan Co., Seoul, Korea).

2.3.2. Surgical Procedure

A full-thickness flap was made on the medial side of both tibiae, followed by exposure of the underlying bone. In each tibia, two holes for implant placement were drilled bicortically using implant surgical drills under copious sterile saline irrigation. The diameter of the drills was increased sequentially, with a final drill diameter of 2.8 mm. Then, the implant with a diameter of 3.3 mm was inserted into the hole and engaged bicortically with sufficient stability. The SLA and M + UV implants were allocated to each hole based on a 2 × 2 Latin square randomization. Following the implant placement, the periosteum and fascia were sutured with 4-0 resorbable polyglactin material (Vicryl, Ethicon, Somerville, MA, USA), and the skin was sutured with 4-0 monofilament nylon (Blue nylon, Ailee, Busan, Korea). Post-operatively, each rabbit was kept in a separate cage and administered with 5 mg/kg of enrofloxacin (Komibiotril, Komipharm International Co., Siheung, Korea) for seven days.

2.3.3. Sacrifice and Microcomputed Tomography (Micro-CT)

Two experimental animals were sacrificed at 10 days and the other two animals at 28 days after the surgery by an intravenous overdose of potassium chloride. The implants were surgically harvested en bloc with the surrounding bone. Then the implant–bone blocks were immediately immersed in a 10% neutral buffered formalin fixative. Micro-CT imaging was performed using a SkyScan 1275 (Bruker microCT, Kontich, Belgium). The X-ray source was set at an acceleration voltage of 100 kV and a pixel size of 10 μm. Each sample was scanned three times, using 360° spiral scanning on the SkyScan 1275 with a scanning time of 2 h. Reconstruction was performed using an NRecon (v. 1.7.3.2, Bruker microCT). The region-of-interest (ROI) was defined as the area within consecutive threads engaged in the upper cortical bone (Figure 1a). The analysis was performed using the CTAn software (v. 1.18.4.0, Bruker microCT; Figure 1b), and it also involved the visualization software DataViewer (v. 1.5.4.0, Bruker microCT) and CTVox (v. 3.3.0, Bruker microCT).

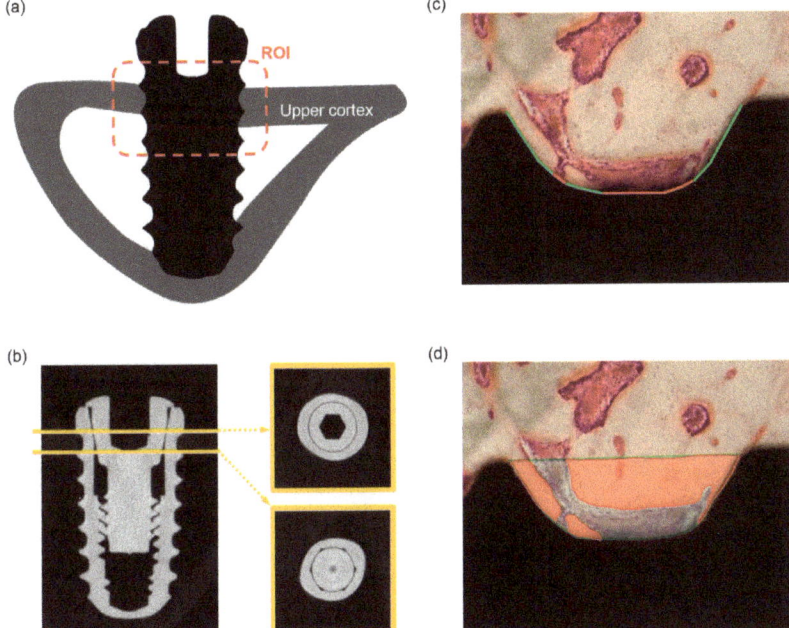

Figure 1. (a) The schematic drawing of the region-of-interest (ROI). The ROI was defined as an area within threads engaged in the upper cortical bone (red dot box); (b) microcomputed tomography (micro-CT) images for the measurement of the bone-to-implant contact ratio (BIC); (c) the bone-to-implant contact ratio (BIC) was calculated by the length of the green line divided by the total length of the well (green and red line); (d) the definition of the bone area (BA) was calculated by the area of red color divided by the total area of the well.

2.3.4. Histological Preparation and Histomorphometric Measurement

After μCT scanning, the implant–bone blocks were dehydrated in a series of ethanol with increasing concentrations and then embedded in light-curing resin (Technovit 7200 VLC, Hereaus Kulzer, Hanau, Germany). The embedded blocks were sectioned perpendicular to the longitudinal axis of the implant using the EXAKT system (EXAKT Apparatebau, Norderstedt, Germany), following the method described by Donath and Breuner [25]. The section was then ground to a thickness of 40 μm and stained with hematoxylin and eosin (H&E) for examination using a light microscope. These undecalcified, ground sections were observed under a light microscope (BX51, Olympus, Tokyo, Japan) to measure the BIC and bone area (BA; Figure 1c,d). The region-of-interest (ROI) was defined as the same area that was used in the micro-CT analysis. The measurement was performed under a ×100 magnification, using the SPOT software version 4.0 (Diagnostic Instruments, Sterling Heights, MI, USA) and Image-Pro Plus (Media Cybernetics, Rockville, MD, USA).

2.4. Statistical Analysis

The Kruskal–Wallis test was used to evaluate statistically significant differences amongst the three groups of discs. If there was a significant difference amongst the three groups, the post-hoc Tukey method was applied. To compare the two groups of implants, the Mann–Whitney U test was performed to determine the statistically significant differences. $p < 0.05$ was set as the statistical significance. All statistical analyses were performed using SPSS 20.0 (IBM Corp., Armonk, NY, USA).

3. Results

3.1. Surface Characteristics

In the overall evaluation of the Ti samples via the FE–SEM images, the M and M + UV surfaces showed similar evidence of machine turning (continuous straight traces) with smooth surfaces, although the SLA surface presented a rougher surface with a typical honeycomb appearance (Figures 2a and 3a).

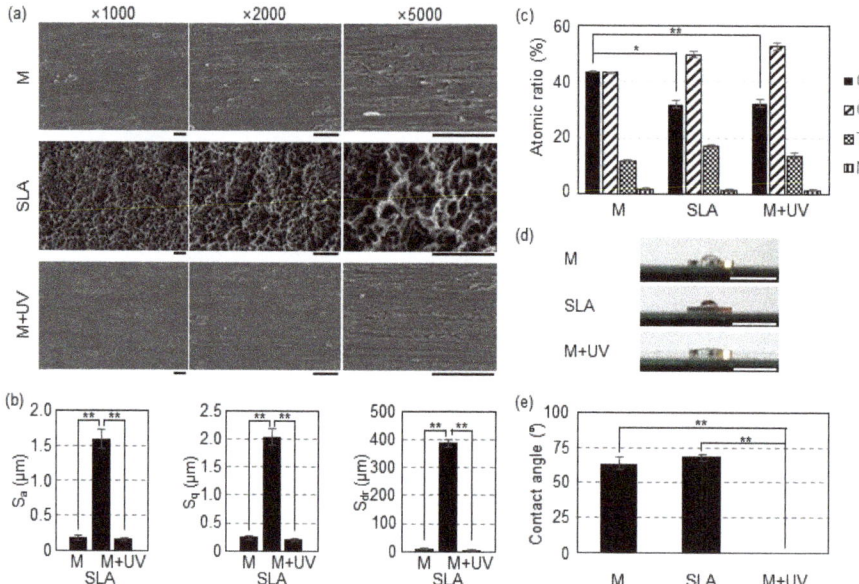

Figure 2. (a) Scanning electron microscopy (SEM) images of the Ti discs. M: Machined surface (top row); SLA: Sandblasted with large grit and acid etched surface (middle row); M + UV: machined surface treated with ultraviolet light (bottom row). Scale bars: 10 μm. (b) Surface roughness (Sa and Sq) and surface area ratio (Sdr) of the Ti discs evaluated by confocal laser scanning microscopy (CLSM) analysis. The values for the M and M + UV surfaces are similar and smaller than the SLA. (c) Element content of the surfaces of the Ti disc according to the energy dispersive x-ray spectrometer (XPS). M + UV shows a significantly lower carbon percentage than the M disc, but there is no significant difference between the SLA and M + UV. (d) The changes in wettability of the Ti discs. Superhydrophilicity after UV treatment for 48 h was observed. Scale bars: 10 mm. (e) The value of the contact angle of the Ti discs. Without UV treatment, the Ti disc was hydrophobic, but became superhydrophilic (zero degree angle) after UV treatment for 48 h. Error bars show the standard deviation. (*) and (**) represents significance compared with each pair, $p < 0.05$ and $p < 0.01$, respectively.

Surface roughness parameters for the samples are shown in Figures 2b and 3b. In the disc specimen, the SLA surface showed higher Sa, Sq, and Sdr values, and these were significantly different from the M and M + UV surfaces ($p < 0.01$). There was no statistical difference between the M and M + UV surfaces ($p > 0.05$). Similarly, for the implant specimen, the M + UV and SLA surfaces were statistically different across all the parameters ($p < 0.05$).

Chemical compositions from the x-ray spectrometer (XPS) revealed that, compared with the 43.42% ± 0.31% carbon in the M disc, the M + UV disc contained about 32.35% ± 1.50% carbon, which was statistically different ($p = 0.000$). The SLA showed 31.87% ± 1.42% carbon, with no significant difference from the M + UV disc ($p = 0.051$). On the other hand, the M + UV implant showed a significantly lower carbon percentage compared to the SLA implant ($p = 0.049$; Figures 2c and 3c).

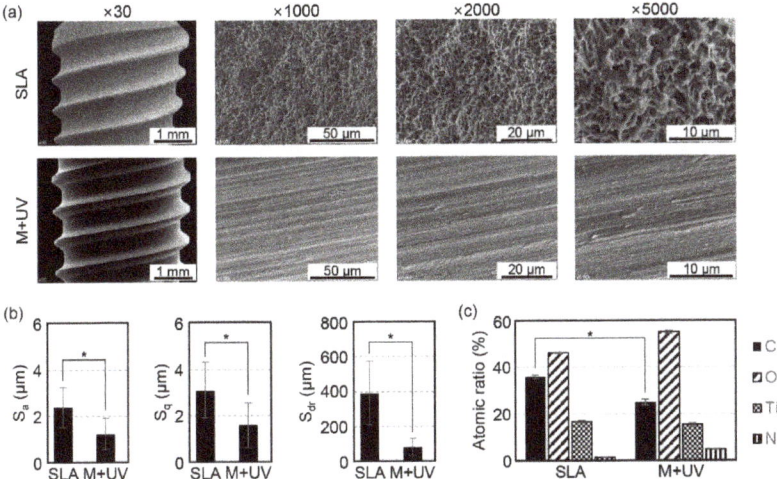

Figure 3. (a) Scanning electron microscopy (SEM) images of the Ti implants. SLA: Sandblasted with large grit and acid etched surface (top row); M + UV: Machined surface treated with ultraviolet light (bottom row). Magnification: ×30, ×1000, ×2000, and ×5000 from the left. (b) Surface roughness (Sa and Sq) and surface area ratio (Sdr) of the Ti discs according to the confocal laser scanning microscopy (CLSM) analysis. The values of the M + UV surfaces are smaller than the SLA, which means it is smoother than the SLA. (c) Element content of the surfaces of the Ti discs according to the energy dispersive x-ray spectrometer (XPS). The M + UV discs contain half of the carbon percentage of the SLA discs. Error bars show the standard deviation. (*) represents the significance compared with each pair, $p < 0.05$.

Contact angle measurement was performed for the disc-type specimens. The M and SLA discs showed a hydrophobic status with angles of 63.6 ± 4.7° and 68.3 ± 2.5°, respectively. On the other hand, 48 h after UV treatment, the M + UV discs showed superhydrophilicity at a 0° contact angle ($p = 0.000$; Figure 2d,e).

3.2. In Vitro Test

3.2.1. Cell Attachment

The CLSM images of the cells are shown in Figure 4a. A wider spread of cells was observed in the M + UV groups compared to the other groups. In the SLA discs, the cells were sharp and needle-like shaped, implying that the osteoblasts were not prone to attach to the SLA surfaces.

3.2.2. Cell Proliferation

The MTT assay showed that the amount of cells increased in a time-dependent way on all the surfaces. On the M + UV surface, the cells proliferated more significantly than the other surfaces at days 1, 3, and 7, with a p-value of less than 0.01, as shown in Figure 4b. On day 7, the amount of cells on the M + UV surface was two times greater than cells on the SLA surface (0.068 ± 0.0005 vs. 0.040 ± 0.001, $p = 0.000$). In particular, at day 3 and 7, the cells proliferated more on the M surface than on the SLA surface.

3.2.3. Quantitative Assessment of the Osteogenic Markers

Figure 4c shows the relative mRNA expression of Col, Alp, and Ocn. The RT–PCR analysis showed that Col was more significantly expressed on the M + UV and SLA surfaces at days 7, 10, and 14 compared to the M surface, although the M + UV and SLA surfaces were not significantly different.

The expression level of Alp on the SLA surface was not different from the M + UV surface at days 1 and 4, but it was significantly higher than the expression levels at days 7, 10, and 14. The M + UV surface expressed the Alp gene more than the M surface at days 4, 7, and 14. The expression level of Ocn on the M + UV surface was significantly higher at day 7, but it was lower at days 10 and 14.

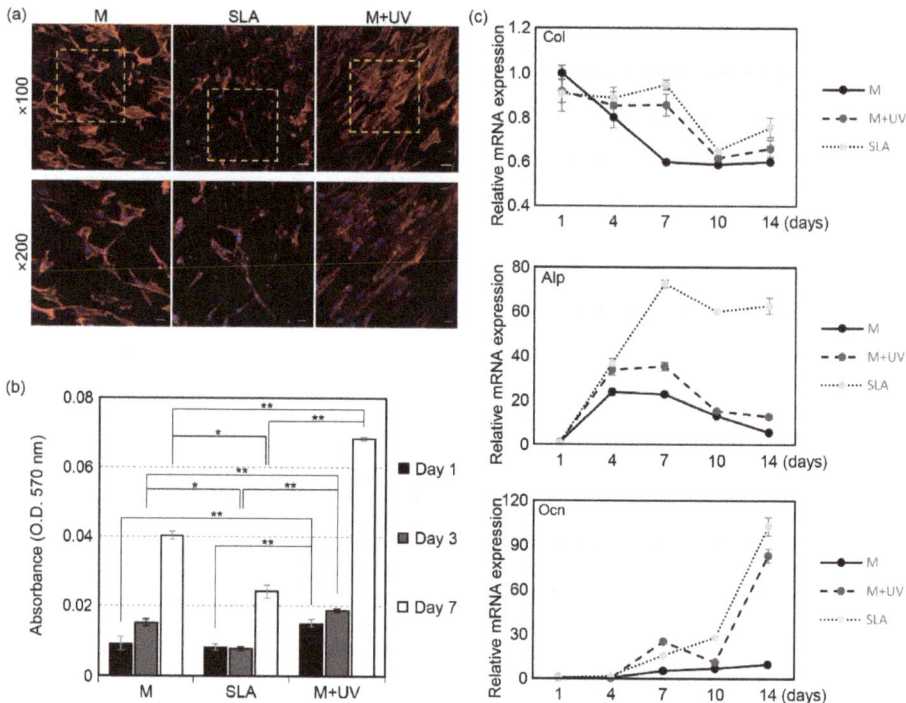

Figure 4. (**a**) Confocal microscopic images of the MC3T3-E1 cells, 24 h after being seeded on the Ti discs. The areas in the dotted box are magnified in the bottom row. Scale bars: 50 μm at ×100 and 20 μm at ×200 magnification. (**b**) Evaluation of cell proliferation of the MC3T3-E1 cells by an MTT assay at 1, 3, and 7 days after being seeded on the Ti discs. (**c**) Evaluation of the cell differentiation of MC3T3-E1 cells by real-time PCR at 1, 4, 7, 10, and 14 days after being seeded on the Ti discs. The relative mRNA expression levels were normalized to glyceraldehyde-3-phosphate dehydrogenase (GAPDH). The osteogenic markers are type I collagen (top), alkaline phosphatase (ALP, middle), and osteocalcin (OCN, bottom). UV photofunctionalization enhanced the osteoblastic gene expression. Error bars show the standard deviation. (*) and (**) represent the significance compared with each pair, $p < 0.05$ and $p < 0.01$, respectively.

3.3. In Vivo Test

3.3.1. Histomorphometry

All the implants were successfully osseointegrated at days 10 and 28 (Figure 5a). At day 10, the BIC ratios of the M + UV implants (55.93% ± 6.19%) were significantly higher than that of the SLA implants (43.38% ± 3.20%, $p = 0.021$). However, at day 28, the BIC ratios of the M + UV implants (64.88% ± 5.35%) were not significantly different from that of the SLA implants (59.93% ± 6.44%, $p = 0.149$; Figure 5b).

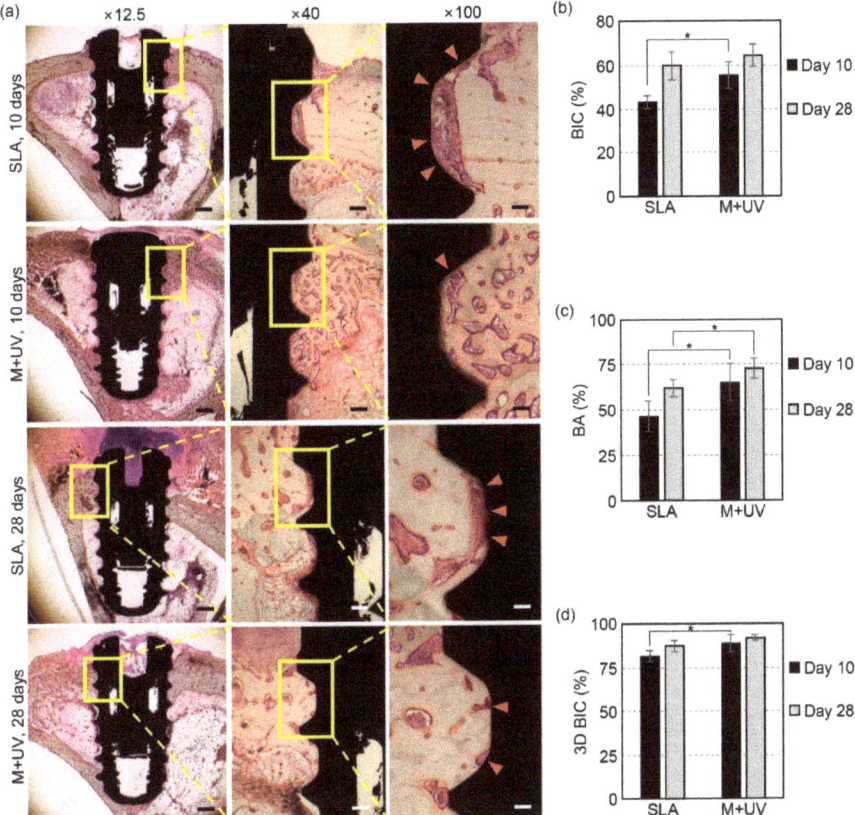

Figure 5. (**a**) Representative histologic sections of the rabbit tibia at 10 and 28 days after the implant placement. In the SLA implant, the osteoblast and organic matrix, which had not mineralized yet, was more observable on the interface between the bone and implant compared to the M + UV implant (red arrow head; magnification ×12.5, ×40, and ×100 from the left, hematoxylin and eosin staining). The scale bars: 1 mm at ×12.5, 200 μm at ×40, and 100 μm at ×100 magnification. (**b**) The bone-to-implant contact ratio (BIC) was evaluated histologically at days 10 and 28. The M + UV implant shows a significantly higher BIC than the SLA at day 10, but there is no significant difference at day 28. (**c**) The bone area ratio (BA) evaluated histologically at 10 and 28 days. The M + UV implants show significantly more BA than the SLA at days 10 and 28. (**d**) The bone-to-implant contact ratio evaluated by micro-CT (3D BIC) at days 10 and 28. The M + UV implants show a significantly higher 3D BIC than the SLA at day 10, but there is no significant difference at day 28. Error bars show the standard deviation. (*) represents the significance compared with each pair, $p < 0.05$.

In terms of the BA, the M + UV implants were significantly higher compared to the SLA implants (46.55% ± 8.59%) at day 10 (65.09% ± 10.42% vs. 46.55% ± 8.59%, $p = 0.042$) and at day 28 (72.70% ± 5.52% vs. 61.83% ± 4.89%, $p = 0.043$; Figure 5c).

3.3.2. Micro-CT

The three-dimensional BIC was evaluated using micro-CT. The micro-CT analysis revealed that the three-dimensional BIC of the M + UV implants was significantly higher than that of the SLA implants at day 10 (88.87% ± 5.1% vs. 81.6% ± 3.28%, $p = 0.046$), but it was not statistically different at day 28 (91.91% ± 1.55% vs. 87.47% ± 2.93%, $p = 0.201$; Figure 5d).

4. Discussion

In this study, we found that UV photofunctionalization on a Ti screw-shaped implant with an M + UV surface showed a higher BIC than the SLA surface at day 10, and there was no significant difference at day 28. This was confirmed in both two-dimensional and three-dimensional measurements. The results indicated that the UV photofunctionalization could accelerate the osseointegration process, and achieve firm fixation between the implant and the surrounding bone earlier. These findings are supported by other studies, where Park et al. found that after four months of healing, the UV-treated implants in rabbits showed a higher BIC than the untreated implants. The authors observed that UV treatment decreased both carbon impurities on the surface and water contact angles [26]. Similarly, Aita et al. showed that the UV-treated acid-etched implants at week two had a push-in value equivalent to the untreated acid-etched implant [18]. Pyo et al. measured the removal torque test in UV-treated implants and showed that it was 50% higher than in untreated implants [27]. Hirota et al. retrospectively studied and found that the use of photofunctionalization reduced the risk of early implant failure with an odds ratio of 0.30 ($p < 0.05$) [28]. Soltanzadeh studied the effect of UV photofunctionalization on immediately loaded implants in a rat model. After the placement, the implants were immediately loaded with 0.46 N of static lateral force. The results showed that osseointegration was successful in 100% of photofunctionalized implants, but 28.6% of untreated ones. The value of the push-in test was 2.4 times higher in photofunctionalized implants [29].

Histologically, the BA was significantly higher at days 10 and 28 in the M + UV compared to the SLA implants, meaning that the M + UV implant had a higher amount of mineralized bones between threads of implants. Pyo et al. evaluated the osteogenic dynamics using fluorescent labeling at four weeks after implant placement; and found that in the UV-treated implant, the interfacial areas between the bone and implant and the areas within the threads were filled with calcein-positive tissues compared to the untreated implants. This meant that UV photofunctionalization could lead to earlier bone deposition [27]. Ueno et al. showed that the UV-treated acid-etched implant had a marked bone formation in a gap healing model without cortical support [20]. Kitajima et al. measured the implant stability quotients (ISQ) for 55 photofunctionalized implants with low and extremely low initial stability at the time of placement and stage-two surgery. Then they calculated the ISQ increase per month, defining the osseointegration speed index (OSI). The OSI ranged 3.9–4.7 substantially higher than the OSIs for untreated implants reported in other literatures (0.36–2.8) [30]. Ijishima et al. evaluated the effect of photofunctionalization on aged rats. The aged rats showed considerably lower biological capabilities (cell attachment, proliferation, and ALP activity) than the young. However, the enhancement of cell attachment and differentiation were observed on the photofunctionalized Ti discs compared with untreated one. Moreover, in the femurs of aged rats, the photofunctionalized mini-implant showed the higher push-in value than untreated one after two weeks of healing. These findings supported that UV photofunctionalization could be also valuable in the compromised sites [31].

Generally, the surface roughness is considered as a main factor for the improvement of osseointegration. However, in this study, UV treatment did not physically change any surface roughness as shown in the SEM and CLSM. Rather, it induced superhydrophilicity (0° angle), reduced the percentage of hydrocarbons, and increased the osteoblast proliferation, attachment, and differentiation, as shown in the in vitro study. This indicated that the only physico-chemical changes in the Ti surface could enhance the biological activities. In the XPS analysis, Roy et al. found that UVC photon energy decreased carbon deposition and the amount of H_2O on Ti surface, and produced many –OH groups (TiOH) without any changes in surface topography. They explained that, through these chemical changes, the UV photofunctionalization could create the superhydrophilicity of Ti. The improvement of biological capabilities by UV photofunctionalization was supported by other studies [32]. Aita et al. showed that Col and osteopontin (Opn) were more expressed in the UV-treated discs [18]. The RT–PCR analysis performed by Zhang et al. showed that the expression of genes encoding Col, Runx2, BMP, and Opn increased in the UV-treated surface [33]. In contrast, Att et al. assessed the RT–PCR of genes for Opn and Ocn in bone marrow cells derived from the femur of eight-week-old male Sprague-Dawley rats, and found that there was no significant difference at days 10 and 20. The differences may have been caused by the kinds of cells, time points, and intensity and wavelength of the UV generator. Further research is required to elucidate this aspect.

With regard to plaque accumulation and peri-implant disease, the implant with a smooth surface is considered to be superior to an implant with a rough surface. Berglundh et al. observed that, at five months after the removal of ligature, bone loss accelerated in the SLA implant but not in the polished implant. Histologically, the size of the inflammatory lesion and the area of plaque were larger in the SLA surface [12]. Additionally, Albouy et al. compared the turned and the roughened implant (Ti-Unite), at six months after the ligature removal, and observed a larger amount of bone loss in the Ti-Unite implant compared to the turned implant (1.47 mm vs. 0.3 mm). This meant that spontaneous progression of peri-implantitis had occurred in the implant with a rough surface [17]. However, the machined implant had a definite drawback in that it had a low BIC level, because the osteoblastic differentiation was lower compared to the smooth surface [34,35]. Therefore, the enhancement of osteoblastic differentiation on the Ti with an M surface by UV photofunctionalization is considered to be inspired. Additionally, UV photofunctionalization itself could decrease plaque formation on Ti surface. De Avila et al. found that after 16 h incubation, there were significantly lower oral bacterial attachment on the UV-treated Ti disc compared to the untreated one [36].

The hydrophilicity of the implant surface can be induced by UV photofunctionalization [18–20,27,37] or preservation in a storage medium [38–40]. Both methods are effective and can improve the bone healing process and attain early osseointegration. However, the latter method has been reported to lead to foreign deposition and little elimination of hydrocarbons on the Ti surface. Moreover, the saline storage method is inferior to UV photofunctionalization in osteoblast spreading and adhesion [41]. Considering this point, UV treatment is considered a safer method to modify the implant surface to make it hydrophilic. On the other hand, Att et al. mentioned that, in UV photofunctionalization, the superhydrophilicity is not a significant factor in explaining the higher BIC in the UV-treated Ti discs compared to the acid-etched ones. The elimination of hydrocarbon on the surface was considered to be a significant factor [19]. The aging of the Ti is related to the contamination and accumulation of the hydrocarbon on the Ti surface, and it can suppress cell recruitment and biologic activity [42,43].

The combination of variables such as duration, intensity, and wavelength can create various modes of UV photofunctionalization, noting that the optimal combination is a controversial issue. The exposure time has been used from 12 min to 48 h [44,45]. However, Aita et al. and Att et al. have shown that, between 24 and 48 h, there was an increase in hydrophilicity and biological effects [18,19]. Additionally, treatment with UVC ($\lambda = 240 \pm 40$ nm) has shown more biological improvements compared to UVA ($\lambda = 360 \pm 40$ nm). Consequently, in our experiment, to maximize the effects of UV, the mode of UV photofunctionalization was determined as UVC treatment for 48 h.

There are still several questions regarding UV photofunctionalization. Strictly, the UV light treatment used in this study may be called physical photo-activation, rather than functionalization, because no chemical application to the surface, which have been shown in the previous studies, were used for enhanced bone response [11,46]. More obvious concept of the UV surface treatment needs to be established in the physical and chemical aspects. Amongst several factors following UV photofunctionalization, we also still need to identify the main factors for the enhancement of biological activity, superhydrophilicity, and removal of hydrocarbon. If they contribute to the improvement, there is a need to understand the mechanism through which they are inter-connected. Therefore, further studies are needed to fully appreciate the effects of UV photofunctionalization.

5. Conclusions

Within the limitations of the present study, UV photofunctionalization of a Ti dental implant with an M surface attained an earlier osseointegration compared to an implant with an SLA surface. The enhancement was considered to result from the superhydrophilicity, the elimination of hydrocarbon on the surface, and the improvement of osteoblastic activities.

Author Contributions: Conceptualization, J.-B.L., I.-C.R., I.-S.L.Y.; methodology, J.-B.L., Y.-H.C., J.-Y.C., I.-S.L.Y.; software, J.-B.L., Y.-H.C.; validation, I.-C.R., I.-S.L.Y.; formal analysis, J.-B.L., I.-S.L.Y.; investigation, J.-B.L., Y.-H.C., J.-Y.C.; resources, J.-Y.C., Y.-J.S., I.-C.R., I.-S.L.Y.; data curation, J.-B.L., J.-Y.C.; writing—original draft preparation, J.-B.L.; writing—review and editing, Y.-H.C., J.-Y.C., Y.-J.S., Y.-M.L., Y.K., I.-C.R., I.-S.L.Y.; visualization, J.-B.L., J.-Y.C.; supervision, I.-C.R., I.-S.L.Y.; project administration, I.-S.L.Y.; funding acquisition, I.-S.L.Y.

Funding: This work was supported by a National Research Foundation of Korea (NRF) grant funded by the Ministry of Science and ICT [NRF-2016R1A2B4014330] and by a Korea Health Industry Development Institute (KHIDI) grant funded by the Ministry of Health & Welfare, Republic of Korea [HI15C1535].

Acknowledgments: The authors greatly thank Kyoung-Hwa Kim and Young-Dan Cho for the material preparation and support.

Conflicts of Interest: The authors declare no potential conflict of interest related to this study. The funders had no role in the design of the study; in the collection, analyses, or interpretation of the data; in the writing of the manuscript, or in the decision to publish the results.

References

1. Henry, P.J. Oral implant restoration for enhanced oral function. *Clin. Exp. Pharmacol. Physiol.* **2005**, *32*, 123–127. [CrossRef] [PubMed]
2. Srinivasan, M.; Meyer, S.; Mombelli, A.; Müller, F. Dental implants in the elderly population: A systematic review and meta-analysis. *Clin. Oral Implant. Res.* **2017**, *28*, 920–930. [CrossRef] [PubMed]
3. De Angelis, F.; Papi, P.; Mencio, F.; Rosella, D.; Di Carlo, S.; Pompa, G. Implant survival and success rates in patients with risk factors: Results from a long-term retrospective study with a 10 to 18 years follow-up. *Eur. Rev. Med. Pharmacol. Sci.* **2017**, *21*, 433–437. [PubMed]
4. Albrektsson, T.; Branemark, P.I.; Hansson, H.A.; Lindstrom, J. Osseointegrated titanium implants. Requirements for ensuring a long-lasting, direct bone-to-implant anchorage in man. *Acta Orthop. Scand.* **1981**, *52*, 155–170. [CrossRef]
5. Gallucci, G.O.; Hamilton, A.; Zhou, W.; Buser, D.; Chen, S. Implant placement and loading protocols in partially edentulous patients: A systematic review. *Clin. Oral Implant. Res.* **2018**, *29*, 106–134. [CrossRef]
6. Chrcanovic, B.R.; Kisch, J.; Albrektsson, T.; Wennerberg, A. Factors Influencing Early Dental Implant Failures. *J. Dent. Res.* **2016**, *95*, 995–1002. [CrossRef] [PubMed]
7. Manzano, G.; Montero, J.; Martin-Vallejo, J.; Del Fabbro, M.; Bravo, M.; Testori, T. Risk Factors in Early Implant Failure: A Meta-Analysis. *Implant Dent.* **2016**, *25*, 272–280. [CrossRef] [PubMed]
8. Diz, P.; Scully, C.; Sanz, M. Dental implants in the medically compromised patient. *J. Dent.* **2013**, *41*, 195–206. [CrossRef]
9. Chrcanovic, B.R.; Albrektsson, T.; Wennerberg, A. Bone Quality and Quantity and Dental Implant Failure: A Systematic Review and Meta-analysis. *Int. J. Prosthodont.* **2017**, *30*, 219–237. [CrossRef]
10. Neves, J.; de Araujo Nobre, M.; Oliveira, P.; Martins Dos Santos, J.; Malo, P. Risk Factors for Implant Failure and Peri-Implant Pathology in Systemic Compromised Patients. *J. Prosthodont.* **2018**, *27*, 409–415. [CrossRef]
11. Albrektsson, T.; Wennerberg, A. Oral implant surfaces: Part 1—Review focusing on topographic and chemical properties of different surfaces and in vivo responses to them. *Int. J. Prosthodont.* **2004**, *17*, 536–543. [PubMed]
12. Berglundh, T.; Gotfredsen, K.; Zitzmann, N.U.; Lang, N.P.; Lindhe, J. Spontaneous progression of ligature induced peri-implantitis at implants with different surface roughness: An experimental study in dogs. *Clin. Oral Implant. Res.* **2007**, *18*, 655–661. [CrossRef] [PubMed]
13. Albouy, J.P.; Abrahamsson, I.; Persson, L.G.; Berglundh, T. Spontaneous progression of peri-implantitis at different types of implants. An experimental study in dogs. I: Clinical and radiographic observations. *Clin. Oral Implant. Res.* **2008**, *19*, 997–1002. [CrossRef] [PubMed]
14. Rasmusson, L.; Kahnberg, K.E.; Tan, A. Effects of implant design and surface on bone regeneration and implant stability: An experimental study in the dog mandible. *Clin. Implant Dent. Relat. Res.* **2001**, *3*, 2–8. [CrossRef] [PubMed]
15. Wennerberg, A.; Albrektsson, T. Effects of titanium surface topography on bone integration: A systematic review. *Clin. Oral Implant. Res.* **2009**, *20*, 172–184. [CrossRef] [PubMed]
16. Feller, L.; Jadwat, Y.; Khammissa, R.A.; Meyerov, R.; Schechter, I.; Lemmer, J. Cellular responses evoked by different surface characteristics of intraosseous titanium implants. *Biomed. Res. Int.* **2015**, *2015*, 171945. [CrossRef]

17. Albouy, J.P.; Abrahamsson, I.; Berglundh, T. Spontaneous progression of experimental peri-implantitis at implants with different surface characteristics: An experimental study in dogs. *J. Clin. Periodontol.* **2012**, *39*, 182–187. [CrossRef]
18. Aita, H.; Hori, N.; Takeuchi, M.; Suzuki, T.; Yamada, M.; Anpo, M.; Ogawa, T. The effect of ultraviolet functionalization of titanium on integration with bone. *Biomaterials* **2009**, *30*, 1015–1025. [CrossRef]
19. Att, W.; Hori, N.; Iwasa, F.; Yamada, M.; Ueno, T.; Ogawa, T. The effect of UV-photofunctionalization on the time-related bioactivity of titanium and chromium-cobalt alloys. *Biomaterials* **2009**, *30*, 4268–4276. [CrossRef]
20. Ueno, T.; Yamada, M.; Suzuki, T.; Minamikawa, H.; Sato, N.; Hori, N.; Takeuchi, K.; Hattori, M.; Ogawa, T. Enhancement of bone-titanium integration profile with UV-photofunctionalized titanium in a gap healing model. *Biomaterials* **2010**, *31*, 1546–1557. [CrossRef]
21. Keleher, J.; Bashant, J.; Heldt, N.; Johnson, L.; Li, Y. Photo-catalytic preparation of silver-coated TiO_2 particles for antibacterial applications. *World J. Microbiol. Biotechnol.* **2002**, *18*, 133–139. [CrossRef]
22. Nakashima, T.; Ohko, Y.; Kubota, Y.; Fujishima, A. Photocatalytic decomposition of estrogens in aquatic environment by reciprocating immersion of TiO_2-modified polytetrafluoroethylene mesh sheets. *J. Photochem. Photobiol. A Chem.* **2003**, *160*, 115–120. [CrossRef]
23. Wang, R.; Hashimoto, K.; Fujishima, A.; Chikuni, M.; Kojima, E.; Kitamura, A.; Shimohigoshi, M.; Watanabe, T. Light-induced amphiphilic surfaces. *Nature* **1997**, *388*, 431. [CrossRef]
24. Chomczynski, P.; Mackey, K. Short technical reports. Modification of the TRI reagent procedure for isolation of RNA from polysaccharide-and proteoglycan-rich sources. *Biotechniques* **1995**, *19*, 942–945. [PubMed]
25. Donath, K.; Breuner, G. A method for the study of undecalcified bones and teeth with attached soft tissues * The Säge-Schliff (sawing and grinding) Technique. *J. Oral Pathol.* **1982**, *11*, 318–326. [CrossRef]
26. Park, K.H.; Koak, J.Y.; Kim, S.K.; Han, C.H.; Heo, S.J. The effect of ultraviolet-C irradiation via a bactericidal ultraviolet sterilizer on an anodized titanium implant: A study in rabbits. *Int. J. Oral Maxillofac. Implant.* **2013**, *28*, 57–66. [CrossRef] [PubMed]
27. Pyo, S.W.; Park, Y.B.; Moon, H.S.; Lee, J.H.; Ogawa, T. Photofunctionalization enhances bone-implant contact, dynamics of interfacial osteogenesis, marginal bone seal, and removal torque value of implants: A dog jawbone study. *Implant Dent.* **2013**, *22*, 666–675. [CrossRef] [PubMed]
28. Hirota, M.; Ozawa, T.; Iwai, T.; Ogawa, T.; Tohnai, I. Effect of Photofunctionalization on Early Implant Failure. *Int. J. Oral Maxillofac. Implant.* **2018**, *33*, 1098–1102. [CrossRef]
29. Soltanzadeh, P.; Ghassemi, A.; Ishijima, M.; Tanaka, M.; Park, W.; Iwasaki, C.; Hirota, M.; Ogawa, T. Success rate and strength of osseointegration of immediately loaded UV-photofunctionalized implants in a rat model. *J. Prosthet. Dent.* **2017**, *118*, 357–362. [CrossRef]
30. Kitajima, H.; Ogawa, T. The Use of Photofunctionalized Implants for Low or Extremely Low Primary Stability Cases. . *Int. J. Oral Maxillofac. Implant.* **2016**, *31*, 439–447. [CrossRef]
31. Ishijima, M.; Ghassemi, A.; Soltanzadeh, P.; Tanaka, M.; Nakhaei, K.; Park, W.; Hirota, M.; Tsukimura, N.; Ogawa, T. Effect of UV Photofunctionalization on Osseointegration in Aged Rats. *Implant Dent.* **2016**, *25*, 744–750. [CrossRef] [PubMed]
32. Roy, M.; Pompella, A.; Kubacki, J.; Szade, J.; Roy, R.A.; Hedzelek, W. Photofunctionalization of Titanium: An Alternative Explanation of Its Chemical-Physical Mechanism. *PLoS ONE* **2016**, *11*, e0157481. [CrossRef] [PubMed]
33. Zhang, H.; Komasa, S.; Mashimo, C.; Sekino, T.; Okazaki, J. Effect of ultraviolet treatment on bacterial attachment and osteogenic activity to alkali-treated titanium with nanonetwork structures. *Int. J. Nanomed.* **2017**, *12*, 4633–4646. [CrossRef] [PubMed]
34. Bowers, K.T.; Keller, J.C.; Randolph, B.A.; Wick, D.G.; Michaels, C.M. Optimization of surface micromorphology for enhanced osteoblast responses in vitro. *Int. J. Oral Maxillofac. Implant.* **1992**, *7*, 302–310. [CrossRef]
35. Keller, J.C.; Schneider, G.B.; Stanford, C.M.; Kellogg, B. Effects of implant microtopography on osteoblast cell attachment. *Implant Dent.* **2003**, *12*, 175–181. [CrossRef] [PubMed]
36. De Avila, E.D.; Lima, B.P.; Sekiya, T.; Torii, Y.; Ogawa, T.; Shi, W.; Lux, R. Effect of UV-photofunctionalization on oral bacterial attachment and biofilm formation to titanium implant material. *Biomaterials* **2015**, *67*, 84–92. [CrossRef] [PubMed]
37. Sawase, T.; Jimbo, R.; Baba, K.; Shibata, Y.; Ikeda, T.; Atsuta, M. Photo-induced hydrophilicity enhances initial cell behavior and early bone apposition. *Clin. Oral Implant. Res.* **2008**, *19*, 491–496. [CrossRef] [PubMed]

38. Buser, D.; Broggini, N.; Wieland, M.; Schenk, R.; Denzer, A.; Cochran, D.L.; Hoffmann, B.; Lussi, A.; Steinemann, S. Enhanced bone apposition to a chemically modified SLA titanium surface. *J. Dent. Res.* **2004**, *83*, 529–533. [CrossRef] [PubMed]
39. Ehlers, H.; Jacobs, F.; Kloppers, H.; Postma, T. The influence of storage media on early osseointegration of titanium implants. *J. Dent. Implant* **2016**, *6*, 3–12. [CrossRef]
40. Wennerberg, A.; Galli, S.; Albrektsson, T. Current knowledge about the hydrophilic and nanostructured SLActive surface. *Clin. Cosmet. Investig. Dent.* **2011**, *3*, 59–67. [CrossRef]
41. Ghassemi, A.; Ishijima, M.; Hasegawa, M.; Mohammadzadeh Rezaei, N.; Nakhaei, K.; Sekiya, T.; Torii, Y.; Hirota, M.; Park, W.; Miley, D.D.; et al. Biological and Physicochemical Characteristics of 2 Different Hydrophilic Surfaces Created by Saline-Storage and Ultraviolet Treatment. *Implant Dent.* **2018**, *27*, 405–414. [CrossRef] [PubMed]
42. Kasemo, B.; Lausmaa, J. Biomaterial and implant surfaces: On the role of cleanliness, contamination, and preparation procedures. *J. Biomed. Mater. Res.* **1988**, *22*, 145–158. [CrossRef] [PubMed]
43. Serro, A.; Saramago, B. Influence of sterilization on the mineralization of titanium implants induced by incubation in various biological model fluids. *Biomaterials* **2003**, *24*, 4749–4760. [CrossRef]
44. Choi, S.H.; Jeong, W.S.; Cha, J.Y.; Lee, J.H.; Lee, K.J.; Yu, H.S.; Choi, E.H.; Kim, K.M.; Hwang, C.J. Overcoming the biological aging of titanium using a wet storage method after ultraviolet treatment. *Sci. Rep.* **2017**, *7*, 3833. [CrossRef] [PubMed]
45. Flanagan, D. Photofunctionalization of Dental Implant. *J. Oral Implantol.* **2016**, *42*, 445–450. [CrossRef] [PubMed]
46. Shayganpour, A.; Rebaudi, A.; Cortella, P.; Diaspro, A.; Salerno, M. Electrochemical coating of dental implants with anodic porous titania for enhanced osseointegration. *Beilstein J. Nanotechnol.* **2015**, *6*, 2183–2192. [CrossRef] [PubMed]

© 2019 by the authors. Licensee MDPI, Basel, Switzerland. This article is an open access article distributed under the terms and conditions of the Creative Commons Attribution (CC BY) license (http://creativecommons.org/licenses/by/4.0/).

Article

A Vitronectin-Derived Bioactive Peptide Improves Bone Healing Capacity of SLA Titanium Surfaces

Chang-Bin Cho [1,†], Sung Youn Jung [2,†], Cho Yeon Park [2], Hyun Ki Kang [2], In-Sung Luke Yeo [1,*] and Byung-Moo Min [2,*]

1. Department of Prosthodontics, Seoul National University School of Dentistry, and Dental Research Institute, 101 Daehak-Ro, Jongno-Gu, Seoul 03080, Korea; istardental@naver.com
2. Department of Oral Biochemistry and Program in Cancer and Developmental Biology, Seoul National University School of Dentistry, and Dental Research Institute, 101 Daehak-Ro, Jongno-Gu, Seoul 03080, Korea; jsy1618@snu.ac.kr (S.Y.J.); choyeon94@snu.ac.kr (C.Y.P.); kang1978@snu.ac.kr (H.K.K.)
* Correspondence: pros53@snu.ac.kr (I.-S.L.Y.); bmmin@snu.ac.kr (B.-M.M.); Tel.: +82-2-2072-2661 (I.-S.L.Y.); +82-2-740-8661 (B.-M.M.); Fax: +82-2-2072-3860 (I.-S.L.Y.); +82-2-740-8665 (B.-M.M.)
† These authors contributed equally to this work.

Received: 13 September 2019; Accepted: 16 October 2019; Published: 17 October 2019

Abstract: In this study, we evaluated early bone responses to a vitronectin-derived, minimal core bioactive peptide, RVYFFKGKQYWE motif (VnP-16), both in vitro and in vivo, when the peptide was treated on sandblasted, large-grit, acid-etched (SLA) titanium surfaces. Four surface types of titanium discs and of titanium screw-shaped implants were prepared: control, SLA, scrambled peptide-treated, and VnP-16-treated surfaces. Cellular responses, such as attachment, spreading, migration, and viability of human osteoblast-like HOS and MG63 cells were evaluated in vitro on the titanium discs. Using the rabbit tibia model with the split plot design, the implants were inserted into the tibiae of four New Zealand white rabbits. After two weeks of implant insertion, the rabbits were sacrificed, the undecalcified specimens were prepared for light microscopy, and the histomorphometric data were measured. Analysis of variance tests were used for the quantitative evaluations in this study. VnP-16 was non-cytotoxic and promoted attachment and spreading of the human osteoblast-like cells. The VnP-16-treated SLA implants showed no antigenic activities at the interfaces between the bones and the implants and indicated excellent bone-to-implant contact ratios, the means of which were significantly higher than those in the SP-treated implants. VnP-16 reinforces the osteogenic potential of the SLA titanium dental implant.

Keywords: vitronectin; RVYFFKGKQYWE motif; cellular responses; dental implants; osseointegration

1. Introduction

Attachment of cells is the first step in cell–biomaterial interactions [1]. The transmembrane proteins on the cell membrane recognize and bind biomacromolecules adsorbed on the surface of the biomaterial [1]. These biomacromolecules are from the extracellular matrix (ECM), controlling cellular behaviors such as attachment, spreading, proliferation, and differentiation depending on the interacting transmembrane receptors [2]. Therefore, these ECM biomolecules, if applied to the titanium dental implant surface, are anticipated to have potential in reinforcing osseointegration [3–5]. One candidate molecule is vitronectin.

Vitronectin, one of the ECM proteins, is an abundant multifunctional glycoprotein found in serum, the extracellular matrix, and bone, and is involved in various physiological processes such as cell attachment, spreading, and migration [6–8]. Vitronectin contributes to healing of the bone surrounding a dental implant by promoting the attachment and spreading of the osteogenic cells [9,10]. This ECM protein appears to play a major role in initial bone healing by reorganizing the intracellular microfilaments and microtubules, which facilitates cell attachment and spreading [11–13]. However, the use of such an original protein has several critical limitations: a high cost for synthesis, antigenicity and instability of the molecule, and steric hindrance of this macromolecule in focal adhesion [5,14,15]. A functional peptide derived from the parent protein is a notable alternative, overcoming these limitations and maintaining the original biological activity [3,5,9,14,16,17]. In addition, bioactive peptides have advantages over larger protein molecules due to their robustness and sterilizability [18]. Recently, a vitronectin-derived functional peptide sequence, RVYFFKGKQYWE (VnP-16), has shown a Janus regulation for bone formation: promotion of osteoblast activity and inhibition of osteoclast activity, which is a desirable effect for osteogenesis that vitronectin does not have [9]. The VnP-16 peptide promoted bone formation by accelerating osteoblast differentiation and activity through direct interaction with β1 integrin followed by focal adhesion kinase (FAK) activation. Concomitantly, the peptide inhibited bone resorption by restraining Janus N-terminal Kinase (JNK)-c-Fos-nuclear factor of activated T cells, cytoplasmic 1 (NFATc1)-induced osteoclast differentiation and αvβ3 integrin-c-Src-proline-rich tyrosine kinase 2 (PYK2)-mediated resorptive function. Moreover, VnP-16 peptide decreased the bone resorbing activity of pre-existing mature osteoclasts without changing their survival rate [9].

The functionalization of a titanium implant via the immobilization of desirable proteins or their bioactive peptides in their native conformations is a promising approach to overcome the bioinertness of the surface, leading to improved osseointegration [18,19]. Several methodologies, including physical adsorption, covalent immobilization via chemical methods, and covalent immobilization via physical methods, have been investigated since the inception of protein functionalization on titanium substrates [19]. Physical adsorption is the simplest method to immobilize proteins on titanium substrates. Chemical immobilization, the most established of the protein functionalization approaches, proved that it was possible to covalently immobilize proteins to titanium substrates, overcoming the unintentional protein release observed in adsorption approaches. Physical covalent immobilization of biomolecules is the most recent approach to dental and orthopedic biomimetic functionalization, and possesses advantages over adsorption and chemical covalent immobilization [19]. Other methods on peptide-functionalization of titanium surfaces are also reported. Control of both peptide orientation and surface concentration is achieved by varying the solution pH or by applying an electric field [18]. In addition, multifunctional coating improves cell adhesion on titanium surfaces by using cooperatively acting peptides [20].

A micro-roughened surface of commercially pure titanium has been clinically used in the field of implant dentistry [10]. A sandblasted large-grit acid-etched (SLA) surface with approximately 1.5 μm of arithmetic mean deviation is known to accelerate osseointegration at the bone-implant interface, compared to the turned surface with no modification at the micro-level [10,21,22]. The application of VnP-16 to the SLA surface increases clinical relevance in the further enhancement of bone formation and makes use of dental implants extended to patients suffering from bone metabolic weaknesses, like osteoporosis. The VnP-16-treated SLA titanium surface has not been investigated yet.

This study aimed to evaluate the early bone response to the VnP-16-treated SLA titanium surface in vivo. In vitro tests were also performed using osteoblast-like cells. The hypothesis underlying this study was that the application of VnP-16 would further reinforce the osteogenic potential of the SLA surface.

2. Materials and Methods

2.1. Cells, Peptides, and Reagents

HOS and MG-63 cells, lines derived from human osteosarcomas, were purchased from the American Type Culture Collection (Rockville, MD, USA) and cultured in Dulbecco's modified Eagle's medium (Gibco BRL, Carlsbad, CA, USA) supplemented with 10% fetal bovine serum. Each peptide was synthesized using the 9-fluorenylmethoxycarbonyl-based solid-phase method with a C-terminal amide using a Pioneer Peptide Synthesizer (Applied Biosystems, Foster City, CA, USA) in the Peptron (Daejeon, Korea). The synthetic peptides used in the study had a purity greater than 95%, as determined using high-performance liquid chromatography. Human plasma vitronectin was obtained from Millipore (Bedford, MA, USA).

2.2. Disc Preparation and Surface Characterization

Titanium disc specimens, which were 0.5 mm thick and 10 mm in radius, were made of commercially pure grade 4 titanium. The discs, serving as control, were prepared by polishing with #600 and #1200 sandpaper. The other discs were subjected to sandblasting with large alumina particles and etched with a hydrochloric acid solution to generate the SLA surface (Deep Implant System, Seongnam, Korea). The SLA titanium discs were rinsed, ultrasonically washed, and dried. One group of the SLA discs was left untreated, another was treated with a scrambled peptide (SP; 10.5 µg/cm^2), and the other was treated with VnP-16 (10.5 µg/cm^2).

The surfaces of the four types of disc were imaged using field emission-scanning electron microscopy (FE-SEM; S-4700, Hitachi, Tokyo, Japan). The element composition of each group was analyzed by electron spectroscopy for the chemical analysis (ESCA; Sigma Probe, Thermo Scientific, Waltham, MA, USA). Confocal laser scanning microscopy (CLSM; LSM 800, Carl Zeiss AG, Oberkochen, Germany) calculated two surface parameters for surface topography of the investigated discs; arithmetic mean deviation (Ra) and the developed surface area ratio (Sdr) [23].

2.3. Cell Attachment and Spreading Assays

The cell attachment assay was performed as described previously [9]. The physical adsorption method was used for the application of peptides. Twenty-four-well culture plates were coated with 0.26 µg/cm^2 human plasma vitronectin for 24 h at 4 °C or 10.5 µg/cm^2 synthetic peptides for 24 h at room temperature, blocked with 1% heat-inactivated bovine serum albumin (BSA) in phosphate-buffered saline (PBS) for 1 h at 37 °C, and then washed with PBS. Cells (1×10^5 cells/500 µL) were added to each plate and incubated in serum-free culture medium for 1 h at 37 °C. After incubation period, unattached cells were removed by rinsing twice with PBS. Attached cells were fixed with 10% formalin for 15 min, stained with 0.5% crystal violet for 1 h, gently washed with distilled water three times, and dissolved with 2% sodium dodecyl sulfate for 5 min. Absorbance was measured at 570 nm using a microplate reader. For cell-spreading assay, cells (7×10^4 cells/500 µL) were added to each substrate-coated plate and incubated for 3 h at 37 °C. To determine cell spreading, formalin-fixed and crystal violet-stained cell surface area was measured with Image-Pro plus software (Version 4.5; Media Cybernetics, Silver Spring, MD, USA).

2.4. Migration Assay

Migration assays were performed using a transwell migration chamber (Corning, Pittston, PA, USA) possessing 8 µm pores as described previously [24]. The lower side of each transwell filter was coated with vitronectin (0.26 µg/cm^2), or synthetic peptides (10.5 µg/cm^2) by drying for 24 h at 4 °C (vitronectin) or for 24 h at room temperature (peptides). Cells (2×10^4 cells/24-well) were seeded in the upper chamber of a transwell filter and allowed to migrate for 24 h at 37 °C. Cells were then fixed with 10% formalin for 15 min and stained with 0.5% crystal violet. Unmigrated cells in the upper side of the transwell filter were removed with a cotton swab, and cell migration was quantified by counting the number of cells that had migrated through the filter. Human placental laminin was used as the positive controls and SP was used as the negative control.

2.5. Cell Viability Assay

The viabilities of cells were investigated using the EZ-Cytox Cell Viability Assay kit (water-soluble tetrazolium salt method; Daeil Lab Service, Seoul, Korea). A 96-well microplate was coated with VnP-16 peptide (0, 10.6, 21.2, or 42.4 µg/cm^2) by drying for 24 h at room temperature. Cells (1.5 ×

10^4 cells/100 µL) were seeded onto a 96-well microplate and then cultured for 24 h or 48 h at 37 °C. The water-soluble tetrazolium salt reagent solution (10 µL) was added to each well, and the plate was incubated for 2 h at 37 °C. The absorbance at 450 nm was then measured using a microplate reader.

2.6. In Vivo Experiment

Sixteen screw-shaped grade 4 titanium implants were made, which were 3.5 mm in major diameter and 11 mm in length (Warantec, Seongnam, Korea). Four implants were used as they were without any surface modification and designated as turned surface (control). The surface of another four implants was SLA (Deep Implant System, Seongnam, Korea). Half of the rest of the eight implants were treated with SP while the other half were treated with VnP-16 (1.0 mg/cm^2). Using the physical adsorption method, the Ti implants were placed on 0.2 ml PCR tubes and coated with the synthetic peptides by drying for 7 d in a vacuum at room temperature.

All the animal experiments performed in this study were approved by the Ethics Committee of Animal Experimentation of the Institutional Animal Care and Use Committee (CRONEX-IACUC 201705001; Cronex, Hwasung, Korea). These experiments were conducted following the Animal Research: Reporting In Vivo Experiments (ARRIVE) guidelines for the care and use of laboratory animals [25]. Four New Zealand white rabbits were used in this study, which were male, approximately five to six months in age and 2.5 to 3.0 kg in weight. The experimental animals were intramuscularly anesthetized with a dose of 15 mg/kg tiletamine hydrochloride and zolazepam hydrochloride (Zoletil, Virbac, Carros, France) and 5 mg/kg xylazine (Rompun, Bayer AG, Leverkusen, Germany). The skin hair was shaved at the tibial area of the rabbits, which was disinfected with aqueous iodine. A full-thickness incision from the skin to the periosteum of the tibiae was made, and the flaps were elevated to expose the medial surfaces of the tibiae. Drilling was conducted for to make the holes on the medial surfaces for implant insertion. The final diameter of the holes was 3.2 mm. Two implants were inserted into each tibia and arranged according to the split plot design. The periosteum and fascia were sutured with 4-0 polyglactin 910 (Vicryl, Ethicon, Somerville, NJ, USA) while the skin was sutured with 4-0 Nylon (Ethilon, Ethicon, Somerville, NJ, USA). Each experimental animal was housed in a separate cage and an antibiotic, enrofloxacin, (Biotril, Komipharm International, Siheung, Korea) was administered to prevent infection.

2.7. Light Microscopic Evaluation

The rabbits were sacrificed under general anesthesia with the intravenous administration of potassium chloride at 14 days after implant insertion. The implants were removed en bloc with the surrounding bones and fixed in 10% neutral buffered formalin for 2 weeks. After formalin fixation, each implant-bone block was dehydrated with ethanol. Then, the blocks were resin-embedded (Technovit 7200, Heraeus Kulzer, Hanau, Germany) and ground for light microscopy using an EXAKT system (EXAKT Apparatebau, Norderstedt, Germany), according to the methods described in the previous studies [23,26]. Sections of the implant-bone blocks were prepared with a final thickness of approximately 50 µm and modified Goldner's Masson trichrome staining [27]. The interfacial areas between the bones and implants were observed and evaluated from the bone crests to 2 mm in depth for histomorphometry, where bone-to-implant contact (BIC) and bone area (BA) ratios were calculated. The image analyses and the histomorphometric calculations were performed on ×100 magnified images using a light microscope (BX51, Olympus, Tokyo, Japan), SPOT version 4.0 software (Diagnostic Instruments, Sterling Heights, MI, USA) and Image-Pro Plus (Media Cybernetics, Rockville, MD, USA).

2.8. Statistics

Descriptive statistics for the data were presented as the mean ± standard deviation (SD). The statistical analyses were performed with R software (version 3.6.1, R Foundation for Statistical Computing, Vienna, Austria). All the data obtained in this study were confirmed to be normally distributed by the Shapiro–Wilk test. Analysis of variance tests were used for comparisons among the groups. When a significant difference was found, Tukey's honestly significant difference test was further applied for pairwise comparison. The level of significance was 0.05 in this study.

3. Results

3.1. Surface Characteristics

The FE-SEM images of the specimens showed very different topographical features between the polished and SLA surfaces (Figure 1A). Some grooves on the overall flat surfaces were found for the polished specimens, while a honeycomb-like irregular topography was observed for the SLA specimens. The treatments of SP and VnP-16 had little effect on the surface physical features of the specimens (Figure 1A). There was no significant difference in either the Ra or Sdr among the SLA titanium discs, regardless of the peptide treatment ($p > 0.05$) (Figure 1B). However, in surface chemistry, the treatments of the functional peptides were confirmed from the results of higher nitrogen contents for the SP- and VnP-16-treated surfaces, compared to those for the other groups (polished and SLA titanium surfaces) ($p < 0.05$) (Figure 1C). The highest content element was carbon for every group.

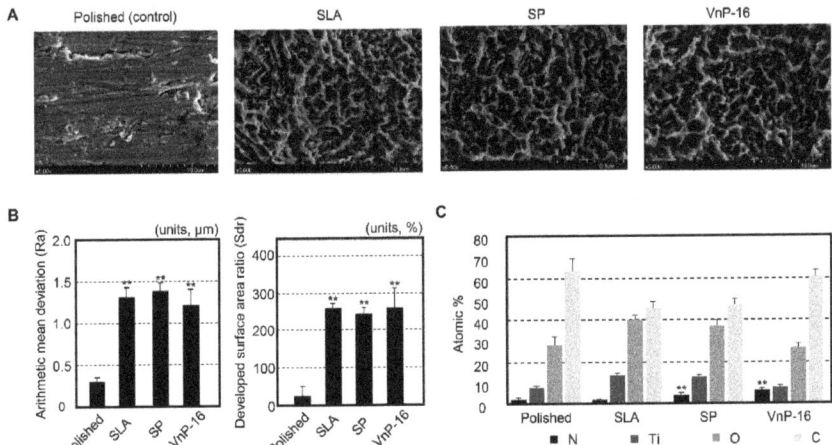

Figure 1. Surface characteristics of the titanium specimens investigated in this study. (**A**) Field emission scanning electron microscopy definitely shows different topographical features between the polished and sandblasted, large-grit, acid-etched (SLA) surfaces. (**B**) The mean values of the measured surface parameters indicated that the peptide treatment did not change the surfaces physically at the micro level. Note the significant differences in the surface parameters between the polished and the other SLA surfaces. ** $p < 0.01$ vs. the polished surface. (**C**) Electron spectroscopy for chemical analysis detected high nitrogen content on the peptide-treated surfaces. Almost no nitrogen was found in the other groups. ** $p < 0.01$ vs. the polished and SLA surfaces (significant differences are marked only for the nitrogen content).

3.2. Effects of VnP-16 Peptide on Cellular Responses of Human Osteoblast-Like Cells

To investigate whether a human vitronectin-derived peptide, VnP-16, could mediate cell behavior of osteoblasts, cell attachment, spreading, and migration of human osteoblast-like cells, including HOS and MG-63, were assayed. The attachment of osteoblast-like cells was evaluated using a cell adhesion assay in a serum-free medium. Human plasma vitronectin strongly promoted cell attachment (Figure 2A upper, B) and spreading (Figure 2A lower, C) in osteoblast-like HOS cells. The VnP-16 peptide also promoted greater cell attachment (Figure 2A upper, B) and spreading (Figure 2A lower, C) than the BSA or SP control, and its attachment and spreading activities were comparable to those of vitronectin (Figure 2A–C). In addition, vehicle and SP did not participate in cell migration in HOS cells. On the other hand, vitronectin and the VnP-16 peptide promoted cell migration in HOS cells, while the VnP-16 peptide was significantly less effective than vitronectin (Figure 2D). The VnP-16 peptide did not affect the proliferation or viability of HOS cells (Figure 2E), indicating that its stimulatory effect on the cell behavior of HOS cells was not due to cytotoxicity or enhanced cell proliferation. These results support that VnP-16 is functionally active in promoting osteoblastic responses.

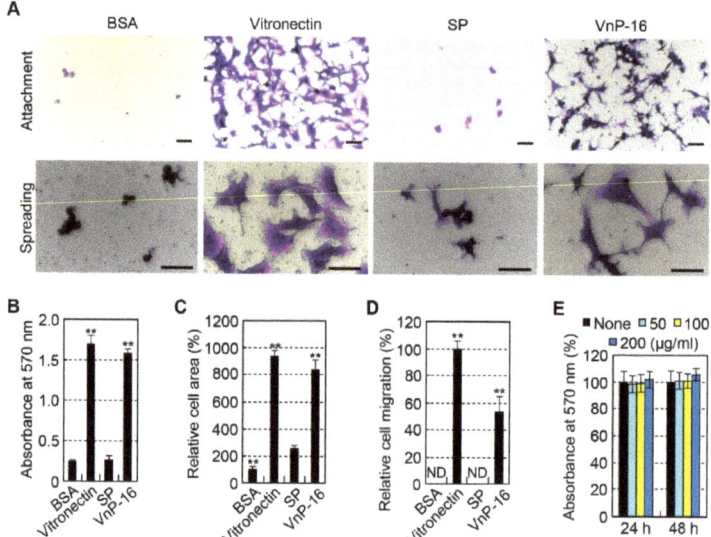

Figure 2. Cell attachment, spreading, and migration of osteoblast-like HOS cells seeded on culture plates treated with vitronectin and synthetic peptides. (**A**) Photographs of osteoblast-like HOS cells adhering (upper panel) and spreading (lower panel) to culture plates treated with 1% bovine serum albumin (BSA), vitronectin (0.26 µg/cm^2), scrambled peptide (SP), and VnP-16 peptide (10.5 µg/cm^2). Bar = 100 µm. (**B**,**C**) Cell attachment (**B**) and spreading (**C**) to immobilized synthetic peptides. HOS cells were allowed to adhere to peptide-treated plates for 1 h (**B**) or 3 h (**C**) in serum-free medium. (**D**) Migration of osteoblast-like HOS cells induced by vitronectin and synthetic peptides. HOS cells were seeded into the upper chambers of transwell filters coated with vitronectin (0.26 µg/cm^2), SP, or VnP-16 (10.5 µg/cm^2) and were incubated for 24 h. ND, not detected. (**E**) The viabilities of osteoblast-like HOS cells treated with VnP-16 for 24 or 48 h. ** $p < 0.01$ vs. the SP-treated control group. Data in (**B**–**E**) ($n = 4$) represent the mean ± SD.

Next, to determine whether the effects of the VnP-16 peptide on the cell behavior of HOS cells were similar to those of other human osteoblast-like cells, we used human osteoblast-like MG-63 cells. Similar, but not identical, results were obtained for cell behavior in MG-63 cells. Human plasma vitronectin and VnP-16 peptide strongly promoted cell attachment (Figure 3A upper, B) and spreading (Figure 3A lower, C) in MG-63 cells compared to the BSA or SP control. In addition, the cell attachment and spreading activities of VnP-16 peptide were comparable to those of vitronectin (Figure 3A–C). Similarly, VnP-16 peptide did not affect the proliferation or viability of MG-63 cells (Figure 3D). However, the cell migration activities of vitronectin and the VnP-16 peptide were different between HOS and MG-63 cells. In other words, vitronectin and the VnP-16 peptide had no effect on cell migration in MG-63 cells (data not shown). Therefore, the cellular responses of the VnP-16 peptide to the human osteoblast-like cells HOS and MG-63 had different effects on cell migration but had similar effects on cell attachment, spreading, and viability.

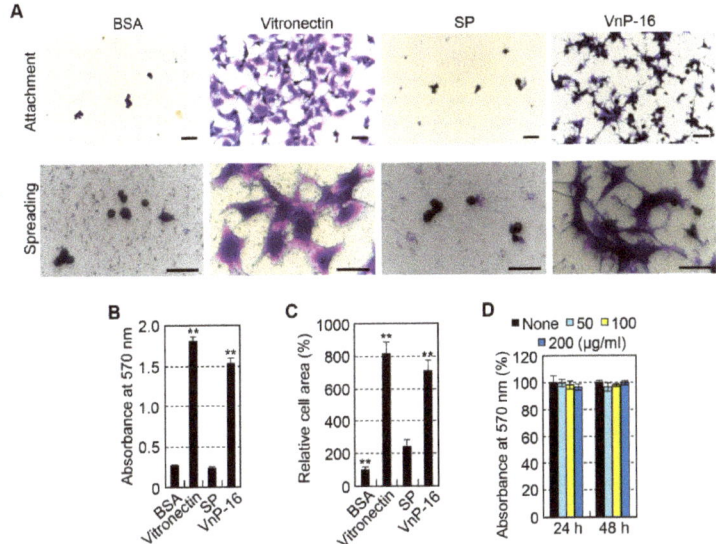

Figure 3. Cell attachment and spreading of osteoblast-like MG-63 cells seeded on culture plates treated with vitronectin and synthetic peptides. (**A**) Photographs of osteoblast-like MG-63 cells adhering (upper panel) and spreading (lower panel) to culture plates treated with 1% bovine serum albumin (BSA), vitronectin (0.26 μg/cm^2), scrambled peptide (SP), and VnP-16 peptide (10.5 μg/cm^2). Bar = 100 μm. (**B–C**) Cell attachment (**B**) and spreading (**C**) to immobilized synthetic peptides. MG-63 cells were allowed to adhere to peptide-treated plates for 1 h (**B**) or 3 h (**C**) in serum-free medium. (**D**) The viabilities of osteoblast-like MG-63 cells treated with VnP-16 for 24 or 48 h. ** $p < 0.01$ vs. the SP-treated control group. Data in (**B–D**) ($n = 4$) represent the mean ± SD.

3.3. Histomorphometry

Every experimental animal was healthy, and no signs of diseases or pathologic states were found until the sacrifice. There were no special inflammatory or immune cells found in the light microscopic view of the specimens. After 14 days of implant insertion, sufficient mineralization was observed in each section (Figure 4A). The mean value and standard deviation of each group were 47.0% ± 7.5% for the turned surface, 64.4% ± 8.6% for SLA, 42.1% ± 18.1% for SP-treated and 65.0% ± 7.2% for the VnP-16-treated surface. The histomorphometric data showed significant differences in BIC ($p = 0.027$). However, the pairwise comparisons between the groups found no significant differences in BIC (Figure 4B). The mean values and standard deviation in the BA were 58.8% ± 6.0% for the turned group, 56.8% ± 6.4% for SLA, 56.6% ± 8.4% for the SP-treated group, and 61.5% ± 10.6% for the VnP-16-treated group (Figure 4C). There were no significant differences in BA among the groups.

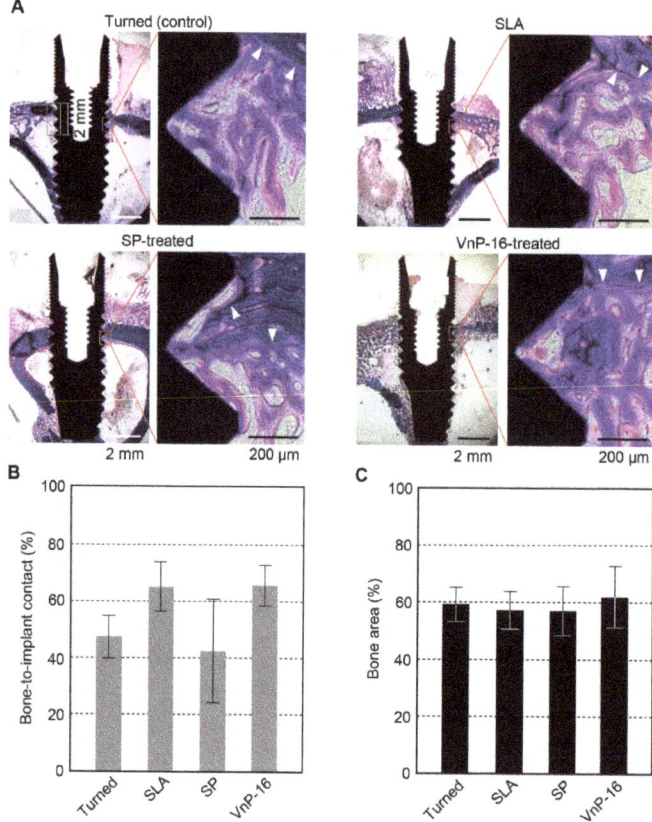

Figure 4. The histologic views and histomorphometric data for bone responses to the turned, SLA, SP-treated SLA, and VnP-16-treated SLA titanium implant surfaces. (**A**) The demarcation lines (white arrowheads), difference in stained colors and maturity of the bone (cancellous or cortical) differentiate the new bone from the existing old bone. Here, the new bone is stained more reddish while the old bone is stained more blueish. (**B**) Bone-to-implant contact ratios were measured, which are defined as the percentage of the implant surface in contact with bone to the total implant surface at the region of interest, which was the area ranging from the bone crest to 2 mm in depth in this study (green edged rectangle in (**A**)). (**C**) The ratio of the area filled with bone to the total area of the region of interest (bone area, or BA ratio) was also measured for each implant.

4. Discussion

The results of these in vitro and in vivo studies indicated that the VnP-16-treated SLA titanium surface augments the initial bone response to a dental implant. The VnP-16 bioactive peptide is expected to accelerate early bone healing further when this peptide is applied to the SLA titanium dental implant. VnP-16 is another candidate biomolecule for stronger osseointegration into a dental implant that is clinically applicable, together with some laminin-derived peptides [3,5,23]. Because VnP-16 has both the upregulation of osteoblast activity and the downregulation of osteoclast activity, different from other peptides, the clinical applicability of this material is considered to be higher [9].

Since VnP-16 downregulates osteoclast activity, this vitronectin-derived peptide is applicable to osteoporotic patients as well as to normal patients. A previous study has already shown the improved bone healing capacity of VnP-16 in an in vivo experiment using ovariectomized rats [9]. The SLA titanium implant is well-known to have long-term clinical performance with high survival rates (higher than 95%) [28,29]. However, this micro-roughened titanium surface has some limitations in the use for patients with a problem in bone metabolism, including osteoporosis [30]. The cell adhesion molecules can potentiate the bone healing capacities of the modified titanium surfaces used in dental clinics without immune responses, which is shown in this study. The effect of VnP-16 on osseointegration of the SLA dental implant needs to be evaluated for normal and osteoporotic patients.

This study used the physical adsorption method for the VnP-16 application to titanium surface. One of the main advantages of this method is the simplicity, that is, an easy process to functionalize the titanium implant surface while this method has several important disadvantages like low effective peptide concentrations, and denaturation of the three dimensional structures of proteins or peptides [19]. The concentration of VnP-16 was determined from a dose-response curve through the cell attachment assay, as in a previous study (data not shown) [9]. The lowest concentration, showing the maximal effect in cell attachment, was used in this study, which solved the problem of low effective peptide concentration. VnP-16 is a peptide composed of 12 amino acids, which are considered to have no specific three dimensional structure [5]. Furthermore, the other approaches to peptide immobilization require additional reactions and costs [19]. Therefore, physical adsorption was the method of choice in this study although more effective approach needs to be investigated continuously.

In the analysis of surface chemistry of this study, higher nitrogen contents on the peptide-treated surfaces imply the applications of the peptides to the titanium implant surfaces. However, the results of the amounts of carbon detected on the surfaces were hard to interpret, despite that carbon was the most abundant on all the surfaces investigated in this study. These unclear results are considered to be from the phenomenon of hydrocarbon contamination on titanium surface, which is usual when titanium is exposed to air [31,32].

Although the analysis of variance test for BIC in this study showed a p-value less than 0.05, the pairwise comparisons found no significant differences between the groups. Perhaps, large standard deviation, especially obtained from the measurements of the turned and SP-treated implants, caused no significant differences in the pairwise comparisons. The high mean BIC ratios in the SLA and VnP-16-treated groups and the low mean BIC in the turned and SP-treated groups were considered to contribute to the significant difference in the analysis of variance test for all the groups. The large standard deviation of the data occurred because of the small sample size in this study. Another reason for the large deviation might be the difficulty in displaying the entire three-dimensional bone-implant interface in light microscopic histology. The selection of one cross-sectional plane for light microscopy is arbitrary, and the data from the cross-section are poorly correlated with the data measured on the whole three dimensional image [33]. In order to obtain the similar data between two- and three-dimensional images, three to four histologic sections for each specimen are needed, which are extremely difficult to prepare from a undecalcified specimen, including a titanium implant [34]. Methodological advancements, like three dimensional imaging analysis by micro-computed tomography, and a technique to make more histologic sections from a hard specimen are needed for more obvious in vivo results.

5. Conclusions

A human vitronectin-derived peptide, VnP-16 (RVYFFKGKQYWE motif), showed excellent histomorphometric osseointegration data without any special antigen–antibody reaction when this peptide was treated on an SLA titanium dental implant, which has been successfully used in clinics. From the in vitro results of this study, VnP-16 promotes the attachment of osteogenic cells and differentiation into osteoblasts, which may increase the bone healing capacity of the SLA's titanium surface. Considering both the in vitro and in vivo results of this study, VnP-16 reinforces the osteogenic potential of the SLA titanium dental implant when this peptide is applied to the SLA surface. In the future, VnP-16 may expand the clinical indications of SLA titanium dental implants.

Author Contributions: Conceptualization, I.-S.L.Y. and B.-M.M.; methodology, I.-S.L.Y. and B.-M.M.; software, C.-B.C., S.Y.J., C.Y.P., and H.K.K.; validation, I.-S.L.Y. and B.-M.M.; formal analysis, C.-B.C. and S.Y.J.; Investigation,

C.-B.C., S.Y.J., C.Y.P., and H.K.K.; resources, I.-S.L.Y. and B.-M.M.; data curation, C.-B.C. and S.Y.J.; writing—original draft preparation, C.-B.C. and S.Y.J.; writing—review and editing, C.Y.P., H.K.K., I.-S.L.Y., and B.-M.M.; visualization, C.-B.C. and S.Y.J.; supervision, I.-S.L.Y. and B.-M.M.; project administration, I.-S.L.Y. and B.-M.M.; funding acquisition, I.-S.L.Y. and B.-M.M.

Funding: This work was supported by the National Research Foundation of Korea (NRF) grant funded by the Korea government (MSIT) (2016R1A2B4014330 and 2019R1F1A1054209).

Conflicts of Interest: The authors declare no conflict of interest.

References

1. Anderson, J.M.; Rodriguez, A.; Chang, D.T. Foreign body reaction to biomaterials. *Semin. Immunol.* **2008**, *20*, 86–100. [CrossRef]
2. Ravanetti, F.; Gazza, F.; D'Arrigo, D.; Graiani, G.; Zamuner, A.; Zedda, M.; Manfredi, E.; Dettin, M.; Cacchioli, A. Enhancement of peri-implant bone osteogenic activity induced by a peptidomimetic functionalization of titanium. *Ann. Anat.* **2018**, *218*, 165–174. [CrossRef]
3. Kang, H.K.; Kim, O.B.; Min, S.K.; Jung, S.Y.; Jang, D.H.; Kwon, T.K.; Min, B.M.; Yeo, I.S. The effect of the DLTIDDSYWYRI motif of the human laminin alpha2 chain on implant osseointegration. *Biomaterials* **2013**, *34*, 4027–4037. [CrossRef]
4. Shekaran, A.; Garcia, A.J. Extracellular matrix-mimetic adhesive biomaterials for bone repair. *J. Biomed. Mater. Res. A* **2011**, *96*, 261–272. [CrossRef]
5. Yeo, I.S.; Min, S.K.; Kang, H.K.; Kwon, T.K.; Jung, S.Y.; Min, B.M. Identification of a bioactive core sequence from human laminin and its applicability to tissue engineering. *Biomaterials* **2015**, *73*, 96–109. [CrossRef]
6. Boron, W.F.; Boulpaep, E.L. *Medical Physiology E-Book*; Elsevier Health Sciences: Kidlington, Oxford, UK, 2016.
7. Cherny, R.C.; Honan, M.A.; Thiagarajan, P. Site-directed mutagenesis of the arginine-glycine-aspartic acid in vitronectin abolishes cell adhesion. *J. Biol. Chem.* **1993**, *268*, 9725–9729.
8. Shin, T.M.; Isas, J.M.; Hsieh, C.L.; Kayed, R.; Glabe, C.G.; Langen, R.; Chen, J. Formation of soluble amyloid oligomers and amyloid fibrils by the multifunctional protein vitronectin. *Mol. Neurodegener.* **2008**, *3*, 16. [CrossRef]
9. Min, S.K.; Kang, H.K.; Jung, S.Y.; Jang, D.H.; Min, B.M. A vitronectin-derived peptide reverses ovariectomy-induced bone loss via regulation of osteoblast and osteoclast differentiation. *Cell Death Differ.* **2018**, *25*, 268–281. [CrossRef]
10. Yeo, I.S. Surface modification of dental biomaterials for controlling bone response. In *Bone Response to Dental Implant Materials*; Elsevier: Kidlington, Oxford, UK, 2017; pp. 43–46.
11. Howlett, C.R.; Evans, M.D.; Walsh, W.R.; Johnson, G.; Steele, J.G. Mechanism of initial attachment of cells derived from human bone to commonly used prosthetic materials during cell culture. *Biomaterials* **1994**, *15*, 213–222. [CrossRef]
12. Rivera-Chacon, D.M.; Alvarado-Velez, M.; Acevedo-Morantes, C.Y.; Singh, S.P.; Gultepe, E.; Nagesha, D.; Sridhar, S.; Ramirez-Vick, J.E. Fibronectin and vitronectin promote human fetal osteoblast cell attachment and proliferation on nanoporous titanium surfaces. *J. Biomed. Nanotechnol.* **2013**, *9*, 1092–1097. [CrossRef]
13. Scotchford, C.A.; Ball, M.; Winkelmann, M.; Voros, J.; Csucs, C.; Brunette, D.M.; Danuser, G.; Textor, M. Chemically patterned, metal-oxide-based surfaces produced by photolithographic techniques for studying protein- and cell-interactions. II: Protein adsorption and early cell interactions. *Biomaterials* **2003**, *24*, 1147–1158. [CrossRef]
14. Cacchioli, A.; Ravanetti, F.; Bagno, A.; Dettin, M.; Gabbi, C. Human Vitronectin-Derived Peptide Covalently Grafted onto Titanium Surface Improves Osteogenic Activity: A Pilot In Vivo Study on Rabbits. *Tissue Eng. Part. A* **2009**, *15*, 2917–2926. [CrossRef]
15. Petrie, T.A.; Raynor, J.E.; Dumbauld, D.W.; Lee, T.T.; Jagtap, S.; Templeman, K.L.; Collard, D.M.; Garcia, A.J. Multivalent integrin-specific ligands enhance tissue healing and biomaterial integration. *Sci. Transl. Med.* **2010**, *2*, 45ra60. [CrossRef]
16. Rezania, A.; Healy, K.E. Biomimetic peptide surfaces that regulate adhesion, spreading, cytoskeletal organization, and mineralization of the matrix deposited by osteoblast-like cells. *Biotechnol. Prog.* **1999**, *15*, 19–32. [CrossRef]

17. Vukicevic, S.; Luyten, F.P.; Kleinman, H.K.; Reddi, A.H. Differentiation of canalicular cell processes in bone cells by basement membrane matrix components: Regulation by discrete domains of laminin. *Cell* **1990**, *63*, 437–445. [CrossRef]
18. Martin, L.J.; Akhavan, B.; Bilek, M.M.M. Electric fields control the orientation of peptides irreversibly immobilized on radical-functionalized surfaces. *Nat. Commun.* **2018**, *9*, 357. [CrossRef]
19. Stewart, C.; Akhavan, B.; Wise, S.G.; Bilek, M.M. A review of biomimetic surface functionalization for bone-integrating orthopedic implants: Mechanisms, current approaches, and future directions. *Prog. Mater. Sci.* **2019**, *106*, 100588. [CrossRef]
20. Pagel, M.; Hassert, R.; John, T.; Braun, K.; Wiessler, M.; Abel, B.; Beck-Sickinger, A.G. Multifunctional Coating Improves Cell Adhesion on Titanium by using Cooperatively Acting Peptides. *Angew Chem. Int. Ed. Engl.* **2016**, *55*, 4826–4830. [CrossRef]
21. Choi, J.Y.; Kang, S.H.; Kim, H.Y.; Yeo, I.L. Control Variable Implants Improve Interpretation of Surface Modification and Implant Design Effects on Early Bone Responses: An In Vivo Study. *Int J. Oral Maxillofac Implant.* **2018**, *33*, 1033–1040. [CrossRef]
22. Wennerberg, A.; Albrektsson, T. On implant surfaces: A review of current knowledge and opinions. *Int. J. Oral Maxillofac. Implant.* **2010**, *25*, 63–74.
23. Kim, S.; Choi, J.Y.; Jung, S.Y.; Kang, H.K.; Min, B.M.; Yeo, I.L. A laminin-derived functional peptide, PPFEGCIWN, promotes bone formation on sandblasted, large-grit, acid-etched titanium implant surfaces. *Int. J. Oral Maxillofac. Implant.* **2019**, *34*, 836–844. [CrossRef] [PubMed]
24. Kim, J.M.; Min, S.K.; Kim, H.; Kang, H.K.; Jung, S.Y.; Lee, S.H.; Choi, Y.; Roh, S.; Jeong, D.; Min, B.M. Vacuolar-type H+-ATPase-mediated acidosis promotes in vitro osteoclastogenesis via modulation of cell migration. *Int. J. Mol. Med.* **2007**, *19*, 393–400. [CrossRef] [PubMed]
25. Kilkenny, C.; Browne, W.J.; Cuthill, I.C.; Emerson, M.; Altman, D.G. Improving bioscience research reporting: The ARRIVE guidelines for reporting animal research. *Osteoarthr. Cartil.* **2012**, *20*, 256–260. [CrossRef] [PubMed]
26. Donath, K.; Breuner, G. A method for the study of undecalcified bones and teeth with attached soft tissues. The Sage-Schliff (sawing and grinding) technique. *J. Oral Pathol.* **1982**, *11*, 318–326. [CrossRef]
27. Gruber, H.E. Adaptations of Goldner's Masson trichrome stain for the study of undecalcified plastic embedded bone. *Biotech. Histochem.* **1992**, *67*, 30–34. [CrossRef]
28. Buser, D.; Janner, S.F.; Wittneben, J.G.; Bragger, U.; Ramseier, C.A.; Salvi, G.E. 10-year survival and success rates of 511 titanium implants with a sandblasted and acid-etched surface: A retrospective study in 303 partially edentulous patients. *Clin. Implant. Dent. Relat Res.* **2012**, *14*, 839–851. [CrossRef]
29. Van Velzen, F.J.; Ofec, R.; Schulten, E.A.; Ten Bruggenkate, C.M. 10-year survival rate and the incidence of peri-implant disease of 374 titanium dental implants with a SLA surface: A prospective cohort study in 177 fully and partially edentulous patients. *Clin. Oral Implant. Res.* **2015**, *26*, 1121–1128. [CrossRef]
30. Tsolaki, I.N.; Madianos, P.N.; Vrotsos, J.A. Outcomes of dental implants in osteoporotic patients. A literature review. *J. Prosthodont.* **2009**, *18*, 309–323. [CrossRef]
31. Choi, S.H.; Jeong, W.S.; Cha, J.Y.; Lee, J.H.; Lee, K.J.; Yu, H.S.; Choi, E.H.; Kim, K.M.; Hwang, C.J. Overcoming the biological aging of titanium using a wet storage method after ultraviolet treatment. *Sci. Rep.* **2017**, *7*, 3833. [CrossRef]
32. Lee, J.B.; Jo, Y.H.; Choi, J.Y.; Seol, Y.J.; Lee, Y.M.; Ku, Y.; Rhyu, I.C.; Yeo, I.L. The Effect of Ultraviolet Photofunctionalization on a Titanium Dental Implant with Machined Surface: An In Vitro and In Vivo Study. *Materials* **2019**, *12*, 2078. [CrossRef]
33. Choi, J.Y.; Park, J.I.; Chae, J.S.; Yeo, I.L. Comparison of micro-computed tomography and histomorphometry in the measurement of bone-implant contact ratios. *Oral Surg. Oral Med. Oral Pathol. Oral Radiol.* **2019**, *128*, 87–95. [CrossRef]
34. Bernhardt, R.; Kuhlisch, E.; Schulz, M.C.; Eckelt, U.; Stadlinger, B. Comparison of bone-implant contact and bone-implant volume between 2D-histological sections and 3D-SRmicroCT slices. *Eur. Cell Mater.* **2012**, *23*, 237–247, discussion 247–248. [CrossRef]

© 2019 by the authors. Licensee MDPI, Basel, Switzerland. This article is an open access article distributed under the terms and conditions of the Creative Commons Attribution (CC BY) license (http://creativecommons.org/licenses/by/4.0/).

Review

Fracture Resistance of Zirconia Oral Implants In Vitro: A Systematic Review and Meta-Analysis

Annalena Bethke [1], Stefano Pieralli [1,2], Ralf-Joachim Kohal [2], Felix Burkhardt [1,2], Manja von Stein-Lausnitz [1], Kirstin Vach [3] and Benedikt Christopher Spies [1,2,*]

[1] Department of Prosthodontics, Geriatric Dentistry and Craniomandibular Disorders, Charité—Universitätsmedizin Berlin, corporate member of Freie Universität Berlin, Humboldt-Universität zu Berlin, and Berlin Institute of Health, Aßmannshauser Str. 4-6, 14197 Berlin, Germany; a.k.bethke@web.de (A.B.); stefano.pieralli@charite.de (S.P.); felix.burkhardt@charite.de (F.B.); manja.von-stein-lausnitz@charite.de (M.v.S.-L.)

[2] Department of Prosthetic Dentistry, Faculty of Medicine, Center for Dental Medicine, Medical Center—University of Freiburg, Hugstetter Str. 55, 79106 Freiburg, Germany; ralf.kohal@uniklinik-freiburg.de

[3] Institute of Medical Biometry and Statistics, Faculty of Medicine, Medical Center—University of Freiburg, University of Freiburg, Stefan-Meier-Str. 26, 79104 Freiburg, Germany; kv@imbi.uni-freiburg.de

* Correspondence: benedikt.spies@charite.de; Tel.: +49-30-450-662546

Received: 20 December 2019; Accepted: 21 January 2020; Published: 24 January 2020

Abstract: Various protocols are available to preclinically assess the fracture resistance of zirconia oral implants. The objective of the present review was to determine the impact of different treatments (dynamic loading, hydrothermal aging) and implant features (e.g., material, design or manufacturing) on the fracture resistance of zirconia implants. An electronic screening of two databases (MEDLINE/Pubmed, Embase) was performed. Investigations including > 5 screw-shaped implants providing information to calculate the bending moment at the time point of static loading to fracture were considered. Data was extracted and meta-analyses were conducted using multilevel mixed-effects generalized linear models (GLMs). The Šidák method was used to correct for multiple testing. The initial search resulted in 1864 articles, and finally 19 investigations loading 731 zirconia implants to fracture were analyzed. In general, fracture resistance was affected by the implant design (1-piece > 2-piece, $p = 0.004$), material (alumina-toughened zirconia/ATZ > yttria-stabilized tetragonal zirconia polycrystal/Y-TZP, $p = 0.002$) and abutment preparation (untouched > modified/grinded, $p < 0.001$). In case of 2-piece implants, the amount of dynamic loading cycles prior to static loading ($p < 0.001$) or anatomical crown supply ($p < 0.001$) negatively affected the outcome. No impact was found for hydrothermal aging. Heterogeneous findings of the present review highlight the importance of thoroughly and individually evaluating the fracture resistance of every zirconia implant system prior to market release.

Keywords: dental implant; zirconia; ceramics; aging; artificial mouth; fracture load; fatigue; chewing simulation; meta-analysis

---

1. Introduction

To date, titanium can be considered the gold standard material in oral implantology [1]. However, due to increasing esthetic standards and a discussed impact of metal/titanium particle release on the pathogenesis of peri-implant bone loss [2,3], a renaissance of ceramic oral implants can be observed in dental media. Nowadays, the market share of zirconia oral implants seems to be increasing, even if still comparatively small compared to conventional titanium implants.

Nonetheless, the superiority of ceramic oral implants regarding esthetics and biocompatibility, or, as an example, the frequently claimed patients' demand for metal-free implantology are still not soundly scientifically evidenced. Nevertheless, the majority of dental experts are of the opinion that zirconia oral implants will be coexistent with titanium implants in the near future [4].

When zirconium dioxide (zirconia, ZrO_2) was introduced as ceramic implant material, research focused to evaluate and improve its osseointegrative potential by creating a microroughened surface topography [5]. In the first instance, parameters like bone-to-implant contact (BIC), push-in values and removal torque were assessed in animal experiments. As a result, zirconia implants with various surface modifications (additive by sintering a porous ceramic layer, subtractive by sandblasting and/or acid-etching or, for example, by texturing the inner surface of a mold in case of an injection-molded implant) can nowadays be considered comparable to titanium implants by means of osseointegration in preclinical studies [6]. This finding was confirmed in clinical trials, however limited to short- and mid-term observation periods and the replacement of up to three adjacent missing teeth (single-tooth restorations and three-unit fixed dental prostheses) using one-piece ceramic implants [7].

From a technical point of view, such a 1-piece design, comprising the abutment and endosseous part in a single piece, might benefit from increased fracture resistance and reduced susceptibility for low-temperature degradation or so-called "aging" (by exposing a reduced total surface area to aging by inducing oral fluids), compared to 2-piece ceramic implants. Furthermore, 1-piece implants do not have a micro-gap in between the assembled implant and abutment. One might consider the absence of such a micro-gap beneficial, since it is capable in hosting bacteria, potentially resulting in marginal inflammation and consecutive bone resorption [8]. However, no advantage of a monobloc design was found for "seamless", 1-piece implants made from titanium [9]. Moreover, from a practitioner's point of view, a 1-piece implant design is associated with several surgical and prosthodontic shortcomings [10]. As an example, submerged implant healing is hardly possible, since the transmucosal part of a 1-piece implant cannot be detached. If no sufficient primary stability can be attained or guided bone regeneration is necessary, a missing option for wound closure might be considered disadvantageous. Furthermore, there is only a limited potential to compensate for mal-positioned implants with the provisional and final restoration. When trying to remove subsections in case of misaligned implants to support a bridge, intra-oral grinding of the zirconia abutment is necessary [11]. This, however, might have an impact upon the osseointegration (due to potential heat development or the displacement of zirconia particles in surrounding tissues) and fracture resistance of the implant [12]. Therefore, a two-piece design represents the favorable option for daily clinical use. Today, several two-piece zirconia implants are available on the market. In these systems, implant-abutment assembly is mostly realized by either luting the abutment to the implant or by screw-retention [13]. Luting the abutment to the implant seals the micro-gap, and allows for initial but irreversible correction of the implant angulation, but misses flexibility for future restorations of the implant. On the other hand, when going for screw-retention, several ceramic implants are still assembled with a titanium screw, and therefore, still not metal-free in the proper sense.

Even if the market share of zirconia dental implants increases, concerns regarding their fracture resistance are still present, and standardized testing protocols for zirconia implants adequately addressing the aging behavior of the final product are still missing [14]. To overcome this, different treatments were proposed to mimic intraoral conditions to the extent possible for the evaluation of ceramic implants. These treatments included thermal aging (high-temperature conditions or thermal cycling) [15,16] and/or dynamic loading procedures (various exposure times and different applied loading modes) [12,17]. Zirconia implants evaluated regarding their fracture resistance in the literature comprised a heterogeneous range of features like material selection (yttria-stabilized tetragonal zirconia polycrystal, Y-TZP or alumina-toughened zirconia, ATZ) [18], design (1- or 2-piece) [13], manufacturing (subtractive or by ceramic injection molding, CIM) [19], restoration (anatomical crown, hemisphere or no restoration) [20,21], abutment preparation (in the case of 1-piece implants) [22], or assembly (in the case of 2-piece implants) [13].

Therefore, the objective of the present systematic review was to evaluate the influence of the aforementioned treatments and features on the fracture resistance of zirconia oral implants in different preclinical studies. The null hypothesis supposed no distinction between treatments and features in relation to bending moment when statically loading the implant to fracture.

2. Materials and Methods

2.1. Study Design

To determine a selection of comparable studies on the question of zirconia implant fracture resistance, the preferred reporting items for systematic reviews and meta-analyses (PRISMA) statement of 2009 was applied [23]. Therefore, this report takes the appropriate Enhancing the Quality and Transparency of health Research (EQUATOR) (http://www.equator-network.org) guidelines into account.

2.2. Focused Question

Is there a variable significantly affecting the fracture resistance of 1- and 2-piece zirconia implants in preclinical in-vitro studies?

2.3. Search Strategy

Two databases, namely the Medical Literature Analysis and Retrieval System Online (MEDLINE) (PubMed) and Embase (accessed via Ovid), were screened for relevant articles. The database specific search strategies consisted of a combination of subject headings and free text words. Data was extracted from the databases on 3rd December 2019 without applying any time restrictions. Thereafter, references of included articles were screened for further records satisfying the inclusion criteria (cross-referencing). In case of the availability of the full methodological procedures in the literature and accessibility of information regarding the included samples, unpublished data of the authors of the present review was likewise included. The resulting studies were imported and stored in a reference managing program (EndNote X9; Clarivate Analytics, Philadelphia, PA, USA). Articles written in English and the German language were considered.

2.4. Screening Process

To build up the search terms, three categories addressing the samples (dental implants), materials (zirconia ceramics) and outcome (fracture load) were combined ("AND"). These categories consisted of combinations ("OR") of free text words and indexed vocabulary (MEDLINE: MeSH terms, Embase: Emtree terms). An asterisk was used in combination with some free text words as a truncation symbol (e. g. "ceramic *") to allow for the so-called "wildcard search".

Pubmed search term:

(((((dental implant [MeSH Terms]) OR (((oral) AND ((implant) OR implants))) OR ((dental) AND ((implant) OR implants))))) AND (((zircon *) OR ceramic *) OR ceramics[MeSH Terms])) AND (((((ageing) OR aging) OR artificial mouth) OR fracture resistance) OR load *)

Embase search term:

('tooth implant'/exp OR (oral AND implant) OR (dental AND implant)) AND (zircon * OR ceramic * OR 'ceramics'/exp) AND (ageing OR aging OR (artificial AND mouth) OR (fracture AND resistance) OR load *)

2.5. Eligibility Criteria

Studies to be included in this systematic review needed to fulfill the following inclusion criteria:

- Language: English or German

- Samples: Screw-shaped, ceramic oral implants containing a minimum of 50% v/v ZrO_2 within the bulk material
- Outcome: Static loading to fracture
- Outcome measure: Bending moment [Ncm or Nmm] or fracture load [N] allowing to calculate the bending moment (e.g., by adopting ISO 14801 or providing data to calculate the lever arm) was provided
- Sample size: Minimum of five samples tested

2.6. Selection of Studies

Concerning the inclusion criteria, both the first author and the senior author of this manuscript (A.B. and B.C.S.) independently screened the titles and abstracts of the extracted data in the reference management program. If sufficient information needed for inclusion or exclusion was not provided within the title or abstract, the corresponding full texts were read. In case of disagreement, a third author (S.P.) was consulted for final decision making.

2.7. Data Extraction

Besides the total number of samples within one study, the number of implants made from different materials (Y-TZP, ATZ), processing routes (subtractive, injection molding), design (1- and 2-piece) and diameters were retrieved. Further features like restoration mode (anatomical crown, hemisphere or no reconstruction), abutment preparation (yes/no in case of 1-piece implants), implant-abutment connection (screwed/bonded in case of 2-piece implants), thermal aging (thermal cycling, high temperature, no aging) or dynamic loading (yes/no), dynamic loading conditions (exerted load and amount of cycles), crosshead speed during static fracture, and angulation, were likewise extracted. This allowed us to group the implants finally subjected to static loading within the included studies in cohorts. For standardization purposes, the bending moment at the time point of fracture [Ncm] was considered the outcome measure of interest, and the corresponding authors of the articles to be included were contacted by email in case of solely providing fracture load values [N] without mentioning the lever arm. Extracted cohorts were subdivided into groups subjected to comparable treatments:

1. No dynamic loading
2. 1–1.2 million loading cycles (50 N)
3. 1–1.2 million loading cycles (100 N)
4. 3.5–5 million loading cycles (100 N)
5. 5 million loading cycles (>500 N)
6. 10 million loading cycles (100 N)

2.8. Statistical Analysis

From the included nineteen studies/datasets, two to twelve observations were extracted each. One observation consisted of the mean bending moment and standard deviation (at the time point of fracture) and/or mean fracture load and standard deviation (including additional information allowing us to calculate the bending moment) of a specific cohort of implants (comprising the same type of implant subjected to the same treatment) extracted from one included study. These observations had sample sizes of 2 to 12 implants. To analyze the effect of specific treatments of features (as indicated in 2.7) on the bending moment, a multilevel mixed-effects generalized linear model was used for each outcome, with each investigation as random effect to cluster observations by the respective studies. The Šidák method was used to correct for multiple testing. The level of significance was set at $p < 0.05$.

In order to compare the aforementioned groups (1–6, depending on load and cycles) for heterogeneity of the data, both inter- and intra-standard deviations with 95% confidence intervals (Cis) were computed. In addition, the cohort-specific standard error of the bending moment was used for weighting. Furthermore, box plots were created for visualization of the data. The data were analyzed with STATA 16.1 (StataCorp LLC, Texas, TX, USA).

3. Results

3.1. Screening Process/Included Data

Screening of two databases using the aforementioned specifically adapted search terms resulted in a total of 1864 records. After the removal of 622 duplicates, another 1202 records were withdrawn for analyses by screening the titles and abstracts. After reading the full texts of the remaining 40 studies, a further 23 manuscripts were excluded (Figure 1). Detailed reasons for exclusion can be found in Table A1. In general, the most frequent reasons for exclusion were the fracture of zirconia abutments assembled with titanium implants (mostly excluded by title and abstract) and the fracture on the restoration level using zirconia one-piece implants as support (mostly excluded during full-text screening). When only the fracture load [N] during static loading was reported, three options allowed for the calculation of the bending moment: (1) embedding was described to fully respect ISO 14801 (prescribing a lever arm of 5.5 mm allowing for the calculation of the bending moment), (2) all details regarding the embedding were provided in the manuscript (e.g., by providing a scheme) or (3) the bending moment and/or lever arm were provided by the authors upon request. As an example, six of the included studies adopted ISO 14801 for embedding [15,17,21,24–26], whereas three provided all necessary information [19,27,28] allowing us to calculate the bending moment (embedding level, angulation, total sample length, point of loading). In the remaining cases the bending moment was reported [13,20,22] or sent by the authors [12,18,29,30]. Finally, 17 full-texts were analyzed in the present systematic review (Table 1). In addition, the datasets of two finalized projects, currently under review and in preparation of the manuscript, were included. Two authors of the present review (R.K. and B.C.S.) were involved in both of these two investigations, and were able to access the full data. The applied materials and methods were already described in detail in precedent publications [21,26]. Since available on the market, the material composition of the included implant systems is likewise available and accessible. In detail, three zirconia implant systems (1-piece: Straumann PURE Ceramic, Straumann AG, Basel, CH; 2-piece: 5s-50-10, Z-Systems AG, Oensingen, CH and Ceralog Hexalobe Implant, Axis biodental, Les Bois, CH) were subjected to identical treatments and fracture load measurements, as described in two of the included studies [21,26]. In the case of Straumann 1-piece (as-received: 609 ± 20 Ncm; loaded/aged: 557 ± 36 Ncm) and Z-Systems 2-piece implants (as-received: 463 ± 21 Ncm; loaded/aged: 443 ± 39 Ncm), aging/loading (as described in [21,26]) did not affect the fracture resistance to a statistically significant level ($p = 0.171$). In contrast, the fracture resistance of 2-piece Ceralog Hexalobe Implants (as-received: 547 ± 89 Ncm; loaded/aged: 413 ± 127 Ncm) was significantly affected ($p = 0.046$) by aging/loading (as described in [21,26]).

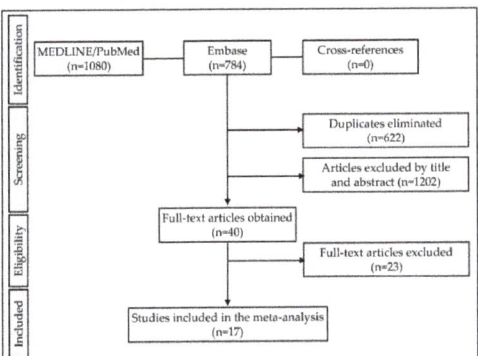

Figure 1. Flowchart according to the preferred reporting items for systematic reviews and meta-analyses (PRISMA) guidelines.

Table 1. A total of 731 one- and two-piece implants made from yttria-stabilized tetragonal zirconia polycrystal (Y-TZP) and alumina-toughened zirconia (ATZ), extracted from 17 studies and two unpublished datasets, subjected to different dynamic loading and thermal aging conditions prior to static loading to fracture, were finally included in meta-analyses.

First Author	Year	Ref.	n	Material	Pieces	Loading Cycles (×10^6)	Thermal Aging
Andreiotelli	2009	[29]	88	Y-TZP	1	0, 1.2	TC, none
Kohal	2009	[30]	32	Y-TZP	2	0, 1.2	TC, none
Kohal	2010	[18]	72	ATZ, Y-TZP	1	0, 1.2, 5	TC, none
Kohal	2011	[12]	48	Y-TZP	1	0, 1.2, 5	TC, none
Rosentritt	2014	[28]	36	Y-TZP	1, 2	1.2	TC
Kohal	2015	[20]	48	Y-TZP	1	0, 5, 10	TC, none
Sanon	2015	[25]	30	Y-TZP	1	0	HT
Spies	2015	[22]	48	ATZ	1	0, 1.2, 5	TC, none
Kammermeier	2016	[27]	30	Y-TZP	1, 2	0, 3.6	TC, none
Preis	2016	[19]	32	ATZ, Y-TZP	2	1	TC, none
Spies	2016	[13]	48	ATZ, Y-TZP	1, 2	0, 10	HT, none
Joda	2017	[24]	11	ATZ	2	0	none
Spies	2017	[21]	28	Y-TZP	2	0, 10	HT, none
Ding	2018	[17]	29	Y-TZP	1	0, 5	none
Spies	2018	[26]	14	ATZ	2	0, 10	HT, none
Monzavi	2019	[15]	60	Y-TZP	1	0	HT, none
Stimmelmayr	2019	[16]	36	Y-TZP	2	1.2	TC
Kohal	2020	*	28	Y-TZP	1, 2	0, 10	HT, none
Zhang	2020	*	13	Y-TZP	2	0, 10	HT, none

* Unpublished data, Ref. = Reference, n = total number of included implants, TC = thermal cycling, HT = high temperature.

3.2. Meta-Analyses

All 17 articles published between 2009 [29] and 2019 [15,16] were included and analyzed in the present meta-analysis. Moreover, unpublished data of two projects currently under review and in preparation of the manuscript were included (Table 1). From the included articles/datasets, 114 observations were extracted or calculated (mean bending moment), comprising different implant features (e.g., diameter, material, crown supply, abutment preparation or implant-abutment-connection) or treatments (e.g., thermal aging or dynamic loading). One observation consisted of the mean bending moment and standard deviation (SD) of up to 12 included implants.

In order to evaluate the impact of different dynamic loading procedures (implants were subjected to prior to fracture loading) on the outcome (bending moment), groups as indicated in Section 2.7 were analyzed for heterogeneity. As a result, standard deviation as a measure of variation within and in between the included studies revealed to be within the same range (Table 2). No heterogeneity of the bending moments for groups 1–6 was found, even if a decreased mean value for group 3 was calculated ($p = 0.612$). This did not change when stratifying the implants according to their design (1-piece: $p = 0.951$; 2-piece: $p = 0.056$).

Table 2. Groups 1–6 (as indicated in 2.7) were tested for heterogeneity regarding the outcome.

Groups	Overall	1	2	3	4	5	6
Effect [1]	395.27	407.42	397.99	262.17	400.73	579.96	448.94
95% CI	330.2–460.3	338.4–476.4	272.1–523.9	195.0–329.3	249.2–552.2	521.8–638.0	373.7–524.1
Intra [2]	103.57	110.07	74.58	100.30	150.33	46.64	57.59
95% CI	89.4–120.0	89.3–135.7	42.4–131.0	61.4–163.8	95.8–235.8	18.2–119.7	28.5–116.4
Inter [3]	126.06	126.92	133.58	$1.670 \times e^{-15}$	146.81	$4.690 \times e^{-18}$	77.72
95% CI	86.5–183.8	82.1–196.2	65.5–272.3	$-\infty$–$+\infty$	59.0–365.5	$-\infty$–$+\infty$	33.9–178.2

[1] Mean bending moment [Ncm], [2] Standard deviation/variation within included studies, [3] Standard deviation/variation in between included studies.

3.3. Outcomes

Outcomes extracted from the 17 included studies and the two unpublished datasets were calculated and stratified for the material selection, manufacturing, implant diameter, anatomical crown supply, abutment preparation (1-piece implants), implant-abutment-connection (IAC; 2-piece implants), thermal aging procedure prior to static loading (none; TC = thermal cycling, mostly in between 5–55 °C; HT = high temperature, mostly in between 60–134 °C) and/or dynamic loading in a chewing simulation device applying different loads (ranging from 50 to > 500 N) for a different amount of cycles (ranging from 1 to 10 millions). In total, 731 implants were available for analyses, revealing a mean bending moment at the time point of fracture of 386.4 ± 167.6 Ncm. Furthermore, the outcome was stratified for 1- and 2-piece implants. Mean bending moments, standard deviations and the included number of implants are listed in Table 3. Significance (linear mixed models, level of significance $p < 0.05$) calculated for differences regarding the implant design, different covariables and treatments can be found in Table 4.

Table 3. Calculated mean bending moment (in Ncm) and standard deviation depending on the implant design, several covariables and treatments.

	Overall [1]			1-Piece			2-Piece		
	n	Mean	SD	n	Mean	SD	n	Mean	SD
Overall	731	386.4	167.6	495	431.9	151.0	236	291.7	162.4
Material									
Y-TZP	577	378.7	160.1	383	422.2	143.4	194	284.3	155.7
ATZ	154	418.7	106.0	112	475.8	180.7	42	318.6	194.0
Manufacturing									
Subtractive	591 [2]	397.5	177.4	417	457.4	154.4	174	260.1	149.6
Injection molded	120 [2]	364.8	116.7	70	329.4	73.7	50	426.8	154.4
Implant diameter									
3.0–3.3 mm	15	207.2	14.3	9	215.0	6.7	6	191.6	-
3.8–4.4 mm	675	394.9	170.4	463	441.3	152.7	212	293.6	165.0
4.5–5.0 mm	41	349.4	125.4	23	388.0	59.4	18	301.2	178.0
Anatomical crown supply									
Yes	209	237.5	96.6	74	327.0	65.4	135	186.9	71.4
No	522	455.2	147.7	421	453.2	154.8	101	463.9	114.2
Abutment preparation									
Yes	-	-	-	112	411.3	126.2	-	-	-
No	-	-	-	383	436.5	156.5	-	-	-
Implant-Abutment-Connection									
Screw-retained	-	-	-	-	-	-	159	327.5	179.0
Bonded	-	-	-	-	-	-	77	217.0	86.0
Thermal aging									
Thermal cycling	310	355.5	171.7	218	426.5	149.4	92	174.6	41.1
High temperature	124	392.9	115.9	75	362.6	96.4	49	453.4	135.1
None	297	406.2	180.4	202	464.0	163.2	95	299.9	164.6
Dynamic loading									
Yes	391	389.4	169.2	250	447.7	146.6	141	279.2	156.5
No/Group 1	340	383.2	166.3	245	417.7	153.6	95	303.5	171.2
Group 2	86	258.1	111.5	66	362.6	59.4	20	174.4	50.2
Group 3	76	394.7	211.2	40	457.3	188.1	36	144.4	15.0
Group 4	132	379.6	159.7	96	437.8	140.5	36	205.1	20.5
Group 5	17	580.8	55.7	17	580.8	55.7	-	-	-
Group 6	80	435.1	108.3	31	420.2	93.0	49	443.6	122.5

n = number of included implants, SD = standard deviation, [1] 1- and 2-piece implants pooled together, [2] the authors of one included study could not provide the manufacturing mode for all included implants [28].

Table 4. Significance (linear mixed models (LMMs), level of significance $p < 0.05$) was calculated for differences regarding the implant design, different covariables and treatments.

Parameter	Options	Significance (p)		
		Overall [1]	1-Piece	2-Piece
Implant design	1-piece, 2-piece	0.004	-	-
Material	Y-TZP, ATZ	0.002	0.001	0.282
Manufacturing	Subtractive, injection-molded	0.749	0.076	0.095
Implant diameter	Range: 3.3–5.0 mm	0.327	0.273	0.191
Anatomical crown	Yes/No	<0.0001	0.080	<0.0001
Abutment preparation	Yes/No	-	<0.0001	-
Connection type	Screw-retained, bonded	-	-	0.584
Thermal aging	TC, HT, none	0.446	0.538	0.776
	Yes/No	0.410	0.559	0.474
Dynamic loading	Applied load [range: 50–500 N]	0.050	0.181	0.202
	Amount of cycles [range: 1–10 × 10^6]	0.238	0.971	<0.0001
	Groups 1–6 [as indicated in 2.7]	0.612	0.951	0.056
Angulation	Range: 30–45°	0.215	0.671	0.003
Crosshead speed	Range: 0.5–10 mm/s	0.261	0.562	<0.0001

[1] 1- and 2-piece implants pooled together, TC = thermal cycling, HT = high temperature.

3.3.1. Implant Design

Eight studies [12,15,17,18,20,22,25,29] focused on 1-piece zirconia implants, whereas six studies solely included 2-piece implants [16,19,21,24,26,30]. The remaining investigations evaluated a mixture of both 1- and 2-piece implants [13,27,28]. Regardless of all other variables, 1-piece implants (431.9 ± 151.0 Ncm) were found to be more fracture resistant than 2-piece implants (291.7 ± 162.4 Ncm, $p = 0.004$; Figure 2).

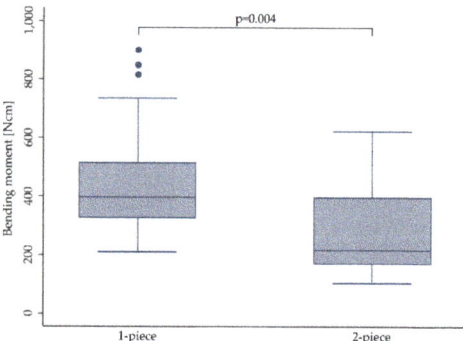

Figure 2. Boxplot showing the bending moment at the time point of fracture for 1- and 2-piece zirconia implants. Whiskers are used to represent all samples lying within 1.5 times the interquartile range (IQR). Dots represent outliers. Detailed data can be found in Tables 3 and 4.

3.3.2. Material

Material selection of the included studies is listed in Table 1. Of the included implants, 577 were made from Y-TZP, whereas 154 were manufactured from ATZ [13,18,19,22,24,26]. When pooling the outcome for 1- and 2-piece zirconia implants, the bending moment at the time point of implant fracture was significantly affected by the material ($p = 0.002$; Table 4). In detail, implants made from alumina-toughened zirconia (ATZ, 418.7 ± 106.0 Ncm) were more fracture-resistant compared to implants made from yttria-stabilized tetragonal zirconia polycrystals (Y-TZP, 378.7 ± 160.1 Ncm,

$p = 0.002$). When stratifying the outcome for 1- and 2-piece implants, however, material selection only affected 1-piece implants ($p = 0.001$, Figure 3a), whereas 2-piece implants performed the same, regardless of the material selection ($p = 0.282$, Figure 3b).

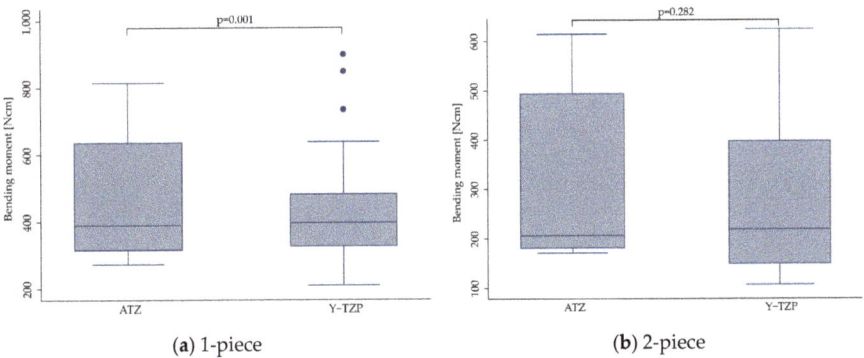

(a) 1-piece

(b) 2-piece

Figure 3. Boxplots showing the bending moment at the time point of fracture depending on the material selection for 1- (a) and 2-piece (b) zirconia implants. Whiskers are used to represent all samples lying within 1.5 times the interquartile range. Dots represent outliers. Detailed data can be found in Tables 3 and 4.

3.3.3. Manufacturing

Manufacturing was mostly subtractive ($n = 591$ implants), but ceramic injection-molding (CIM) was likewise used for the production ($n = 120$ implants) [15,19,21,25]. There was no statistically significant difference in the fracture resistance of implants when manufacturing method (subtractive: 397.5 ± 177.4 Ncm, CIM: 364.8 ± 116.7 Ncm) was regarded ($p > 0.095$). Boxplots can be seen in Figure 4.

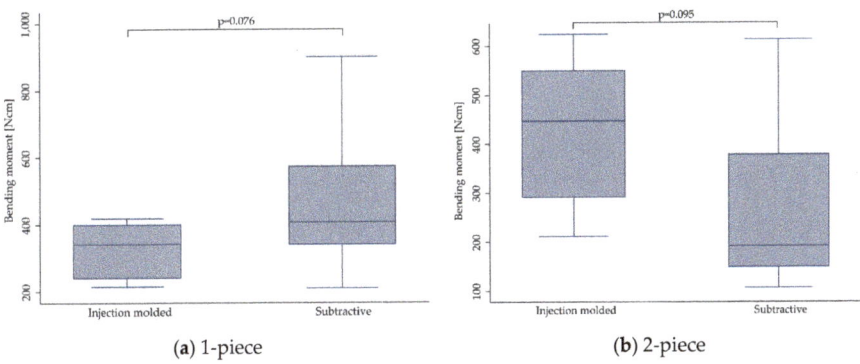

(a) 1-piece

(b) 2-piece

Figure 4. Boxplots showing the bending moment at the time point of fracture depending on the manufacturing method for 1- (a) and 2-piece (b) zirconia implants. Whiskers are used to represent all samples lying within 1.5 times the interquartile range. Detailed data can be found in Tables 3 and 4.

3.3.4. Implant Diameter

No statistically significant difference could be calculated for the bending moment at the time point of fracture regarding the implant diameter ranging from 3 to 5 mm ($p = 0.327$). This did not change when stratifying the outcome for 1- ($p = 0.273$) and 2-piece ($p = 0.191$) implants. However, the included studies evaluated only very few implants in the range of 3 mm (range: 3.0–3.3 mm; $n = 15$, 207.2 ± 14.3 Ncm) [24,27] and 5 mm (range: 4.5–5.0 mm; $n = 41$, 349.4 ± 125.4 Ncm) [15,19,27,28], whereas the majority of implants had a diameter in the range of 4 mm (range: 3.8–4.4 mm; n = 675, 394.9 ± 170.4 Ncm). Boxplots can be seen in Figure 5.

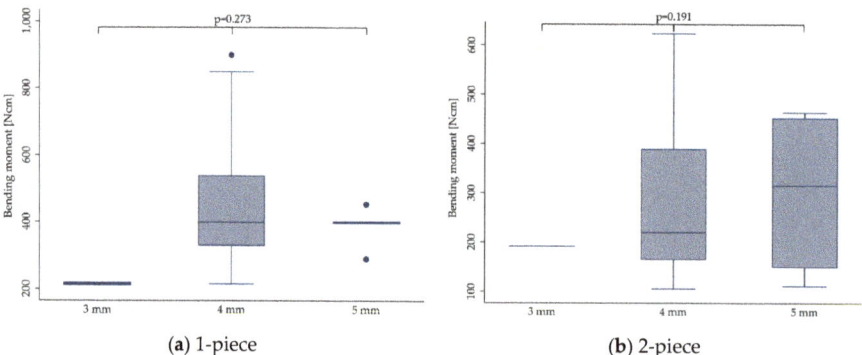

(a) 1-piece   (b) 2-piece

Figure 5. Boxplots showing the bending moment at the time point of fracture depending on the implant diameter for 1- (**a**) and 2-piece (**b**) zirconia implants. Whiskers are used to represent all samples lying within 1.5 times the interquartile range. Dots represent outliers. Detailed data can be found in Tables 3 and 4.

3.3.5. Anatomical Crown Supply

Of the included 731 implants, 209 were restored with an anatomically shaped crown, mostly made from ceramic materials. Most of the crowns were designed to replace maxillary central incisors but also some premolar reconstructions were included. The remaining 522 implants did not receive any reconstruction and were directly loaded to the abutment or were equipped with a non-anatomical stainless-steel hemisphere according to ISO 14801. When pooling the data for 1- and 2-piece implants, anatomical crown supply (237.5 ± 96.6 Ncm) negatively affected the outcome compared to implants with no crowns or equipped with a hemisphere (455.2 ± 147.7 Ncm, $p < 0.0001$). When stratifying for 1- and 2-piece implants (Figure 6), statistical significance was only reached for the group of 2-piece implants ($p < 0.0001$), likewise revealing an inferior outcome for implants restored with anatomical crowns. Fracture resistance of 1-piece implants was not affected by crown supply ($p = 0.080$).

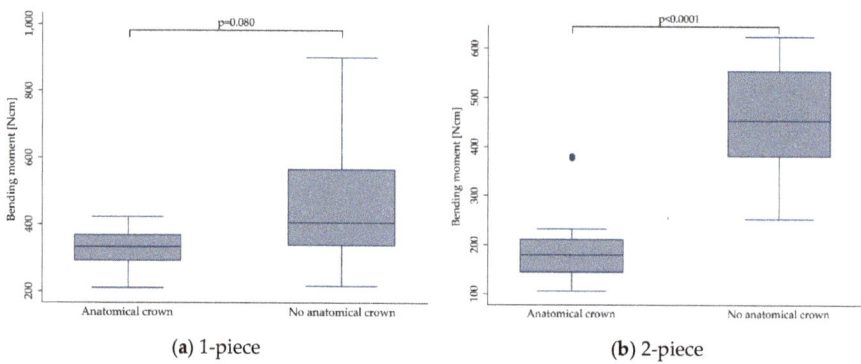

(a) 1-piece   (b) 2-piece

Figure 6. Boxplots showing the bending moment at the time point of fracture depending on the crown supply for 1- (**a**) and 2-piece (**b**) zirconia implants. Whiskers are used to represent all samples lying within 1.5 times the interquartile range. Dots represent outliers. Detailed data can be found in Tables 3 and 4.

3.3.6. Abutment Preparation and Implant-Abutment-Connection (IAC)

Of the 1-piece implants (n = 495), 112 abutments were prepared/modified by grinding [12,18,22,29], whereas 383 abutments remained untouched until fracture. In most cases, abutment preparation should simulate a clinically relevant situation of a 1-piece implant installed in anterior regions of the mouth. In both groups, some implants were restored with anatomically shaped incisor crowns, and some did not receive any reconstruction. Grinding of the abutment (411.3 ± 126.2 Ncm) resulted in a significantly reduced bending moment at the time point of fracture compared to non-grinded implants (436.5 ± 156.5 Ncm, $p < 0.0001$; Figure 7a).

Of the two-piece implants included in the present review (n = 236), 159 abutments were assembled by screw retention [13,16,19,21,24,26]. Most screws were made from titanium, but also gold and polyetheretherketone (PEEK; in one study, carbon-fiber-reinforced [26]) were used. The remaining 77 two-piece implants were irreversibly assembled by adhesive bonding [13,19,27,28,30]. The type of abutment retention (screw-retained: 327.5 ± 179.0 Ncm, bonded: 217.0 ± 86.0 Ncm) did not affect the fracture resistance ($p = 0.584$; Figure 7b).

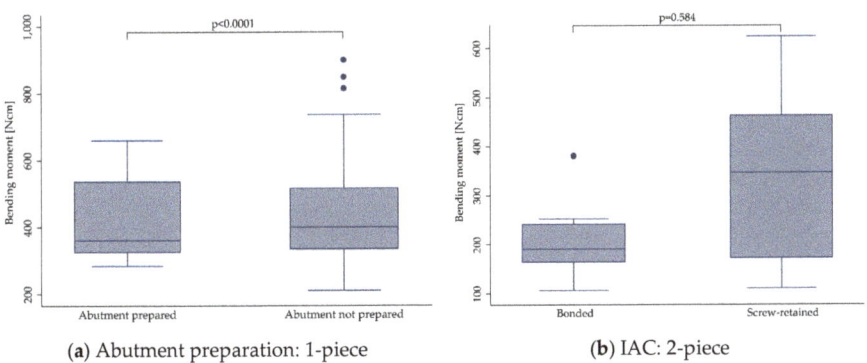

(a) Abutment preparation: 1-piece   (b) IAC: 2-piece

Figure 7. Boxplots showing the bending moment at the time point of fracture depending on the abutment preparation for 1-piece (a) and depending on the implant-abutment-connection (IAC) for 2-piece (b) zirconia implants. Whiskers are used to represent all samples lying within 1.5 times the interquartile range. Dots represent outliers. Detailed data can be found in Tables 3 and 4.

3.3.7. Thermal Aging

Regardless of the implant design, in 297 implants, no aging was induced prior to static loading to fracture, whereas 124 implants were subjected to a high temperature (HT) treatment in a humid environment, ranging from 60 up to 134 °C for different time periods lasting from 5–30 h (134 °C) [15,25] to 60 days (85 °C) [21,26]. High temperature treatment was applied in combination or during dynamic loading or alone. The remaining 310 implants were subjected to a thermal cycling (TC) procedure, exposing the samples to a changing water bath set at 5 and 55 °C [12,16,18–20,22,27–30]. The latter was mostly performed during dynamic loading in a chewing simulation device. Compared to untreated implants (406.2 ± 180.4 Ncm), neither HT treatment (392.9 ± 115.9 Ncm) nor TC (355.5 ± 171.7 Ncm) did affect the fracture resistance ($p = 0.446$). This did not change when calculating the outcome for 1- ($p = 0.538$) and 2-piece implants ($p = 0.776$) separately (Figure 8).

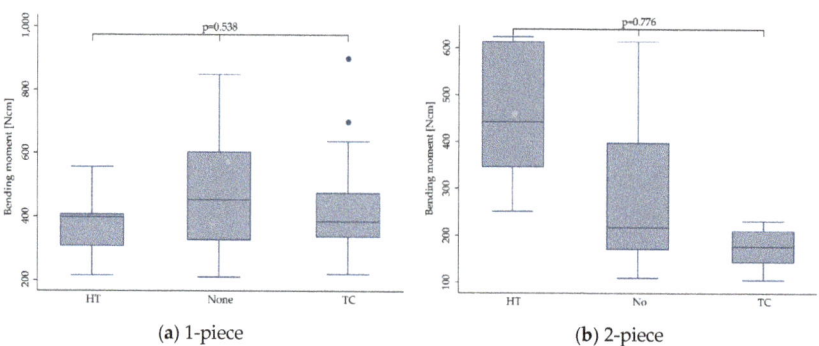

(a) 1-piece   (b) 2-piece

Figure 8. Boxplots showing the bending moment at the time point of fracture, depending on the thermal aging conditions (none, HT = high temperature, TC = thermal cycling) for 1- (**a**) and 2-piece (**b**) zirconia implants. Whiskers are used to represent all samples lying within 1.5 times the interquartile range. Dots represent outliers. Detailed data can be found in Tables 3 and 4.

3.3.8. Dynamic Loading

The effect of dynamic loading was evaluated from different perspectives. The simplest one assigned the included implants to two categories subjected to either no dynamic loading procedure ("No") or those being subjected to dynamic loading ("Yes"; Figure 9, Table 4). Furthermore, the effect of dynamic loading was evaluated regarding the dynamically "applied load", ranging from 45 [30] up to more than 500 N [17], or regarding the "amount of cycles" ranging from 1.2 [12,16,18,22,28–30] to 10 million [13,20,21,26] loading cycles. Finally, a combination of "applied load" and "amount of cycles" was used to from six groups, as mentioned in Section 2.7 (Figure 10).

When pooling the extracted data for 1- and 2-piece implants, dynamic loading did not affect the fracture resistance (dynamically-loaded implants showed a mean bending moment at the time point of fracture of 389.4 ± 169.2 Ncm compared to 383.2 ± 166.3 Ncm calculated for non-loaded implants ($p = 0.410$)). This did not change when evaluating 1- and 2-piece implants separately ($p > 0.474$). Solely the category "applied load" was close to statistical significance ($p = 0.05$). However, none of the multiple pairwise comparisons comparing different dynamically applied loads showed a statistically significant difference ($p > 0.07$). When solely evaluating 2-piece implants, "amount of cycles" significantly affected the fracture resistance ($p < 0.0001$), whereas "applied load" ($p = 0.202$) and groups 1–6 respecting the applied load and the amount of cycles ($p = 0.056$) did not affect the outcome.

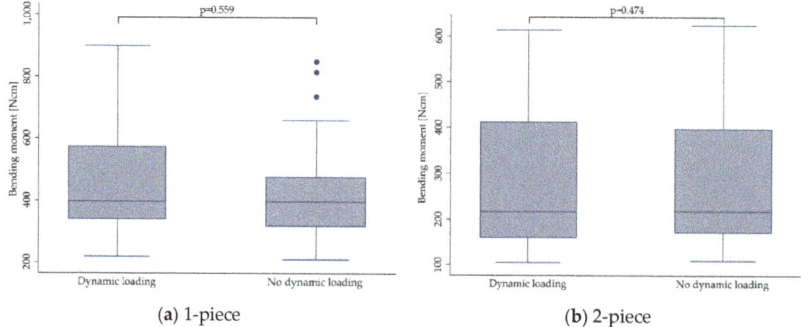

(a) 1-piece   (b) 2-piece

Figure 9. Boxplots showing the bending moment at the time point of fracture depending on dynamic loading (Yes: Implants were subjected to dynamic loading, No: Implants were not dynamically loaded) for 1- (**a**) and 2-piece (**b**) zirconia implants. Whiskers are used to represent all samples lying within 1.5 times the interquartile range. Dots represent outliers. Detailed data can be found in Tables 3 and 4.

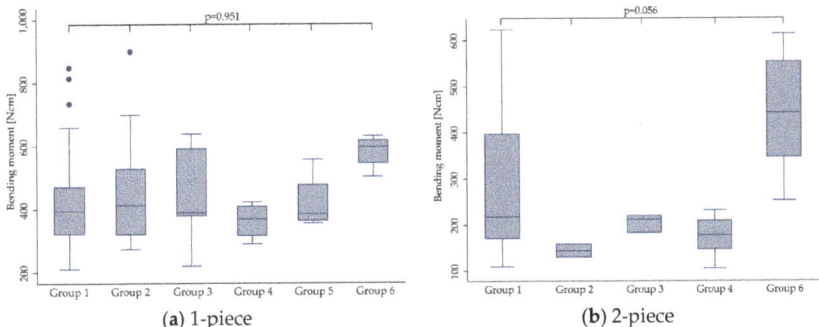

(a) 1-piece

(b) 2-piece

Figure 10. Boxplots showing the bending moment at the time point of fracture depending on dynamic loading conditions respecting the applied load and amount of cycles (as categorized in Section 2.7) for 1- (**a**) and 2-piece (**b**) zirconia implants. Whiskers are used to represent all samples lying within 1.5 times the interquartile range. Dots represent outliers. Detailed data can be found in Tables 3 and 4. No 2-piece implants were allocated to group 5.

4. Discussion

The present systematic review and meta-analysis included the data of 17 studies and two unpublished datasets. To be finally able to compare the outcomes of the included data, it was necessary to extract or calculate the bending moment at the time point of implant fracture [Ncm], since the mostly reported fracture load values [N] do not respect the leverage (length of the lever arm) and are therefore, if not considering a rigorously standardized embedding procedure as described in ISO 14801, not comparable to each other. Of the included 19 investigations/datasets, three studies reported the bending moment individually calculated for each included implant [13,20,22], whereas six studies [15,17,21,24–26] and the two included unpublished datasets fully respected ISO 14801 for embedding. Fully respecting this ISO implies the fixation of the endosseous part in a rigid clamping device or embedding in a material with a modulus of elasticity higher than 3 GPa. Moreover, the embedding/clamping level should respect a distance of 3.0 ± 0.5 mm apically from the nominal bone level, as specified in the manufacturer's instructions for use. Furthermore, implant abutments need to be equipped with a non-anatomical hemisphere designed to realize a distance of $l = 11.0 \pm 0.5$ mm from the center of the hemisphere to the embedding level (Figure 11).

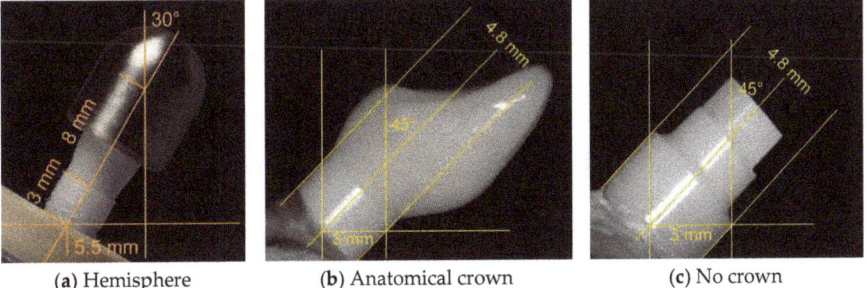

(a) Hemisphere (b) Anatomical crown (c) No crown

Figure 11. Exemplary schemes of embedded implants according to ISO 14801 (**a**) [21], equipped with an anatomically shaped incisor crown (**b**) or without any restorative supply (**c**) [20]. When embedding the samples according to ISO 14801, the lever arm measures 5.5 mm. In the latter two cases, the lever arm needs to be individually calculated and reported.

When loading such samples with an angle of α = 30° to the vertical, the lever arm (y) or bending moment (M) for this configuration can be calculated with the reported fracture load (F) by using Equation (1).

$$M = y \cdot F = \sin \alpha \cdot l \cdot F \tag{1}$$

This results in y = 0.55 cm when embedding according to ISO 14801. For the aforementioned publications/datasets fully respecting ISO 14801 for embedding and reporting the fracture load values [N], the bending moment was therefore calculated by multiplying the fracture load with 0.55. Interestingly, some of the included investigations reported embedding according to ISO 14801, but solely adopted the embedding level (simulation of a bony recession of 3 mm), and sometimes the angulation (30°), but did not use a loading hemisphere, finally resulting in a lever arm different to 5.5 mm, as proposed by the ISO standard [16,19,27,28]. In most cases, anatomical crowns (maxillary premolars or incisors) made from ceramic materials were used instead of the hemisphere, finally resulting in altered lever arms and loading conditions. In the investigations of one group, the crown design and embedding procedure were described in detail (l, α and F were reported), allowing us to calculate y and M [19,27,28]. To calculate the bending moment for the remaining studies, authors needed to provide the necessary data upon request or standardized photographs provided in the publications, or by the authors needed to allow the approximation of the lever arm by using an image analysis software (ImageJ, National Institutes of Health, Bethesda, MD, USA) [12,16,18,29,30]. In order to be able to compare the outcome of preclinical studies evaluating the fracture resistance of dental implants, it is therefore recommended to either fully adopt an ISO standard for the embedding procedure or to provide the bending moment additionally to the fracture load. Considering different lever arms due to different embedding procedures for the implants included in this systematic review and meta-analysis, one needs to keep in mind that dynamic loading prior to static loading to fracture can result in altered fatigue, even if the applied load was the same.

The heterogeneity of the included samples comprising a mixture of market-available products (finally sterilized and incorporating a micro-roughened surface) [15,16,22,24,26] but also prototype implants (e.g., with or without any surface post-processing) [13,19,21,25,28,30] represents a major limitation of the present systematic review and meta-analysis. However, it was shown that, for example, surface modifications like micro-roughening to enhance osseointegration or steam-sterilization can significantly compromise fracture strength and ageing kinetics [31,32].

Another shortcoming of this systematic review presents the fact that of the 19 included datasets, more than half (nine published and two unpublished studies) were at least partially authored by the collaborates of the current paper. This might be considered a reasonable risk of bias. However, the present review was conducted according to standardized guidelines, and the available literature was systematically screened on the basis of predefined search terms and inclusion criteria. Modifying the search strategy, outcome measure or inclusion criteria in consequence of unexpected or homogeneously authored findings would likewise present a source of bias.

Regarding the treatments, the included samples have been subjected to prior to loading, and six groups (representing different categories of loading conditions as indicated in Section 2.7) have been evaluated for heterogeneity of the outcome. As a result, no heterogeneity of the bending moments for groups 1–6 was found ($p = 0.612$). This did not change when stratifying the implants according to their design (1-piece: $p = 0.951$; 2-piece: $p = 0.056$). Therefore, it was decided to pool the data of all groups for any further calculations, and yet still, one can hardly generalize the present findings and apply them to a specific zirconia implant system.

No statistically significant influence of hydrothermal aging on the fracture resistance of zirconia implants was calculated in the present review. It is important to note that aging or so-called low-temperature degradation (LTD) can, depending upon the sample quality and surface conditions, result in both increased [21,25] and decreased [33] fracture load. This might be explained by the following: Assuming a zirconia sample surface with various process-related defects/impurities, the largest defects/impurities are thought to act as "locus minoris resistentiae", and can thereby be considered representative for the fracture resistance of this sample. Increased fracture load of such zirconia samples after a hydrothermal aging procedure is thought to be attributed to a transformed layer at the sample surface, inducing a compressive stress on the surface, tending to close a potential advancing crack at such existing defects/impurities located on the surface. This phenomenon is liable to cause an increase in the strength of the material, and was described for the first time three decades ago [34]. On the other side, at some point when the degradation process penetrates deeper into the material, the contribution from the aging may instead cause the strength of the same sample to be decreased, since once transformed to the monoclinic, zirconia grains cannot exhibit stress-induced phase transformation toughening anymore [33]. As an example, in the included investigation of Monzavi and co-workers [15] the effect of artificial aging on the mechanical resistance and micromechanical properties of commercially- and noncommercially-available zirconia dental implants was evaluated. In this study, the bending moment was significantly increased after aging for three of six groups, whereas two groups showed no influence of the aging procedure, and one group was negatively affected in terms of fracture resistance by the treatment [15]. When pooling the outcomes of the included studies showing positive, negative or no effects of LTD on the fracture resistance of zirconia implants in one dataset, as happened in the present meta-analyses, no effect of hydrothermal aging on the bending moment at the time point of fracture was calculated ($p > 0.446$). This, however, might be misleading, since several of the included studies indeed showed that aging can significantly affect the fracture resistance. However, due to the explanation given at the beginning of this paragraph, both in a negative or positive way. Therefore, missing significance, as calculated for pooled data in this review, should not be interpreted as an argument to refrain from aging tests of a zirconia implant system prior to market release. Therefore, pooling the data from different studies using the different conditions of thermal aging needs to be considered a limitation of the present review. It is discussed in the literature that the present amount of transformation to the monoclinic on the surface of as-delivered zirconia implants can be decisive for the ongoing fracture resistance after further hydrothermal aging procedures. In detail, implants showing no or very limited transformation to the monoclinic when released to the market (e.g., due to final temperature annealing [35] or manufacturing by ceramic injection-molding [21,25]) were observed to be less fracture-resistant in the original as-delivered state, but significantly gained fracture resistance due to increasing compressive stress at the sample surface after transformation to the monoclinic occurred. In contrast, samples already revealing a transformed layer of several micrometers (e.g., due to subtractive manufacturing or post-processing steps like sandblasting in order to roughen the surface to enhance osseointegration [26]) mostly do not benefit from further aging by means of an increased fracture resistance. Besides the amount of already transformed grains, implant surface topography showed to have a significant impact on aging susceptibility and its impact on fracture resistance [32,36]. As an example, implants structured with porous or alveolar surfaces were more likely to be negatively affected by aging procedures due to interconnected porosities in the surface layer, offering a path for the transformation to start at every surface accessible by water [25]. Finally, a layer structured in this way can be transformed in a shorter period of time.

Of the implants included in the present investigation, 209 of 731 were restored with anatomically-shaped crowns [16,19,20,27–30]. Most of these crowns were designed as maxillary central incisors, and were manufactured from: lithium disilicate [20], veneered [29] or monolithic [19,27,28] zirconia, or porcelain fused to metal [30]. Another included study restored the implants with maxillary first premolar restorations made from lithium disilicate [16], whereas Joda and collaborates restored the implants with non-anatomical hemispheres likewise made from lithium disilicate [24].

Most of the included studies not restoring the implants with anatomically-shaped crowns were conducted by adopting ISO 14801. According to this standard, the loading force shall be applied to the hemispherical loading surface, by a loading device with a plane surface normal to the loading direction of the machine, without additional horizontal loading forces. In contrast, especially incisor crowns present an inclined plane when loaded during the dynamic and finally static loading procedure, resulting in an increased shear force. Additionally, some investigations applied horizontal forces during the dynamic loading procedure (as it happens in the oral cavity), causing further fatigue of the sample [20,29,30]. Therefore, not the restoration itself, but the altered investigational setup, resulting in increased shear forces and fatigue during static loading, and in some cases, precedent chewing simulation might be considered responsible for decreased fracture resistance. Nonetheless, this finding should be taken into account when drafting international standards in order to guarantee clinical safety, since the anatomical reconstruction of zirconia oral implants and horizontal shear forces during loading represent clinical reality. Regarding the nature or location of failure, 1-piece implants mostly fractured at the embedding level or slightly below, with crack initiation on the tensile side of the implant. As described in the included studies, it seems that the fracture mode was not affected by crown supply. In 2-piece implants, fracture modes were generally observed to be highly heterogeneous, depending on the mode of assembly and the materials used.

When it comes to clinical reality, the fracture resistance of a zirconia implant should finally withstand the maximum voluntary bite forces of the patients. Nonetheless, one cannot find the definition of any indication specific (e.g., for implants installed in anterior or posterior regions) minimum value for the fracture strength of a zirconia implant in ISO 14801. This, as an example, is provided in detail in ISO 6872 for ceramic materials used for reconstructions (e.g., crowns, bridges) in dentistry [37]. Taking the highest bending moment measured in vivo (95 Ncm) with the help of strain gauge abutments into account [38], and applying a safety buffer of 100%, one might consider a minimum fracture resistance of 200 Ncm sufficient to guarantee clinical safety. When applying this requirement to the included studies, mostly 2-piece prototype implants and implants with a reduced diameter ($\leq$ 3.3 mm) did not meet this demand [19,24,27,28,30].

Of the zirconia implants included in the present investigation, 577 were manufactured from Y-TZP and 154 from ATZ. Overall, implant stability was significantly affected by the material, in favor of ATZ ($p = 0.002$). When evaluating 1- and 2-piece implants separately, however, only 1-piece implants made from ATZ performed better ($p = 0.001$), whereas 2-piece implants performed the same, regardless of the material selection ($p = 0.282$). This might be explained by the fact that 1-piece zirconia implants or even, as an example, 2-piece titanium implants are mostly made from one single material (in the case of titanium: the implant, the abutment and the abutment screw are mostly fabricated from titanium). In contrast, most of the available 2-piece zirconia systems represent a multi-material complex comprising at least two or sometimes even three different materials. In some cases, only the implant body is manufactured from zirconia, whereas the screw (e.g., titanium or PEEK) and/or abutment (e.g., glass-fiber or polyetherketoneketone/PEKK) might be manufactured from different materials revealing different aging or degradation behavior during treatments (hydrothermal aging, dynamic loading), precedent to final static loading to fracture. To date, sound correlations to approximate intraoral aging conditions in an accelerated way in the dental laboratory are mostly available for zirconia ceramics, but missing for screw and abutment materials prone to degradation in aqueous environments, like e.g., polyetherketones [39,40]. In consequence, no standardized testing procedures were proposed to the present date, sufficiently evaluating multi-material, 2-piece implants regarding their fracture resistance, and individually respecting the degradation behavior of several included components. Regrettably, the sample size and heterogeneity of the extracted data gathered from 2-piece implants included in the present review did not allow for the statistical evaluation of a potential impact of the screw or abutment material on the fracture resistance of 2-piece zirconia implants. In one of the included studies, the aim was to measure the abutment rotation and fracture load of 2-piece zirconia implants screwed with three different abutment screw materials [16]. Implants and abutments of the included system were assembled with screws made from gold, titanium and PEEK.

As a result, no significant differences were found for these three materials, even if PEEK screws showed inferior results. When choosing PEEK as an abutment screw material, the incorporation of continuous carbon fibers proved to positively affect the maximum tensile strength of the screw [41]. However, a strengthening effect on the entire implant-abutment complex in case of zirconia implants still needs to be evidenced. In one of the included studies [26], a 2-piece ATZ implant system assembled with a carbon-fiber-reinforced abutment screw showed to be non-inferior compared to a market-established 2-piece titanium implant of a highly comparable design regarding its fracture resistance.

5. Conclusions

The null hypothesis of the present review, supposing no distinction between treatments and features in relation to bending moment when statically loading a zirconia implant to fracture, needs to be partially rejected. The focused question can be answered as follows: In general, 1-piece implants can be considered more fracture resistant than 2-piece implants, even if some of the included studies showed very promising results for 2-piece zirconia implants. When focusing on 1-piece implants, implants made from ATZ are more fracture resistant than implants made from Y-TZP. Due to its negative impact on fracture resistance, abutment preparation of 1-piece zirconia implants should be avoided. When drafting international standards to guarantee clinical safety, one should keep in mind that the loading of anatomically shaped crowns might result in the decreased fracture resistance of zirconia implants compared to non-anatomical loading hemispheres, as mentioned in ISO 14801. Further research is needed to define adequate hydrothermal aging and dynamic loading conditions for 2-piece ceramic implants, nowadays mostly comprising a multi-material complex.

Author Contributions: Conceptualization, B.C.S.; methodology, B.C.S. and A.B.; validation, S.P., M.v.S.-L., F.B. and K.V.; statistical analysis, K.V.; writing—original draft preparation, B.C.S. and A.B.; writing—review and editing, S.P., K.V., R.-J.K., F.B. and M.v.S.-L. All authors have read and agreed to the published version of the manuscript.

Funding: This research received no external funding.

Acknowledgments: We acknowledge support from the German Research Foundation (DFG) and the Open Access Publication Fund of Charité—Universitätsmedizin Berlin.

Conflicts of Interest: The authors declare no conflict of interest.

Appendix A

Table A1. Articles excluded after screening of the full-texts.

First Author	Year	Ref.	Reason for Exclusion
Young	1972	[42]	Narrative review
Kohal	2006	[43]	Root analogue implants
Silva	2009	[44]	No static loading to fracture (step-stress fatigue)
Silva	2011	[45]	No static loading to fracture (pendulum impact tester)
Van Dooren	2012	[46]	Narrative review and case report
Iijima	2013	[47]	No evaluation of dental implants (discs)
Mobilio	2013	[48]	No static loading to fracture (strain measurements), <5 samples
Sanon	2013	[36]	No static loading to fracture (step-stress fatigue)
Cattani-Lorente	2014	[31]	No evaluation of dental implants (bar-shaped samples)
Rohr	2015	[49]	Fracture on the restoration level
Kamel	2017	[50]	Calculation of bending moment not possible, no author response
Karl	2017	[51]	No static loading to fracture (insertion torque measurements)
Korabi	2017	[52]	No static loading to fracture (finite element analysis)
Monzavi	2017	[53]	No static loading to fracture (accelerated aging only)
Zietz	2017	[54]	No zirconia implants
Baumgart	2018	[55]	No static loading to fracture
Rohr	2018	[56]	Fracture on the restoration level
Zaugg	2018	[57]	Fracture on the restoration level
Faria	2019	[58]	No evaluation of dental implants (discs)
Nueesch	2019	[59]	Fracture on the restoration level
Rohr	2019	[60]	Fracture on the restoration level
Scherrer	2019	[61]	Fractographic analysis of clinically fractured zirconia implants
Siddiqui	2019	[62]	Cyclic fatigue w/o subsequent fracture loading

References

1. Bosshardt, D.D.; Chappuis, V.; Buser, D. Osseointegration of titanium, titanium alloy and zirconia dental implants: Current knowledge and open questions. *Periodontol. 2000* **2017**, *73*, 22–40. [CrossRef] [PubMed]
2. Kniha, K.; Kniha, H.; Grunert, I.; Edelhoff, D.; Holzle, F.; Modabber, A. Esthetic Evaluation of Maxillary Single-Tooth Zirconia Implants in the Esthetic Zone. *Int. J. Periodontics Restor. Dent.* **2019**, *39*, e195–e201. [CrossRef] [PubMed]
3. Fretwurst, T.; Nelson, K.; Tarnow, D.P.; Wang, H.L.; Giannobile, W.V. Is Metal Particle Release Associated with Peri-implant Bone Destruction? An Emerging Concept. *J. Dent. Res.* **2018**, *97*, 259–265. [CrossRef] [PubMed]
4. Sanz, M.; Noguerol, B.; Sanz-Sanchez, I.; Hammerle, C.H.F.; Schliephake, H.; Renouard, F.; Sicilia, A.; Cordaro, L.; Jung, R.; Klinge, B.; et al. European Association for Osseointegration Delphi study on the trends in Implant Dentistry in Europe for the year 2030. *Clin. Oral Implant. Res.* **2019**, *30*, 476–486. [CrossRef]
5. Sennerby, L.; Dasmah, A.; Larsson, B.; Iverhed, M. Bone tissue responses to surface-modified zirconia implants: A histomorphometric and removal torque study in the rabbit. *Clin. Implant. Dent. Relat. Res.* **2005**, *7*, S13–S20. [CrossRef]
6. Pieralli, S.; Kohal, R.-J.; Lopez Hernandez, E.; Doerken, S.; Spies, B.C. Osseointegration of zirconia dental implants in animal investigations: A systematic review and meta-analysis. *Dent. Mater.* **2018**, *34*, 171–182. [CrossRef]
7. Pieralli, S.; Kohal, R.J.; Jung, R.E.; Vach, K.; Spies, B.C. Clinical Outcomes of Zirconia Dental Implants: A Systematic Review. *J. Dent. Res.* **2017**, *96*, 38–46. [CrossRef]
8. Prithviraj, D.R.; Gupta, V.; Muley, N.; Sandhu, P. One-piece implants: Placement timing, surgical technique, loading protocol, and marginal bone loss. *J. Prosthodont.* **2013**, *22*, 237–244. [CrossRef]
9. Östman, P.O.; Hellman, M.; Albrektsson, T.; Sennerby, L. Direct loading of Nobel Direct and Nobel Perfect one-piece implants: A 1-year prospective clinical and radiographic study. *Clin. Oral Implant. Res.* **2007**, *18*, 409–418. [CrossRef]
10. Cionca, N.; Hashim, D.; Mombelli, A. Zirconia dental implants: Where are we now, and where are we heading? *Periodontol. 2000* **2017**, *73*, 241–258. [CrossRef]
11. Spies, B.C.; Witkowski, S.; Vach, K.; Kohal, R.J. Clinical and patient-reported outcomes of zirconia-based implant fixed dental prostheses: Results of a prospective case series 5 years after implant placement. *Clin. Oral Implant. Res.* **2018**, *29*, 91–99. [CrossRef] [PubMed]
12. Kohal, R.J.; Wolkewitz, M.; Tsakona, A. The effects of cyclic loading and preparation on the fracture strength of zirconium-dioxide implants: An in vitro investigation. *Clin. Oral Implant. Res.* **2011**, *22*, 808–814. [CrossRef] [PubMed]
13. Spies, B.C.; Nold, J.; Vach, K.; Kohal, R.J. Two-piece zirconia oral implants withstand masticatory loads: An investigation in the artificial mouth. *J. Mech. Behav. Biomed. Mater.* **2016**, *53*, 1–10. [CrossRef] [PubMed]
14. Frigan, K.; Chevalier, J.; Zhang, F.; Spies, B.C. Is a Zirconia Dental Implant Safe When It Is Available on the Market? *Ceramics* **2019**, *2*, 44. [CrossRef]
15. Monzavi, M.; Zhang, F.; Meille, S.; Douillard, T.; Adrien, J.; Noumbissi, S.; Nowzari, H.; Chevalier, J. Influence of artificial aging on mechanical properties of commercially and non-commercially available zirconia dental implants. *J. Mech. Behav. Biomed. Mater.* **2020**, *101*, 103423. [CrossRef]
16. Stimmelmayr, M.; Lang, A.; Beuer, F.; Mansour, S.; Erdelt, K.; Krennmair, G.; Guth, J.F. Mechanical stability of all-ceramic abutments retained with three different screw materials in two-piece zirconia implants-an in vitro study. *Clin. Oral Investig.* **2019**. [CrossRef]
17. Ding, Q.; Zhang, L.; Bao, R.; Zheng, G.; Sun, Y.; Xie, Q. Effects of different surface treatments on the cyclic fatigue strength of one-piece CAD/CAM zirconia implants. *J. Mech. Behav. Biomed. Mater.* **2018**, *84*, 249–257. [CrossRef]
18. Kohal, R.J.; Wolkewitz, M.; Mueller, C. Alumina-reinforced zirconia implants: Survival rate and fracture strength in a masticatory simulation trial. *Clin. Oral Implant. Res.* **2010**, *21*, 1345–1352. [CrossRef]
19. Preis, V.; Kammermeier, A.; Handel, G.; Rosentritt, M. In vitro performance of two-piece zirconia implant systems for anterior application. *Dent. Mater.* **2016**, *32*, 765–774. [CrossRef]

20. Kohal, R.J.; Kilian, J.B.; Stampf, S.; Spies, B.C. All-Ceramic Single Crown Restauration of Zirconia Oral Implants and Its Influence on Fracture Resistance: An Investigation in the Artificial Mouth. *Materials* **2015**, *8*, 1577–1589. [CrossRef]
21. Spies, B.C.; Maass, M.E.; Adolfsson, E.; Sergo, V.; Kiemle, T.; Berthold, C.; Gurian, E.; Fornasaro, S.; Vach, K.; Kohal, R.J. Long-term stability of an injection-molded zirconia bone-level implant: A testing protocol considering aging kinetics and dynamic fatigue. *Dent. Mater.* **2017**, *33*, 954–965. [CrossRef] [PubMed]
22. Spies, B.C.; Sauter, C.; Wolkewitz, M.; Kohal, R.J. Alumina reinforced zirconia implants: Effects of cyclic loading and abutment modification on fracture resistance. *Dent. Mater.* **2015**, *31*, 262–272. [CrossRef] [PubMed]
23. Moher, D.; Liberati, A.; Tetzlaff, J.; Altman, D.G. Preferred Reporting Items for Systematic Reviews and Meta-Analyses: The PRISMA Statement. *PLoS Med.* **2009**, *6*, e1000097. [CrossRef] [PubMed]
24. Joda, T.; Voumard, B.; Zysset, P.K.; Bragger, U.; Ferrari, M. Ultimate force and stiffness of 2-piece zirconium dioxide implants with screw-retained monolithic lithium-disilicate reconstructions. *J. Prosthodont. Res.* **2018**, *62*, 258–263. [CrossRef] [PubMed]
25. Sanon, C.; Chevalier, J.; Douillard, T.; Cattani-Lorente, M.; Scherrer, S.S.; Gremillard, L. A new testing protocol for zirconia dental implants. *Dent. Mater.* **2015**, *31*, 15–25. [CrossRef]
26. Spies, B.C.; Fross, A.; Adolfsson, E.; Bagegni, A.; Doerken, S.; Kohal, R.J. Stability and aging resistance of a zirconia oral implant using a carbon fiber-reinforced screw for implant-abutment connection. *Dent. Mater.* **2018**, *34*, 1585–1595. [CrossRef]
27. Kammermeier, A.; Rosentritt, M.; Behr, M.; Schneider-Feyrer, S.; Preis, V. In vitro performance of one- and two-piece zirconia implant systems for anterior application. *J. Dent.* **2016**, *53*, 94–101. [CrossRef]
28. Rosentritt, M.; Hagemann, A.; Hahnel, S.; Behr, M.; Preis, V. In vitro performance of zirconia and titanium implant/abutment systems for anterior application. *J. Dent.* **2014**, *42*, 1019–1026. [CrossRef]
29. Andreiotelli, M.; Kohal, R.-J. Fracture Strength of Zirconia Implants after Artificial Aging. *Clin. Implant. Dent. Relat. Res.* **2009**, *11*, 158–166. [CrossRef]
30. Kohal, R.; Finke, H.C.; Klaus, G. Stability of Prototype Two-Piece Zirconia and Titanium Implants after Artificial Aging: An In Vitro Pilot Study. *Clin. Implant. Dent. Relat. Res.* **2009**, *11*, 323–329. [CrossRef]
31. Cattani-Lorente, M.; Scherrer, S.S.; Durual, S.; Sanon, C.; Douillard, T.; Gremillard, L.; Chevalier, J.; Wiskott, A. Effect of different surface treatments on the hydrothermal degradation of a 3Y-TZP ceramic for dental implants. *Dent. Mater.* **2014**, *30*, 1136–1146. [CrossRef] [PubMed]
32. Chevalier, J.; Loh, J.; Gremillard, L.; Meille, S.; Adolfson, E. Low-temperature degradation in zirconia with a porous surface. *Acta Biomater.* **2011**, *7*, 2986–2993. [CrossRef] [PubMed]
33. Kim, H.T.; Han, J.S.; Yang, J.H.; Lee, J.B.; Kim, S.H. The effect of low temperature aging on the mechanical property & phase stability of Y-TZP ceramics. *J. Adv. Prosthodont.* **2009**, *1*, 113–117. [CrossRef] [PubMed]
34. Virkar, A.V.; Huang, J.L.; Cutler, R.A. Strengthening of oxide ceramics by transformation-induced stress. *J. Am. Ceram. Soc.* **1987**, *70*, 164–170. [CrossRef]
35. Fischer, J.; Schott, A.; Martin, S. Surface micro-structuring of zirconia dental implants. *Clin. Oral Implants Res.* **2016**, *27*, 162–166. [CrossRef]
36. Sanon, C.; Chevalier, J.; Douillard, T.; Kohal, R.J.; Coelho, P.G.; Hjerppe, J.; Silva, N.R. Low temperature degradation and reliability of one-piece ceramic oral implants with a porous surface. *Dent. Mater.* **2013**, *29*, 389–397. [CrossRef]
37. Peixoto, H.E.; Bordin, D.; Cury, A.A.D.B.; Da Silva, W.J.; Faot, F. The role of prosthetic abutment material on the stress distribution in a maxillary single implant-supported fixed prosthesis. *Mater. Sci. Eng. C* **2016**, *65*, 90–96. [CrossRef]
38. Morneburg, T.R.; Pröschel, P.A. In vivo forces on implants influenced by occlusal scheme and food consistency. *Int. J. Prosthodont.* **2003**, *16*, 481–486.
39. Aversa, R.; Apicella, A. Osmotic Tension, Plasticization and Viscoelastic response of amorphous Poly-Ether-Ether-Ketone (PEEK) equilibrated in humid environments. *Am. J. Eng. Appl. Sci.* **2016**, *9*, 565–573. [CrossRef]

40. Chevalier, J.; Cales, B.; Drouin, J.M. Low-Temperature Aging of Y-TZP Ceramics. *J. Am. Ceram. Soc.* **1999**, *82*, 2150–2154. [CrossRef]
41. Schwitalla, A.D.; Abou-Emara, M.; Zimmermann, T.; Spintig, T.; Beuer, F.; Lackmann, J.; Muller, W.D. The applicability of PEEK-based abutment screws. *J. Mech. Behav. Biomed. Mater.* **2016**, *63*, 244–251. [CrossRef] [PubMed]
42. Young, F.A., Jr. Ceramic tooth implants. *J. Biomed. Mater. Res.* **1972**, *6*, 281–296. [CrossRef] [PubMed]
43. Kohal, R.J.; Klaus, G.; Strub, J.R. Zirconia-implant-supported all-ceramic crowns withstand long-term load: A pilot investigation. *Clin. Oral Implant. Res.* **2006**, *17*, 565–571. [CrossRef] [PubMed]
44. Silva, N.R.; Coelho, P.G.; Fernandes, C.A.; Navarro, J.M.; Dias, R.A.; Thompson, V.P. Reliability of one-piece ceramic implant. *J. Biomed. Mater. Res. B Appl. Biomater.* **2009**, *88*, 419–426. [CrossRef]
45. Silva, N.R.; Nourian, P.; Coelho, P.G.; Rekow, E.D.; Thompson, V.P. Impact Fracture Resistance of Two Titanium-Abutment Systems Versus a Single-Piece Ceramic Implant. *Clin. Implant. Dent. Relat. Res.* **2011**, *13*, 168–173. [CrossRef]
46. Van Dooren, E.; Calamita, M.; Calgaro, M.; Coachman, C.; Ferencz, J.L.; Pinho, C.; Silva, N.R. Mechanical, biological and clinical aspects of zirconia implants. *Eur. J. Esthet. Dent.* **2012**, *7*, 396–417. [PubMed]
47. Iijima, T.; Homma, S.; Sekine, H.; Sasaki, H.; Yajima, Y.; Yoshinari, M. Influence of surface treatment of yttria-stabilized tetragonal zirconia polycrystal with hot isostatic pressing on cyclic fatigue strength. *Dent. Mater. J.* **2013**, *32*, 274–280. [CrossRef] [PubMed]
48. Mobilio, N.; Stefanoni, F.; Contiero, P.; Mollica, F.; Catapano, S. Experimental and numeric stress analysis of titanium and zirconia one-piece dental implants. *Int. J. Oral Maxillofac. Implant.* **2013**, *28*, e135–e142. [CrossRef] [PubMed]
49. Rohr, N.; Coldea, A.; Zitzmann, N.U.; Fischer, J. Loading capacity of zirconia implant supported hybrid ceramic crowns. *Dent. Mater.* **2015**, *31*, e279–e288. [CrossRef]
50. Kamel, M.; Vaidyanathan, T.K.; Flinton, R. Effect of Abutment Preparation and Fatigue Loading in a Moist Environment on the Fracture Resistance of the One-Piece Zirconia Dental Implant. *Int. J. Oral Maxillofac. Implant.* **2017**, *32*, 533–540. [CrossRef]
51. Karl, M.; Scherg, S.; Grobecker-Karl, T. Fracture of Reduced-Diameter Zirconia Dental Implants Following Repeated Insertion. *Int. J. Oral Maxillofac. Implant.* **2017**, *32*, 971–975. [CrossRef] [PubMed]
52. Korabi, R.; Shemtov-Yona, K.; Rittel, D. On stress/strain shielding and the material stiffness paradigm for dental implants. *Clin. Implant. Dent. Relat. Res.* **2017**, *19*, 935–943. [CrossRef] [PubMed]
53. Monzavi, M.; Noumbissi, S.; Nowzari, H. The Impact of In Vitro Accelerated Aging, Approximating 30 and 60 Years In Vivo, on Commercially Available Zirconia Dental Implants. *Clin. Implant. Dent. Relat. Res.* **2017**, *19*, 245–252. [CrossRef] [PubMed]
54. Zietz, C.; Vogel, D.; Mitrovic, A.; Bader, R. Mechanical investigation of newly hybrid dental implants. *Biomedizinische Technik* **2017**, *62* (Suppl. 1), S447.
55. Baumgart, P.; Kirsten, H.; Haak, R.; Olms, C. Biomechanical properties of polymer-infiltrated ceramic crowns on one-piece zirconia implants after long-term chewing simulation. *Int. J. Implant. Dent.* **2018**, *4*, 16. [CrossRef] [PubMed]
56. Rohr, N.; Martin, S.; Fischer, J. Correlations between fracture load of zirconia implant supported single crowns and mechanical properties of restorative material and cement. *Dent. Mater. J.* **2018**, *37*, 222–228. [CrossRef]
57. Zaugg, L.K.; Meyer, S.; Rohr, N.; Zehnder, I.; Zitzmann, N.U. Fracture behavior, marginal gap width, and marginal quality of vented or pre-cemented CAD/CAM all-ceramic crowns luted on Y-TZP implants. *Clin. Oral Implant. Res.* **2018**, *29*, 175–184. [CrossRef]
58. Faria, D.; Pires, J.M.; Boccaccini, A.R.; Carvalho, O.; Silva, F.S.; Mesquita-Guimaraes, J. Development of novel zirconia implant's materials gradated design with improved bioactive surface. *J. Mech. Behav. Biomed. Mater.* **2019**, *94*, 110–125. [CrossRef]
59. Nueesch, R.; Conejo, J.; Mante, F.; Fischer, J.; Martin, S.; Rohr, N.; Blatz, M.B. Loading capacity of CAD/CAM-fabricated anterior feldspathic ceramic crowns bonded to one-piece zirconia implants with different cements. *Clin. Oral Implant. Res.* **2019**, *30*, 178–186. [CrossRef]
60. Rohr, N.; Balmer, M.; Muller, J.A.; Martin, S.; Fischer, J. Chewing simulation of zirconia implant supported restorations. *J. Prosthodont. Res.* **2019**, *63*, 361–367. [CrossRef]

61. Scherrer, S.S.; Mekki, M.; Crottaz, C.; Gahlert, M.; Romelli, E.; Marger, L.; Durual, S.; Vittecoq, E. Translational research on clinically failed zirconia implants. *Dent. Mater.* **2019**, *35*, 368–388. [CrossRef] [PubMed]
62. Siddiqui, D.A.; Sridhar, S.; Wang, F.; Jacob, J.J.; Rodrigues, D.C. Can Oral Bacteria and Mechanical Fatigue Degrade Zirconia Dental Implants in Vitro? *ACS Biomater. Sci. Eng.* **2019**, *5*, 2821–2833. [CrossRef]

© 2020 by the authors. Licensee MDPI, Basel, Switzerland. This article is an open access article distributed under the terms and conditions of the Creative Commons Attribution (CC BY) license (http://creativecommons.org/licenses/by/4.0/).

Review

Biological Responses to the Transitional Area of Dental Implants: Material- and Structure-Dependent Responses of Peri-Implant Tissue to Abutments

Jung-Ju Kim [1,†], Jae-Hyun Lee [2,†], Jeong Chan Kim [1], Jun-Beom Lee [1] and In-Sung Luke Yeo [3,*]

1. Department of Periodontology, Seoul National University School of Dentistry, Seoul 03080, Korea; freetist@gmail.com (J.-J.K.); perio_kjc@naver.com (J.C.K.); dent.jblee@gmail.com (J.-B.L.)
2. Department of Prosthodontics, One-Stop Specialty Center, Seoul National University Dental Hospital, Seoul 03080, Korea; jhlee.snudh@gmail.com
3. Department of Prosthodontics, School of Dentistry and Dental Research Institute, Seoul National University, Seoul 03080, Korea
* Correspondence: pros53@snu.ac.kr; Tel.: +82-2-2072-2661
† Jung-Ju Kim and Jae-Hyun Lee contributed equally to this work.

Received: 21 November 2019; Accepted: 18 December 2019; Published: 22 December 2019

Abstract: The stability of peri-implant tissue is essential for the long-term success of dental implants. Although various types of implant connections are used, little is known about the effects of the physical mechanisms of dental implants on the stability of peri-implant tissue. This review summarizes the relevant literature to establish guidelines regarding the effects of connection type between abutments and implants in soft and hard tissues. Soft tissue seals can affect soft tissue around implants. In external connections, micromobility between the abutment and the hex component of the implant, resulting from machining tolerance, can destroy the soft tissue seal, potentially leading to microbial invasion. Internal friction connection implants induce strain on the surrounding bone via implant wall expansion that translates into masticatory force. This strain is advantageous because it increases the amount and quality of peri-implant bone. The comparison of internal and external connections, the two most commonly used connection types, reveals that internal friction has a positive influence on both soft and hard tissues.

Keywords: abutment; dental implant; implant connection; marginal bone; peri-implantitis

---

1. Introduction

A dental implant is an artificial organ that replaces a missing natural tooth. Implants should be able to function properly in the human body [1]. Long-term stability should be predictable and in line with current trends toward increased life expectancy [2–6]. For dental implants to function properly for a long time in the oral cavity, they should experience neither mechanical nor biological complications, including those related to soft and hard tissues surrounding the implant [7–11].

The stability of the implant–abutment connection is an important factor affecting the long-term success of dental implants in clinical practice [12,13]. To prevent complications resulting from unstable implant–abutment connections, implant–abutment biomechanics are investigated theoretically and experimentally. In this review, internal friction connection and external hex-type connection implant–abutment joints are explored. Dental implants belonging to these two systems are currently the most widely used in clinical practice [13–15]. The nature of these connections is summarized, and their effects on tissues surrounding implants are analyzed for both soft and hard tissues.

2. Typical Dental Implant Connection Types

2.1. External Hex Connection: Butt-Jointed Interface

In external hex-type connections, the abutment is connected to the implant by an abutment screw. This connection is also called a butt joint interface because the flat surfaces on the top of the implant and the bottom of the abutment are in direct contact with each other. This external connection is stabilized entirely by fastening of the abutment screw. When torque is applied to the abutment screw connecting the implant and the abutment, the abutment screw is elongated, generating a preload. Screw dynamics play an important role in this interaction because the mechanism linking this connection is fundamentally dependent on the preload of the screw [16]. The preload acts as a clamping force on the implant–abutment complex, and provides stability to the connection [17].

The Brånemark implant (NobelBiocare, Zurich, Switzerland), the first commercially available screw-shaped implant, is a representative external connection-type implant. When masticatory force is applied to this type of implant, the vertical component of the masticatory force is supported by the top platform of the implant. The lateral component of the masticatory force is placed on the hex structure of the top of the implant and the abutment screw, and the rotational component of this force is resisted by the hex structure.

Previous biomechanical analyses indicate that the manufacturer's recommended torque of 30–35 Ncm is not sufficient to prevent screw loosening [17,18]. Repeated tightening of the abutment screw is considered essential because of the preload loss in screw mechanics, which is the cause of screw loosening [17]. It is also helpful to use other methods in addition to screw tightening to maintain the stability of the implant–abutment connection. One option that makes this possible is the use of the frictional force generated between the interfaces.

2.2. Internal Friction Connection: Frictional Interface

In internal friction connections, stability is maintained by close contact between the inner surface of the implant and the outer surface of the abutment, in addition to the preload applied to the abutment screw. The tight contact between the abutment and implant creates frictional force, and this plays a major role in supporting the stability of the connection. Therefore, this type of connection is also called a friction-screw-retained connection.

The degree of tapering at the interface between the implant and the abutment is an important factor to consider when determining the abutment mobility of the internal conical connection. A wider tapering angle will result in a more unstable connection in this type of internal connection, although the possibility of implant fracture decreases [16,19]. In addition, the role of frictional forces in maintaining the stability of the connection is decreased, while the burden on the abutment screws is increased (Figure 1).

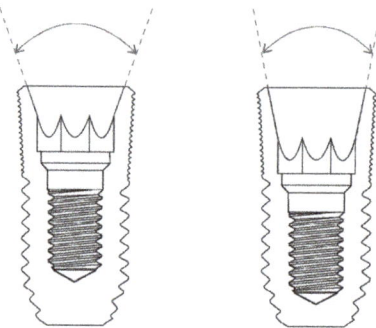

Figure 1. The degree of conical connection. A wider tapering angle (left) results in a more unstable connection for abutment mobility in internal connection-type implants.

A systematic review reported that internal friction connection provides less microleakage at the implant–abutment interface than external hex connection type in both static and dynamic loading tests [20]. The presence of gap could lead to accumulation of bacteria and the phenomenon might affect the success rate of implants.

When an abutment is connected to an internal conical connection-type implant, the abutment is first brought into contact with the upper part of the implant inner wall [19,21]. As the implant at the top is slightly flared, it comes into greater contact with the abutment as the implant is combined. The abutment screw becomes slightly loosened when the abutment is completely combined with the implant by opening the implant wall. Because of the preload loss from such an abutment sinking and settling effect, repeated screw tightening is essential [18,22].

3. Soft Tissue Responses to Different Implant System Materials and Structures

3.1. The Soft Tissue Seal Theory

Humans are exposed to a variety of external environments involving external forces, ultraviolet rays, and microorganisms. Human skin is the first line of defense that protects the human body against these external stresses. When human skin is pierced, the resulting hole must be closed by the immune system and healing mechanisms.

Teeth are one of the few organs in the human body that are located across the skin. The root of the tooth is surrounded by alveolar bone, while the part that penetrates through the soft tissue is in contact with epithelial tissue and/or the connective tissue of the mucosa [23]. Holes in the mucosa created by teeth are sealed by a special structure composed of epithelium and connective tissue [23]. An internal basal lamina and the hemi-desmosomes of epithelial tissue are attached to teeth, and a combination of dento-gingival fiber and cementum links teeth to the surrounding connective tissue. The holes connecting the inside and outside of the human body that contain teeth are secured by soft tissue attachments [23].

Like natural teeth, dental implants are an artificial organ located in a hole on the surface of the body, and the hole is sealed via a mechanism similar to the one that seals the holes around natural teeth using soft tissues. However, these sealing mechanisms are not identical to each other. Dental implants are in contact with the alveolar bone or soft tissue (epithelial and connective tissue) of the transition area (Figure 2) [24]. Fibers in the connective tissue attached to the abutment mainly run parallel to the surface and are circular in shape, whereas dento-gingival fibers, such as Sharpey's fibers, are attached vertically to the cementum in natural teeth (Figure 1) [24].

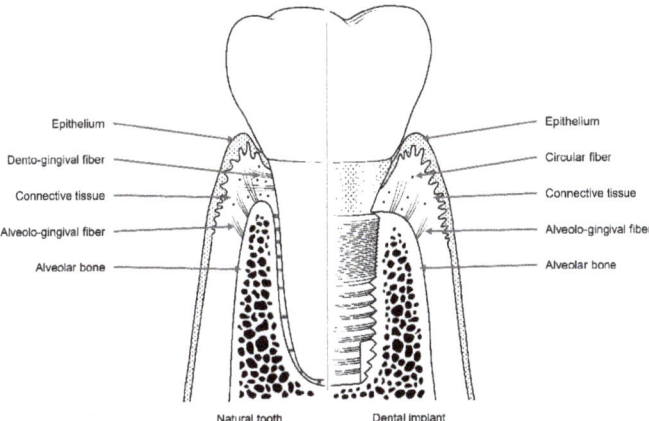

Figure 2. Soft and hard tissues with collagen fibers surrounding a natural tooth (**left**) and a dental implant (**right**).

A scanning electron microscopy study showed that there are two distinct layers in the area within 200 µm of where the implant abutment comes into contact with the connective tissue. The inner 40 µm of this layer contains multiple fibroblasts that are attached directly to the surface of the abutment, while the outer 160 µm of this layer contains numerous collagen fibers [25]. Hence, the attachment of connective tissue to the abutment is maintained by fibroblasts attached to the titanium surface of the implant, and the elasticity of circular collagen fibers. For this reason, compared with the attachments of natural teeth, which involve direct attachment to connective tissue, attachments formed by connective tissue around an implant abutment are weaker.

Epithelial tissue is attached to the implant abutment via an internal basal lamina and hemi-desmosomes in a manner similar to the attachment of natural teeth [26–29]. However, epithelium attachment includes the internal basal lamina and hemi-desmosomes formed only in the lower part of the peri-implant epithelium around the implant abutment, whereas in natural teeth, these attachments are widely distributed throughout the junctional epithelium–tooth interface [30–32]. Thus, epithelial adhesions that form around the implant abutment are more vulnerable than those that form around natural teeth owing to their limited area of distribution.

In summary, both epithelial and connective tissues are more weakly attached to implant abutments than to natural teeth. Hence, holes on the surface of the human body that contain implants are more vulnerable to containment failure. This mechanism of blockade by soft tissue is called a 'soft tissue seal' [30,33].

3.2. Attachment of Soft Tissue

The stability and immobility of soft tissue attachments in contact with the implant abutment are important factors that affect the long-term prognosis of the implant [30,33]. If the soft tissues around the implant are deformed owing to the movement of the lip, cheek, tongue, or jaw, the weak soft tissue seal surrounding the implant may be destroyed. This allows microbes to penetrate through the damaged mucosal seal, increasing the likelihood of disease around the implant [30,33]. Therefore, a stable soft tissue seal is essential to prevent microbial invasion and peri-implant disease [33–35].

The soft tissues around natural teeth are separated into the lining mucosa and masticatory mucosa, of which the masticatory mucosa is composed of the free and attached gingiva (Figure 3). The attached gingiva is in permanent and intimate contact with the surface of the enamel, the cementum, and the alveolar bone, thereby immobilizing the soft tissue [23]. It is, therefore, possible to retain the firmness and health of the mucosal seal by preventing it from detaching.

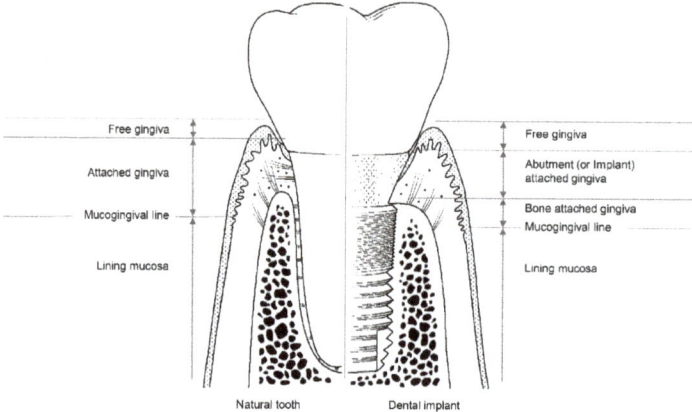

Figure 3. The structure of soft tissue around a natural tooth (**left**) and a dental implant (**right**).

Among the soft tissues surrounding an implant, the attached gingiva differs from those surrounding natural teeth [24]. The gingiva attached to the peri-implant can be divided into two parts: bone-attached and abutment (or implant)-attached gingiva (Figure 3). The bone-attached gingiva is identical to the corresponding tissue in natural gums, and rigidly immobilizes the soft tissue. However, as mentioned above, the abutment- (or implant)-attached gingiva more weakly adheres than the attachment gingiva that surrounds natural teeth [24]. Owing to the weak structure of the abutment- (or implant)-attached gingiva, the width of the bone-attached gingiva around the implant should be wide enough to prevent mucosal mobility, and thereby prevent the occurrence of peri-implant disease due to the collapse of the mucosal seal [24,33].

3.3. Disruption of the Soft Tissue Seal

The soft tissue seal is mainly destroyed via two mechanisms; instability of the peri-implant mucosa or the implant-abutment assembly. If there is no bone-attached gingiva around the implant, the soft tissue can become mobile, and the mucosal seal will inevitably rupture. When the mucosal seal is destroyed, bacteria can penetrate the internal environment through the transmucosal rupture site, potentially leading to peri-implant disease [36]. Therefore, because of the weakness of the soft tissue attachment, it is important to ensure there is sufficient bone-attached gingiva around the implant. Thus, a plan for implant surgery should be established to maintain the bone-attached gingiva following implant placement.

Instability in the connection between the implant and the abutment can also lead to the disruption of the soft tissue seal, which may present as mobility of the abutment or the implant. Mobility of the implant occurs only when osseointegration fails, but mobility of the abutment can occur as a result of various causes, including fractures of the abutment or implant, even following successful osseointegration. The most common cause of abutment mobility is the loosening of the abutment screw that connects it to the implant. Elongation and loosening of the screw caused by the lateral force of the masticatory load are more frequently observed in external connection-type implants than in internal connection-type implants [15].

In external-type implants, there is slight machining tolerance around the hex component, resulting in micromobility in the abutment. This external hex connection is, therefore, a mobile structure. Thus, most occlusal forces are concentrated on the abutment screw. This micromobility is likely to cause disruption of the soft tissue seal, bacterial infiltration, and peri-implant disease [37] (Figure 4). Implant systems with an external hex connection were originally developed for the mandibular restoration of a completely edentulous patient wearing a maxillary complete denture. The occlusal force in such patients is weak, and this mobile connection is able to bear this weak masticatory force with few screw loosening events or breakdown of the soft tissue seal. However, such unstable implant–abutment connections can provoke severe peri-implant problems for a partially edentulous patient with antagonistic natural teeth and a strong occlusal force. To reduce the micromobility of the abutment, the external hex must be lengthened, and the machining tolerance between the male and female components must be minimized. Furthermore, it may be possible to fabricate an abutment screw using a stronger material, or by widening the platform size of the top of the implant to minimize the force concentrated on the abutment screw. In addition, when the abutment is fastened to the implant, applying higher torque may also be employed to increase the stability of the connection. However, none of these methods can completely eliminate the micromobility of the abutment in external connection-type implants.

In internal connection-type implants, the abutment–implant connection is firm owing to a process similar to cold-welding. Unlike external connections, in which the occlusal force is concentrated mainly on the abutment screw, the occlusal force is transmitted to the implant inner wall through the abutment–implant connection in internal connection-type implants. Therefore, less screw loosening occurs in this type of implant. In addition, the abutment–implant contact area is wider internal connection-type implants, and this prevents stress from concentrating at specific sites such as the abutment screw, and contributes to the stability of the soft tissue seal.

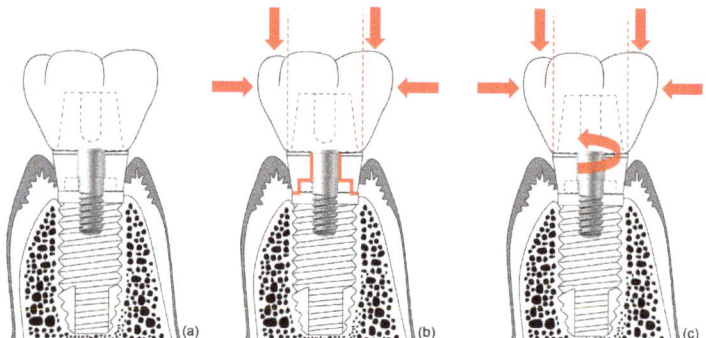

Figure 4. Schematic illustration of the 'soft tissue seal' theory. Micromobility of the abutment in external connection-type implants disrupts the surrounding soft tissue seal, which plays an important role in preventing external irritants from penetrating into the body. (**a**) External connection-type implant. (**b**) Parts experiencing the greatest stress (red lines) from eccentric forces (red arrows). (**c**) Abutment screw-loosening (red rotated arrow) by eccentric forces (red arrows).

3.4. Submerged and Nonsubmerged Implants

Depending on whether the top of the implant is at the alveolar bone or gingival level, implants can be divided into submerged-type and nonsubmerged-type. For submerged implants, the soft tissue seal is formed at the abutment area, rather than at the neck part for the nonsubmerged type. Histologically, there are no significant differences in the degree and pattern of soft tissue attachment to implants between submerged and nonsubmerged types [38]. When nonsubmerged implants are placed at the proper vertical position of the alveolar bone, the mobility of the abutment does not interfere with the soft tissue seal, or cause marginal bone resorption, because the interface between the abutment and the implant is positioned outside the soft tissue [34,35].

When this design was first developed, microgap bacteria residing between the abutment and implant were considered to be the cause of marginal bone resorption [39–41]. Therefore, a design was developed to export the position of this microgap outside the body [42]. However, such a design limits the customization of the patient's emergence profile, resulting in difficulties in terms of long-term clinical results for aesthetically acceptable implant prostheses.

To maintain an adequate soft tissue seal, there must be a proper foundation of underlying bone. However, nonsubmerged-type implants can be susceptible to marginal bone resorption due to the concentration of stress at the top of the implant [43].

3.5. Materials for Abutment

Achieving soft tissue seals may also depend on the type of material used for the abutment. Abrahamsson et al. (1998) investigated the stability of soft tissue seals attached via gold alloy, dental porcelain, titanium, and aluminum oxide. In this study, marginal bone resorption occurred when seals were attached to surfaces composed of gold alloy and dental porcelain, because no soft tissue seal formed. However, when titanium and aluminum oxide were used, a soft tissue seal was achieved, and marginal bone was not absorbed [44]. Welander et al. (2008) also reported that epithelial adhesions receded around abutments attached via gold alloy [45]. Therefore, when using a UCLA abutment, the transmucosal area of the abutment is made of gold or dental porcelain, and a soft tissue seal may not be properly achieved, resulting in marginal bone resorption.

More recently, titanium and zirconia have been used as materials for abutments. One advantage of zirconia is that it is more aesthetic than titanium. While several studies show that there are no significant differences in the soft tissue seal between these two materials, zirconia has a lower fracture strength than titanium, and is thus more likely to be associated with mechanical complications [45–47]. In addition, adhesion of bacteria to titanium occurs less readily than to zirconia [48].

Recent advances in digital dentistry have made it possible to fabricate and restore implant-supported restorations in just a single visit. To this end, lithium disilicate and polymer-infiltrated ceramic network (PICN), which are easy to process on the day, are used as abutment material. These materials are used as implant abutments after cementing to the prefabricated titanium base with adhesive resin cement. Although little research has been conducted on the soft tissue seal of these new materials, Smallidge et al. have reported favorable epithelial cell growth on the PICN surface when PICN has a relatively low surface roughness (Ra < 0.254 μm) [49]. Another recent study reported that the surface roughness of lithium disilicate was smoother than that of PICN after the same polishing process with 6 μm diamond slurry, and the surface roughness of both of materials was unaffected by the ultrasonic scaling procedure [50]. Mehl et al. found that the adhesive resin joint connecting the titanium base and the abutment materials had neither influence on soft tissue anatomy nor on bone loss in the animal study [51]. In zirconia abutments, however, junctional epithelium was significantly shorter than titanium one-piece abutments [51]. Ongun et al. measured the mechanical properties of these hybrid-type abutments including titanium bases, showing that PICN had lower fracture resistance and weaker adhesion to resin cements compared with lithium disilicate ceramic [52]. Such mechanical failure of PICN hybrid-type abutments could also cause the soft tissue seal to break.

3.6. Detachment of Abutments

When the abutment becomes disconnected from a submerged-type implant, the soft tissue seal is broken, and microorganisms in the oral cavity can then penetrate into the tissues surrounding the implant [53]. This may result in the loss of marginal bone. Abrahamsson et al. reported that the absorption of marginal bone is doubled when the abutment is detached five times [54]. For these reasons, some clinicians have proposed the 'one abutment-one time (OAOT)' concept to prevent marginal bone loss [55,56]. Clinicians should be aware that during a prosthetic restoration procedure, or during the postrestoration maintenance period, it may be advantageous to minimize the process of removing the abutment.

A systematic review suggested that the OAOT protocol might preserve peri-implant bone loss and soft tissue changes for two reasons [57]. First of all, the micro-gap between abutment and implant could cause to bacterial leakage and micromotion, and this might lead to the inflammation of peri-implant soft tissue and bone resorption. The OAOT protocol could provide less micro-gap because healing or temporary abutments were installed by less preloading force (<10 Ncm) than final abutments (about 30 Ncm) [57]. Second, the OAOT protocol could reduce the disruption of the soft tissue seal around the implant abutment complex to avoid repeated the dis/reconnection of abutments [57].

3.7. Surface Modification of Abutments

Various technologies have been explored for improving the soft tissue seal at the transmucosal part of the abutment, including surface treatments, such as coating, machining, blasting, plasma spraying, etching, and laser processing [58].

Yang et al. reported that, when ultraviolet light is applied to surfaces, gingival fibroblasts proliferate more readily on surfaces made of zirconia [59]. Several studies showed that, when an abutment is laser-treated, the connective tissue is directly attached to the surface of the abutment by perpendicular fibers [60–62]. In addition, as the surface roughness of the transmucosal part of the abutment increases, the soft tissue seal improves. This may be because a rougher surface has a larger surface area to which the soft tissue can attach [63]. A previous study also suggested that hydrothermal treatment on titanium alloy could change minimally surface topography to enhance hydrophilicity [64].

Subsequently, the treatment contributed to the integration of epithelial cells to the surface and might facilitate the healthy epithelial tissue sealing around the transmucosal part of the implant [64].

The surface of the abutment can be modified to change the surface roughness or the free energy. While this may enhance the soft tissue seal, it can also increase plaque accumulation and the tendency for bacterial colonization to occur [65,66]. Thus, surface treatments applied to abutments should be carefully selected.

4. Hard Tissue Responses to Implant System Materials and Structures

4.1. The Bone Stimulation Theory

The soft and hard tissues surrounding the implant play complementary roles. Alveolar bone is a hard tissue that withstands the masticatory force applied to the implant, and serves to transmit it to the jawbone. The soft tissue, gingiva, and mucosa protect the alveolar bone from external irritants such as bacteria. The condition of the soft tissue is maintained by the underlying alveolar bone, and the alveolar bone is protected by the overlying soft tissue.

In general, after the restoration of an implant, between 1 and 1.5 mm of the marginal bone around the implant is absorbed during the first year owing to the application of occlusal force, and 0.2 mm is absorbed per year thereafter [67]. On the basis of the results of studies showing that marginal bone is steadily absorbed every year, it was thought that long-length implants should be beneficial for long-term predictability [67]. This was observed in long-term studies of the Brånemark system using an external connection-type implant, which is limited and inapplicable to some internal friction connections [14,68,69]. By contrast, several studies on a certain implant connection system reported that peri-implant marginal bone is increased by occlusal loading [68,69]. Because the use of long implants is likely to cause damage to important anatomical structures such as nerves, it may be more advantageous to insert short implants if warranted for long-term prognosis without marginal bone resorption.

In natural dentition, when periodontal disease occurs or a natural tooth is lost, the alveolar bone of the corresponding region is resorbed. However, when the natural tooth and the periodontal tissue are healthy, the alveolar bone is well-maintained. Except for a small amount of loss due to mechanical degeneration or aging, the alveolar bone should be well-maintained throughout life, as periodontal ligaments have been shown to transfer adequate stimulation to alveolar bone. Otherwise, disuse atrophy of the bone can occur. From this point of view, if the peri-implant bone is appropriately stimulated by the implant, the peri-implant marginal bone will not be resorbed, and the quantity and quality of the bone are likely to be preserved.

The Astra implant system (Dentsply Sirona, Charlotte, NC, United States) involves an internal friction connection type, and was the first connection system to embody this concept. The implant–abutment connection of this system improves the amount and quality of peri-implant bone by properly transmitting the masticatory force to the surrounding alveolar bone. This hypothesis was reported by Frost [70] and can be applied to alveolar bone as well as to other bones in the human body [70]. Hence, under conditions involving the application of appropriate strain to human bone, osteoblasts are activated, which increases the amount and quality of the bone.

Before existing implant systems were widely used in clinical practice, various types and shapes of implants were developed, and most disappeared without displaying any ability to withstand mechanical or biological complications [71,72]. The Brånemark system was the first to achieve successful long-term predictable prognosis, and the consequential popularization of implants. Unlike previous plain-shaped systems, this system can deliver occlusal force to the alveolar bone because it has a thread on the implant surface [71,72]. The thread of the implant transforms the shear stress generated at the interface between the implant and the bone into compressive stress so that the appropriate stimulus can be transmitted to the bone (Figure 5). This stimulation is one of the factors that allows the bone to remain stable over the long term.

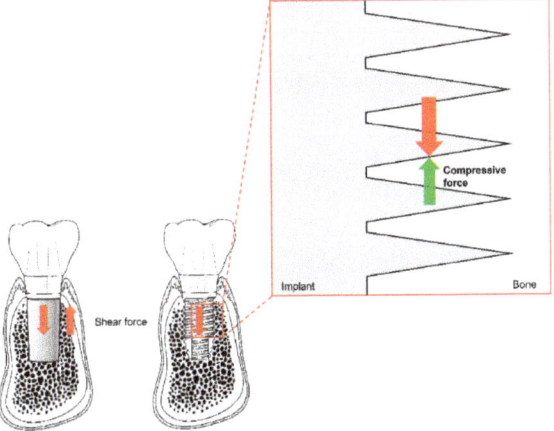

Figure 5. The thread function of non-threaded (**left**) and threaded (**right**) dental implants. Shear force (red arrows) is transformed into compressive force (green arrow) by the threads. Note the reduction of shear force due to the partial switch to compressive force in the magnified diagram (vertically and horizontally; not marked).

A disadvantage of these external connection-type implants is that the soft tissue seal formed around the abutment is destroyed by micromobility at the connection. The Brånemark implant represents the beginning of modern implants, and was the first system to perform well in clinical practice. However, marginal bone resorption was found to be caused by the destruction of the soft-tissue seal. To prevent such marginal bone loss, an Astra internal connection implant system was developed. This internal friction system maintains bone by applying appropriate stimulation to the bone around the implant without loss of the soft tissue seal around the implant.

4.2. Mechanism of Bone Stimulation

The connection of the Astra implant system has an internal conical shape with a slope of 11 degrees. The occlusal force applied to the abutment is transmitted to the implant through this conical connection. The masticatory force delivered to the implant thus becomes the source of the stain to stimulate the alveolar bone [73]. Thus, when the abutment receives occlusal force and sinks downward, the conical opening of the implant is opened wider, and the bone around the implant is consequentially stimulated (Figure 6). This stimulation activates osteoblasts in the alveolar bone, thereby increasing the amount and quality of alveolar bone. This increase in alveolar bone can lead to a positive change in the results of clinical procedures. This reduces the need for invasive bone grafting or the placement of excessively long implants during implant surgery. This represents a positive result for both the patient and the surgeon.

4.3. Prevention of Hard Tissue Loss

The marginal bone loss that occurs in the mobile connection of external connection-type implants does not lead directly to the failure of the implant. However, it can cause considerable complications in the tissue surrounding the implant. In general, these external connection implants are known to lose marginal bone up to the second or third thread level of the implant [74]. This bone loss can alter the properties of the overlying soft tissue, resulting in a reduction in the attached gingiva. This weakens the soft tissue seal and increases the likelihood of bacterial invasion and peri-implantitis. A detailed periodical examination of the condition of tissues around the implant is thus essential for long-term success when external connection-type implants are used.

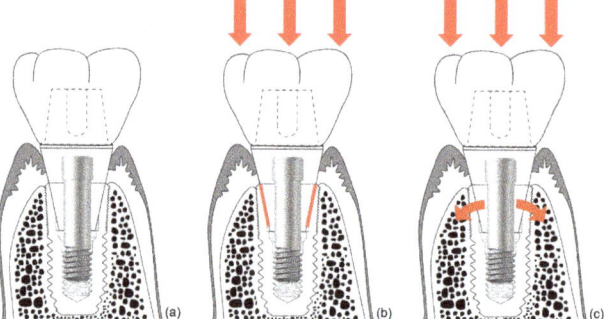

Figure 6. Schematic drawing of bone stimulation. When occlusal force is applied to an internal connection-type implant, the implant around the connection expands and stimulates the surrounding bone to induce bone proliferation. (**a**) Internal friction connection-type implant. (**b**) The occlusal force is transmitted to the implant through the conical connection parts (red lines). (**c**) The coronal expansion of the implant (curved red arrows) becomes the source of the strain that stimulates the alveolar bone.

The issue of whether gingiva must be present for the health of natural teeth has been investigated. Clinically, even if the gingiva cannot completely wrap the tooth, it should be able to prevent the movement of the surrounding soft tissue from being transmitted to the free gingiva [75]. In other words, cheek, lip, and mandibular movements should not be transferred to the marginal gingiva. To achieve this, a certain amount of attached gingiva must be present to prevent the penetration of bacteria into the periodontal tissue [76].

This principle of natural dentition should also be borne in mind when considering implant restorations [77]. An appropriate amount of attached gingiva should be present, rather than free gingiva alone, to prevent microbial invasion into tissue surrounding the implant. According to the systematic review in 2018, if there is the insufficient attached gingiva around implant, the apically positioned flap with autogenous graft is suggested to increase the width of the gingiva [78]. However, it is not enough simply to increase the width of attached gingiva, because this surgical procedure is unable to alter the level of mucogingival junction [79]. Additional vestibuloplasty is, therefore, recommended to create "new" mucogingival junction by preventing the movement of the lips and cheeks from affecting the free gingiva.

In addition, abutment and restoration contour have been recently reported to have effects on marginal bone loss. An emergence angle of more than 30 degrees and a convex emergence profile may increase the risk of peri-implantitis [80]. The association between the restoration contour and the soft tissue seal remains uncertain. However, in order to prevent the bone loss, it is important to maintain the favorable soft tissue environment by adjusting the emergence angle, the emergence profile, the position of a restoration margin, and the removal of excess cement [80–82].

5. Summary

Dental implants should be capable of performing dental functions for a prolonged period of time. For this to be possible, the health of the surrounding soft and hard tissues is essential. Alveolar bone must be able to withstand masticatory force, and the overlying soft tissue should be able to protect the alveolar bone from external irritants.

The soft tissue seal is weaker around implants than around natural teeth. Thus, to avoid damage, the peri-implant soft tissue seal should not be mobile. Free and abutment-attached gingiva cannot prevent the destruction of the soft tissue seal that results from the movements of the lips and cheeks. Additionally, it is impossible to prevent all micromobility of the abutment in external connection-type implants. Hence, it is necessary to secure sufficient bone-attached gingiva, or use an implant system with extensive and deep connections, such as internal connection-type implants. The resulting immobility contributes to the maintenance of a healthy soft tissue seal, and prevents bacterial invasion into the peri-implant tissue. In addition, internal connection-type implants can transform occlusal load into strain that stimulates the surrounding bone through the abutment–implant connection. This strain enhances the amount and quality of peri-implant bone.

Taking these findings into account, clinicians may find it advantageous to select internal conical connection-type implants rather than external connection-type implants, to allow dental implants to function properly as artificial organs with long-term predictability.

Author Contributions: All authors have read and agree to the published version of the manuscript. Conceptualization, J.-J.K., J.-H.L., and I.-S.L.Y.; data curation, J.-J.K. and J.-H.L.; formal analysis, J.-J.K. and J.-H.L.; funding acquisition, I.-S.L.Y.; investigation, J.-J.K., J.-H.L., and J.-B.L.; methodology, J.-J.K., J.-H.L., J.C.K., J.-B.L., and I.-S.L.Y.; project administration, I.-S.L.Y.; resources, J.C.K. and I.-S.L.Y.; software, J.-H.L.; supervision, I.-S.L.Y.; validation, J.-J.K. and J.C.K.; visualization, J.-H.L. and J.C.K.; writing—original draft, J.-H.L., J.-J.K., and I.-S.L.Y.; writing—review and editing, J.-H.L., J.-J.K., J.C.K., J.-B.L., and I.-S.L.Y. All authors have read and agreed to the published version of the manuscript.

Funding: This work was supported by a grant from the Korea Health Technology R&D Project through the Korea Health Industry Development Institute (KHIDI), which is funded by the Ministry of Health and Welfare of the Republic of Korea (grant number: HI15C1535).

Conflicts of Interest: The authors declare no conflict of interest.

References

1. Pjetursson, B.E.; Heimisdottir, K. Dental implants—Are they better than natural teeth? *Eur. J. Oral Sci.* **2018**, *126*, 81–87. [CrossRef] [PubMed]
2. Albrektsson, T.; Chrcanovic, B.; Ostman, P.O.; Sennerby, L. Initial and long-term crestal bone responses to modern dental implants. *Periodontology 2000* **2017**, *73*, 41–50. [CrossRef] [PubMed]
3. De Bruyn, H.; Christiaens, V.; Doornewaard, R.; Jacobsson, M.; Cosyn, J.; Jacquet, W.; Vervaeke, S. Implant surface roughness and patient factors on long-term peri-implant bone loss. *Periodontology 2000* **2017**, *73*, 218–227. [CrossRef] [PubMed]
4. Howe, M.S.; Keys, W.; Richards, D. Long-term (10-year) dental implant survival: A systematic review and sensitivity meta-analysis. *J. Dent.* **2019**, *84*, 9–21. [CrossRef] [PubMed]
5. Srinivasan, M.; Meyer, S.; Mombelli, A.; Muller, F. Dental implants in the elderly population: A systematic review and meta-analysis. *Clin. Oral Implant. Res.* **2017**, *28*, 920–930. [CrossRef] [PubMed]
6. Starch-Jensen, T.; Aludden, H.; Hallman, M.; Dahlin, C.; Christensen, A.E.; Mordenfeld, A. A systematic review and meta-analysis of long-term studies (five or more years) assessing maxillary sinus floor augmentation. *Int. J. Oral Maxillofac. Surg.* **2018**, *47*, 103–116. [CrossRef] [PubMed]
7. De Kok, I.J.; Duqum, I.S.; Katz, L.H.; Cooper, L.F. Management of Implant/Prosthodontic Complications. *Dent. Clin. North Am.* **2019**, *63*, 217–231. [CrossRef]
8. Esposito, M.; Hirsch, J.; Lekholm, U.; Thomsen, P. Differential diagnosis and treatment strategies for biologic complications and failing oral implants: A review of the literature. *Int. J. Oral Maxillofac. Implant.* **1999**, *14*, 473–490.
9. Hsu, Y.T.; Mason, S.A.; Wang, H.L. Biological implant complications and their management. *J. Int. Acad. Periodontol.* **2014**, *16*, 9–18.
10. Lang, N.P.; Wilson, T.G.; Corbet, E.F. Biological complications with dental implants: Their prevention, diagnosis and treatment. *Clin. Oral Implant. Res.* **2000**, *11*, 146–155. [CrossRef]
11. Sadid-Zadeh, R.; Kutkut, A.; Kim, H. Prosthetic failure in implant dentistry. *Dent. Clin. North Am.* **2015**, *59*, 195–214. [CrossRef] [PubMed]
12. Vetromilla, B.M.; Brondani, L.P.; Pereira-Cenci, T.; Bergoli, C.D. Influence of different implant-abutment connection designs on the mechanical and biological behavior of single-tooth implants in the maxillary esthetic zone: A systematic review. *J. Prosthet. Dent.* **2019**, *121*, 398–403.e3. [CrossRef] [PubMed]
13. Lemos, C.A.A.; Verri, F.R.; Bonfante, E.A.; Santiago Junior, J.F.; Pellizzer, E.P. Comparison of external and internal implant-abutment connections for implant supported prostheses. A systematic review and meta-analysis. *J. Dent.* **2018**, *70*, 14–22. [CrossRef] [PubMed]

14. de Medeiros, R.A.; Pellizzer, E.P.; Vechiato Filho, A.J.; Dos Santos, D.M.; da Silva, E.V.; Goiato, M.C. Evaluation of marginal bone loss of dental implants with internal or external connections and its association with other variables: A systematic review. *J. Prosthet. Dent.* **2016**, *116*, 501–506. [CrossRef]
15. Gracis, S.; Michalakis, K.; Vigolo, P.; Vult von Steyern, P.; Zwahlen, M.; Sailer, I. Internal vs. external connections for abutments/reconstructions: A systematic review. *Clin. Oral Implant. Res.* **2012**, *23*, 202–216. [CrossRef]
16. Bozkaya, D.; Muftu, S. Mechanics of the tapered interference fit in dental implants. *J. Biomech.* **2003**, *36*, 1649–1658. [CrossRef]
17. Jeong, C.G.; Kim, S.K.; Lee, J.H.; Kim, J.W.; Yeo, I.S.L. Clinically available preload prediction based on a mechanical analysis. *Arch. Appl. Mech.* **2017**, *87*, 2003–2009. [CrossRef]
18. Michalakis, K.X.; Calvani, P.L.; Muftu, S.; Pissiotis, A.; Hirayama, H. The effect of different implant-abutment connections on screw joint stability. *J. Oral Implantol.* **2014**, *40*, 146–152. [CrossRef]
19. Lee, J.H.; Huh, Y.H.; Park, C.J.; Cho, L.R. Effect of the Coronal Wall Thickness of Dental Implants on the Screw Joint Stability in the Internal Implant-Abutment Connection. *Int. J. Oral Maxillofac. Implant.* **2016**, *31*, 1058–1065. [CrossRef]
20. Mishra, S.K.; Chowdhary, R.; Kumari, S. Microleakage at the Different Implant Abutment Interface: A Systematic Review. *J. Clin. Diagn. Res.* **2017**, *11*, ZE10–ZE15. [CrossRef]
21. Yilmaz, B.; Gilbert, A.B.; Seidt, J.D.; McGlumphy, E.A.; Clelland, N.L. Displacement of Implant Abutments Following Initial and Repeated Torqueing. *Int. J. Oral Maxillofac. Implant.* **2015**, *30*, 1011–1018. [CrossRef] [PubMed]
22. Saleh Saber, F.; Abolfazli, N.; Jannatii Ataei, S.; Taghizade Motlagh, M.; Gharekhani, V. The effect of repeated torque tightening on total lengths of implant abutments in different internal implantabutment connections. *J. Dent. Res. Dent. Clin. Dent. Prospect.* **2017**, *11*, 110–116. [CrossRef] [PubMed]
23. Bartold, P.M.; Walsh, L.J.; Narayanan, A.S. Molecular and cell biology of the gingiva. *Periodontology 2000* **2000**, *24*, 28–55. [CrossRef] [PubMed]
24. Berglundh, T.; Lindhe, J.; Ericsson, I.; Marinello, C.P.; Liljenberg, B.; Thomsen, P. The soft tissue barrier at implants and teeth. *Clin. Oral Implant. Res.* **1991**, *2*, 81–90. [CrossRef] [PubMed]
25. Moon, I.S.; Berglundh, T.; Abrahamsson, I.; Linder, E.; Lindhe, J. The barrier between the keratinized mucosa and the dental implant. An experimental study in the dog. *J. Clin. Periodontol.* **1999**, *26*, 658–663. [CrossRef] [PubMed]
26. Atsuta, I.; Ayukawa, Y.; Furuhashi, A.; Yamaza, T.; Tsukiyama, Y.; Koyano, K. Promotive effect of insulin-like growth factor-1 for epithelial sealing to titanium implants. *J. Biomed. Mater. Res. A* **2013**, *101*, 2896–2904. [CrossRef] [PubMed]
27. Donley, T.G.; Gillette, W.B. Titanium endosseous implant-soft tissue interface: A literature review. *J. Periodontol.* **1991**, *62*, 153–160. [CrossRef]
28. Gould, T.R.; Westbury, L.; Brunette, D.M. Ultrastructural study of the attachment of human gingiva to titanium in vivo. *J. Prosthet. Dent.* **1984**, *52*, 418–420. [CrossRef]
29. Ikeda, H.; Shiraiwa, M.; Yamaza, T.; Yoshinari, M.; Kido, M.A.; Ayukawa, Y.; Inoue, T.; Koyano, K.; Tanaka, T. Difference in penetration of horseradish peroxidase tracer as a foreign substance into the peri-implant or junctional epithelium of rat gingivae. *Clin. Oral Implant. Res.* **2002**, *13*, 243–251. [CrossRef]
30. Atsuta, I.; Ayukawa, Y.; Kondo, R.; Oshiro, W.; Matsuura, Y.; Furuhashi, A.; Tsukiyama, Y.; Koyano, K. Soft tissue sealing around dental implants based on histological interpretation. *J. Prosthodont. Res.* **2016**, *60*, 3–11. [CrossRef]
31. Atsuta, I.; Yamaza, T.; Yoshinari, M.; Goto, T.; Kido, M.A.; Kagiya, T.; Mino, S.; Shimono, M.; Tanaka, T. Ultrastructural localization of laminin-5 (gamma2 chain) in the rat peri-implant oral mucosa around a titanium-dental implant by immuno-electron microscopy. *Biomaterials* **2005**, *26*, 6280–6287. [CrossRef]
32. Ikeda, H.; Yamaza, T.; Yoshinari, M.; Ohsaki, Y.; Ayukawa, Y.; Kido, M.A.; Inoue, T.; Shimono, M.; Koyano, K.; Tanaka, T. Ultrastructural and immunoelectron microscopic studies of the peri-implant epithelium-implant (Ti-6Al-4V) interface of rat maxilla. *J. Periodontol.* **2000**, *71*, 961–973. [CrossRef] [PubMed]
33. Abdallah, M.N.; Badran, Z.; Ciobanu, O.; Hamdan, N.; Tamimi, F. Strategies for Optimizing the Soft Tissue Seal around Osseointegrated Implants. *Adv. Healthc. Mater.* **2017**, *6*, 1700549. [CrossRef] [PubMed]

34. Hermann, J.S.; Buser, D.; Schenk, R.K.; Cochran, D.L. Crestal bone changes around titanium implants. A histometric evaluation of unloaded non-submerged and submerged implants in the canine mandible. *J. Periodontol.* **2000**, *71*, 1412–1424. [CrossRef] [PubMed]
35. Hermann, J.S.; Schoolfield, J.D.; Schenk, R.K.; Buser, D.; Cochran, D.L. Influence of the size of the microgap on crestal bone changes around titanium implants. A histometric evaluation of unloaded non-submerged implants in the canine mandible. *J. Periodontol.* **2001**, *72*, 1372–1383. [CrossRef] [PubMed]
36. Monje, A.; Insua, A.; Wang, H.L. Understanding Peri-Implantitis as a Plaque-Associated and Site-Specific Entity: On the Local Predisposing Factors. *J. Clin. Med.* **2019**, *8*, 279. [CrossRef]
37. Liu, Y.; Wang, J. Influences of microgap and micromotion of implant-abutment interface on marginal bone loss around implant neck. *Arch. Oral Biol.* **2017**, *83*, 153–160. [CrossRef]
38. Abrahamsson, I.; Berglundh, T.; Wennstrom, J.; Lindhe, J. The peri-implant hard and soft tissues at different implant systems. A comparative study in the dog. *Clin. Oral Implant. Res.* **1996**, *7*, 212–219. [CrossRef]
39. D'Ercole, S.; Scarano, A.; Perrotti, V.; Mulatinho, J.; Piattelli, A.; Iezzi, G.; Tripodi, D. Implants with internal hexagon and conical implant-abutment connections: An in vitro study of the bacterial contamination. *J. Oral Implantol.* **2014**, *40*, 30–36. [CrossRef]
40. Koutouzis, T.; Mesia, R.; Calderon, N.; Wong, F.; Wallet, S. The effect of dynamic loading on bacterial colonization of the dental implant fixture-abutment interface: An in vitro study. *J. Oral Implantol.* **2014**, *40*, 432–437. [CrossRef]
41. Piattelli, A.; Vrespa, G.; Petrone, G.; Iezzi, G.; Annibali, S.; Scarano, A. Role of the microgap between implant and abutment: A retrospective histologic evaluation in monkeys. *J. Periodontol.* **2003**, *74*, 346–352. [CrossRef] [PubMed]
42. Buser, D.; Mericske-Stern, R.; Dula, K.; Lang, N.P. Clinical experience with one-stage, non-submerged dental implants. *Adv. Dent. Res.* **1999**, *13*, 153–161. [CrossRef] [PubMed]
43. Hansson, S. A conical implant-abutment interface at the level of the marginal bone improves the distribution of stresses in the supporting bone. An axisymmetric finite element analysis. *Clin. Oral Implant. Res.* **2003**, *14*, 286–293. [CrossRef] [PubMed]
44. Abrahamsson, I.; Berglundh, T.; Lindhe, J. Soft tissue response to plaque formation at different implant systems. A comparative study in the dog. *Clin. Oral Implant. Res.* **1998**, *9*, 73–79. [CrossRef] [PubMed]
45. Welander, M.; Abrahamsson, I.; Berglundh, T. The mucosal barrier at implant abutments of different materials. *Clin. Oral Implant. Res.* **2008**, *19*, 635–641.
46. Kohal, R.J.; Weng, D.; Bachle, M.; Strub, J.R. Loaded custom-made zirconia and titanium implants show similar osseointegration: An animal experiment. *J. Periodontol.* **2004**, *75*, 1262–1268. [CrossRef] [PubMed]
47. Soo, S.Y.; Silikas, N.; Satterthwaite, J. Measurement of Fracture Strength of Zirconia Dental Implant Abutments with Internal and External Connections Using Acoustic Emission. *Materials* **2019**, *12*, 2009. [CrossRef]
48. de Avila, E.D.; de Molon, R.S.; Lima, B.P.; Lux, R.; Shi, W.; Junior, M.J.; Spolidorio, D.M.; Vergani, C.E.; de Assis Mollo Junior, F. Impact of Physical Chemical Characteristics of Abutment Implant Surfaces on Bacteria Adhesion. *J. Oral Implantol.* **2016**, *42*, 153–158. [CrossRef]
49. Smallidge, M.J.; Sabol, J.V.; Aita-Holmes, C.; Chuang, H.; Dickinson, D.P. Human Gingival Epithelial Growth In Vitro on a Polymer-Infiltrated Ceramic Network Restorative Material. *J. Prosthodont.* **2019**, *28*, 541–546. [CrossRef]
50. Lee, J.H.; Kim, S.H.; Han, J.S.; Yeo, I.S.L.; Yoon, H.I.; Lee, J. Effects of ultrasonic scaling on the optical properties and surface characteristics of highly translucent CAD/CAM ceramic restorative materials: An in vitro study. *Ceram. Int.* **2019**, *45*, 14594–14601. [CrossRef]
51. Mehl, C.; Gassling, V.; Schultz-Langerhans, S.; Acil, Y.; Bahr, T.; Wiltfang, J.; Kern, M. Influence of Four Different Abutment Materials and the Adhesive Joint of Two-Piece Abutments on Cervical Implant Bone and Soft Tissue. *Int. J. Oral Maxillofac. Implant.* **2016**, *31*, 1264–1272. [CrossRef] [PubMed]
52. Ongun, S.; Kurtulmus-Yilmaz, S.; Meric, G.; Ulusoy, M. A Comparative Study on the Mechanical Properties of a Polymer-Infiltrated Ceramic-Network Material Used for the Fabrication of Hybrid Abutment. *Materials* **2018**, *11*, 1681. [CrossRef] [PubMed]
53. Rodriguez, X.; Vela, X.; Mendez, V.; Segala, M.; Calvo-Guirado, J.L.; Tarnow, D.P. The effect of abutment dis/reconnections on peri-implant bone resorption: A radiologic study of platform-switched and non-platform-switched implants placed in animals. *Clin. Oral Implant. Res.* **2013**, *24*, 305–311. [CrossRef] [PubMed]

54. Abrahamsson, I.; Berglundh, T.; Lindhe, J. The mucosal barrier following abutment dis/reconnection. An experimental study in dogs. *J. Clin. Periodontol.* **1997**, *24*, 568–572. [CrossRef] [PubMed]
55. Canullo, L.; Bignozzi, I.; Cocchetto, R.; Cristalli, M.P.; Iannello, G. Immediate positioning of a definitive abutment versus repeated abutment replacements in post-extractive implants: 3-year follow-up of a randomised multicentre clinical trial. *Eur. J. Oral Implantol.* **2010**, *3*, 285–296.
56. Degidi, M.; Nardi, D.; Piattelli, A. One abutment at one time: Non-removal of an immediate abutment and its effect on bone healing around subcrestal tapered implants. *Clin. Oral Implant. Res.* **2011**, *22*, 1303–1307. [CrossRef]
57. Wang, Q.Q.; Dai, R.; Cao, C.Y.; Fang, H.; Han, M.; Li, Q.L. One-time versus repeated abutment connection for platform-switched implant: A systematic review and meta-analysis. *PLoS ONE* **2017**, *12*, e0186385. [CrossRef]
58. Ghensi, P.; Bettio, E.; Maniglio, D.; Bonomi, E.; Piccoli, F.; Gross, S.; Caciagli, P.; Segata, N.; Nollo, G.; Tessarolo, F. Dental Implants with Anti-Biofilm Properties: A Pilot Study for Developing a New Sericin-Based Coating. *Materials* **2019**, *12*, 2429. [CrossRef]
59. Yang, Y.; Zhou, J.; Liu, X.; Zheng, M.; Yang, J.; Tan, J. Ultraviolet light-treated zirconia with different roughness affects function of human gingival fibroblasts in vitro: The potential surface modification developed from implant to abutment. *J. Biomed. Mater. Res. B Appl. Biomater.* **2015**, *103*, 116–124. [CrossRef]
60. Iglhaut, G.; Becker, K.; Golubovic, V.; Schliephake, H.; Mihatovic, I. The impact of dis-/reconnection of laser microgrooved and machined implant abutments on soft- and hard-tissue healing. *Clin. Oral Implant. Res.* **2013**, *24*, 391–397. [CrossRef]
61. Nevins, M.; Camelo, M.; Nevins, M.L.; Schupbach, P.; Kim, D.M. Connective tissue attachment to laser-microgrooved abutments: A human histologic case report. *Int. J. Periodontics Restor. Dent.* **2012**, *32*, 385–392.
62. Nevins, M.; Kim, D.M.; Jun, S.H.; Guze, K.; Schupbach, P.; Nevins, M.L. Histologic evidence of a connective tissue attachment to laser microgrooved abutments: A canine study. *Int. J. Periodontics Restor. Dent.* **2010**, *30*, 245–255.
63. Guida, L.; Oliva, A.; Basile, M.A.; Giordano, M.; Nastri, L.; Annunziata, M. Human gingival fibroblast functions are stimulated by oxidized nano-structured titanium surfaces. *J. Dent.* **2013**, *41*, 900–907. [CrossRef] [PubMed]
64. Sakamoto, Y.; Ayukawa, Y.; Furuhashi, A.; Kamo, M.; Ikeda, J.; Atsuta, I.; Haraguchi, T.; Koyano, K. Effect of Hydrothermal Treatment with Distilled Water on Titanium Alloy for Epithelial Cellular Attachment. *Materials* **2019**, *12*, 2748. [CrossRef]
65. Albouy, J.P.; Abrahamsson, I.; Berglundh, T. Spontaneous progression of experimental peri-implantitis at implants with different surface characteristics: An experimental study in dogs. *J. Clin. Periodontol.* **2012**, *39*, 182–187. [CrossRef]
66. Yeo, I.S.; Kim, H.Y.; Lim, K.S.; Han, J.S. Implant surface factors and bacterial adhesion: A review of the literature. *Int. J. Artif. Organs* **2012**, *35*, 762–772. [CrossRef]
67. Albrektsson, T.; Zarb, G.; Worthington, P.; Eriksson, A.R. The long-term efficacy of currently used dental implants: A review and proposed criteria of success. *Int. J. Oral Maxillofac. Implant.* **1986**, *1*, 11–25.
68. Palmer, R.M.; Palmer, P.J.; Smith, B.J. A 5-year prospective study of Astra single tooth implants. *Clin. Oral Implant. Res.* **2000**, *11*, 179–182. [CrossRef]
69. Puchades-Roman, L.; Palmer, R.M.; Palmer, P.J.; Howe, L.C.; Ide, M.; Wilson, R.F. A clinical, radiographic, and microbiologic comparison of Astra Tech and Branemark single tooth implants. *Clin. Implant. Dent. Relat. Res.* **2000**, *2*, 78–84. [CrossRef]
70. Frost, H.M. Perspectives: bone's mechanical usage windows. *Bone Miner.* **1992**, *19*, 257–271. [CrossRef]
71. Buser, D.; Sennerby, L.; De Bruyn, H. Modern implant dentistry based on osseointegration: 50 years of progress, current trends and open questions. *Periodontology 2000* **2017**, *73*, 7–21. [CrossRef] [PubMed]
72. Taylor, T.D.; Agar, J.R. Twenty years of progress in implant prosthodontics. *J. Prosthet. Dent.* **2002**, *88*, 89–95. [CrossRef] [PubMed]
73. de Vasconcellos, L.G.; Kojima, A.N.; Nishioka, R.S.; de Vasconcellos, L.M.; Balducci, I. Axial loads on implant-supported partial fixed prostheses for external and internal hex connections and machined and plastic copings: Strain gauge analysis. *J. Oral Implantol.* **2015**, *41*, 149–154. [CrossRef] [PubMed]

74. Tarnow, D.P.; Cho, S.C.; Wallace, S.S. The effect of inter-implant distance on the height of inter-implant bone crest. *J. Periodontol.* **2000**, *71*, 546–549. [CrossRef]
75. Friedman, M.T.; Barber, P.M.; Mordan, N.J.; Newman, H.N. The "plaque-free zone" in health and disease: A scanning electron microscope study. *J. Periodontol.* **1992**, *63*, 890–896. [CrossRef]
76. Lang, N.P.; Loe, H. The relationship between the width of keratinized gingiva and gingival health. *J. Periodontol.* **1972**, *43*, 623–627. [CrossRef]
77. Thoma, D.S.; Muhlemann, S.; Jung, R.E. Critical soft-tissue dimensions with dental implants and treatment concepts. *Periodontology 2000* **2014**, *66*, 106–118. [CrossRef]
78. Giannobile, W.V.; Jung, R.E.; Schwarz, F.; Groups of the 2nd Osteology Foundation Consensus, M. Evidence-based knowledge on the aesthetics and maintenance of peri-implant soft tissues: Osteology Foundation Consensus Report Part 1-Effects of soft tissue augmentation procedures on the maintenance of peri-implant soft tissue health. *Clin. Oral Implant. Res.* **2018**, *29*, 7–10. [CrossRef]
79. Ainamo, A.; Bergenholtz, A.; Hugoson, A.; Ainamo, J. Location of the mucogingival junction 18 years after apically repositioned flap surgery. *J. Clin Periodontol.* **1992**, *19*, 49–52.
80. Katafuchi, M.; Weinstein, B.F.; Leroux, B.G.; Chen, Y.W.; Daubert, D.M. Restoration contour is a risk indicator for peri-implantitis: A cross-sectional radiographic analysis. *J. Clin Periodontol.* **2018**, *45*, 225–232. [CrossRef]
81. Canullo, L.; Schlee, M.; Wagner, W.; Covani, U.; Montegrotto Group for the Study of Peri-implant, D. International Brainstorming Meeting on Etiologic and Risk Factors of Peri-implantitis, Montegrotto (Padua, Italy), August 2014. *Int. J. Oral Maxillofac. Implant.* **2015**, *30*, 1093–1104. [CrossRef] [PubMed]
82. Derks, J.; Schaller, D.; Hakansson, J.; Wennstrom, J.L.; Tomasi, C.; Berglundh, T. Effectiveness of Implant Therapy Analyzed in a Swedish Population: Prevalence of Peri-implantitis. *J. Dent. Res.* **2016**, *95*, 43–49. [CrossRef] [PubMed]

© 2019 by the authors. Licensee MDPI, Basel, Switzerland. This article is an open access article distributed under the terms and conditions of the Creative Commons Attribution (CC BY) license (http://creativecommons.org/licenses/by/4.0/).

Article

Improvement in Fatigue Behavior of Dental Implant Fixtures by Changing Internal Connection Design: An In Vitro Pilot Study

Nak-Hyun Choi [1,†], Hyung-In Yoon [2,†], Tae-Hyung Kim [3] and Eun-Jin Park [1,*]

1. Department of Prosthodontics, School of Medicine, Ewha Womans University, Seoul 07985, Korea; hyun_bc@naver.com
2. Department of Prosthodontics, School of Dentistry and Dental Research Institute, Seoul National University, Seoul 03080, Korea; prosyhi@naver.com
3. Kim and Lee Dental Clinic, Seoul 06626, Korea; kthrock@nate.com
* Correspondence: prosth@ewha.ac.kr; Tel.: +82-2-2650-5042
† These authors contributed equally to this work.

Received: 9 September 2019; Accepted: 4 October 2019; Published: 7 October 2019

Abstract: (1) Background: The stability of the dental implant–abutment complex is necessary to minimize mechanical complications. The purpose of this study was to compare the behaviors of two internal connection type fixtures, manufactured by the same company, with different connection designs. (2) Methods: 15 implant–abutment complexes were prepared for each group of Osseospeed® TX (TX) and Osseospeed® EV (EV): 3 for single-load fracture tests and 12 for cyclic-loaded fatigue tests (nominal peak values as 80%, 60%, 50%, and 40% of the maximum breaking load) according to international standards (UNI EN ISO 14801:2013). They were assessed with micro-computed tomography (CT), and failure modes were analyzed by scanning electron microscope (SEM) images. (3) Results: The maximum breaking load [TX: 711 ± 36 N (95% CI; 670–752), EV: 791 ± 58 N (95% CI; 725–857)] and fatigue limit (TX: 285 N, EV: 316 N) were higher in EV than those in TX. There was no statistical difference in the fracture areas ($P > 0.99$). All specimens with 40% nominal peak value survived 5×10^6 cycles, while 50% specimens failed before 10^5 cycles. (4) Conclusions: EV has improved mechanical properties compared with TX. A loading regimen with a nominal peak value between 40% and 50% is ideal for future tests of implant cyclic loading.

Keywords: dental implants; fracture strength; mechanical stress; fatigue; dental implant–abutment connection; dental implant–abutment design

1. Introduction

Dental implants have been a fairly reliable and predictable treatment option for edentulous patients since their introduction [1]. In previous systematic reviews, a 5-year survival rate of 95.6–97.2% and 10-year survival rate of 93.1% in implants supporting fixed partial dentures were reported, implying that implants have a high survival rate [2–4]. Although dental implants have been clinically and scientifically studied as a viable treatment option to restore the edentulous area [5–7], complications remain a big concern for clinicians. Complication of dental implants can be mainly classified as either biological or mechanical. Biological complications include early loss of osseointegration, marginal bone loss, and peri-implantitis, eventually leading to the implants failing and falling out. Mechanical complications include loose abutment or screw, veneer or ceramic fracture, loss of retention, sinking down of abutment, and fracture of implant fixture, abutment, or screw [2]. Previous reports have demonstrated incidence rates of implant fixture fracture of 0.2–1.1% and abutment or screw fracture of 0.7–2.3% [2]. In particular, fracture of implant fixture is a catastrophic complication, which requires extensive surgical treatments. To overcome mechanical failure of dental implants and guarantee long-term clinical success, the stability of the implant–abutment connection to withstand a masticatory load is important [4,8]. The mechanical stability of the connection may be affected by modifying the implant–abutment connection design as well as by improving material properties of the components [9]. Recently, manufacturers of dental implant systems such as Astra Tech Dental have introduced a modified connection design of the dental implant fixture to improve its mechanical properties. To date, however, only a few studies have demonstrated the effect of different connection designs on the mechanical properties of the implant fixtures [10].

Fatigue is the process of localized, permanent structural change of a material under fluctuating stress [11]. Mechanical complications of implants are generally caused by fatigue stress related to mechanical overload [12]. The interpretation of fatigue limit in implants is slightly different from general mechanics. The fatigue limit of dental implants is defined as "the maximum loading value that can withstand 5×10^6 cycles", contrary to the general definition in mechanics: "the maximum loading value that can withstand infinite cycles" [11,13]. To evaluate fatigue stress in the laboratory, finite element analysis and cyclic loading can be utilized [14–16]. While finite element analysis is considered to simulate fairly reliable results, cyclic loading is used as a method to observe the mechanical properties of actual specimens [16]. To standardize the testing method in the laboratory, ISO 14801 was suggested to simulate a "worst-case scenario" applied on an implant–abutment assembly and consists of sinusoidally curved cyclic loading [13]. These methods can be utilized to substitute in vivo tests, and while a generalized clinical conclusion may not be drawn, a tendency can be observed to provide insight to researchers and clinicians.

The purpose of this study was to compare the mechanical behaviors of two internal connection type dental implant fixtures with different connection designs manufactured by a single manufacturer. By strictly adhering to the procedures considered as the norm, it may provide data on implant systems widely used on the market, and moreover, supply background for additional protocols to the universal standard. The null hypothesis was that the fatigue behavior including the mode of failure of the dental implant–abutment complex is not affected by the modification of the connection design of the dental implant fixture.

2. Materials and Methods

2.1. Preparation of Specimens

The fixtures and abutments tested in this study are listed in Table 1 and the flow of the experiment is shown in Figure 1. A total of 30 implant–abutment assemblies were prepared for test and control groups (n = 15 per group): Osseospeed® EV (EV) and Osseospeed® TX (TX) (Astra Tech Dental, Dentsply Sirona Implants, Mölndal, Sweden). Among the 15 specimens, 3 were tested for single-load fracture tests to identify the maximum fracture load and the other 12 specimens were divided into 4 groups (n = 3 each) for fatigue test under cyclic loading. Each specimen was marked with an indelible marker indicating where the load would be applied to analyze the fractured surface with a scanning electron microscope (SEM). Each implant and abutment was connected with a torque of 25 N cm, as recommended by the manufacturer.

2.2. Micro-CT Image Observation

The implant–abutment assemblies were scanned with micro-computed tomography (CT) scanner (SkyScan 1172, Bruker, Kontich, Belgium) to obtain a series of detailed structural images prior to cyclic loading. The samples were firmly fixed to a full 360° rotational inspection jig. The frame rate was four frames per rotational step of 0.5°, for a total of 2880 images per specimen.

Table 1. Materials used in this study. All fixtures and abutments were composed of commercially pure grade 4 titanium.

Components	Test Group	Control Group
Implant Fixture	OsseoSpeed® EV (4.2 mm × 11 mm)	OsseoSpeed® TX (4.0 mm × 11 mm)
Abutment	TiDesign® EV Abutment height: 5.5 mm Gingival height: 2.5 mm	TiDesign® Abutment height: 5.5 mm Gingival height: 1.5 mm
	Manufacturer: Dentsply Sirona Implants, Mölndal, Sweden.	

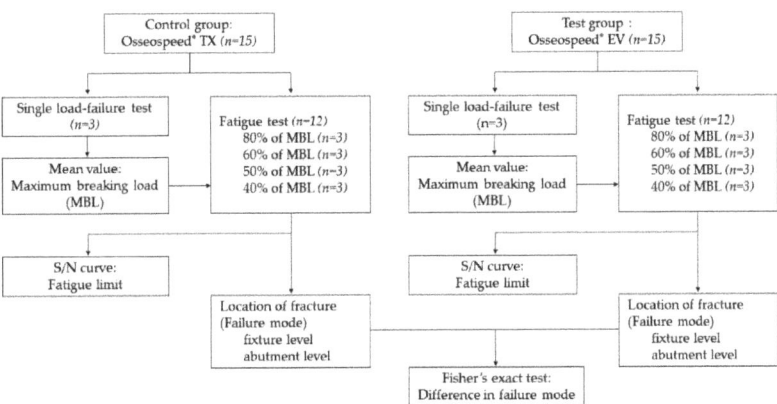

Figure 1. Overall flow of the experiment.

2.3. Single-Load Failure Test and Fatigue

All mechanical tests were performed according to ISO 14801:2013 (Figure 2). The testing apparatus should impose a force within ± 5% of the maximum error range of the nominal peak value with constant frequency. The testing apparatus should also be able to monitor maximum and minimum load values and to stop when the specimen fractures. A servo-hydraulic test system (MTS Landmark, Minneapolis, MN, USA) under load control was used. Single-load failure tests and fatigue tests were conducted in an atmospheric environment of 20 °C ± 5 °C. Each implant–abutment assembly was inserted in a custom stainless-steel jig and collet up to the first thread of the implant fixture (approximately 3.0 mm). The collets were then held at a 30° off-axis angle and fixed to the jig and testing machine. A hemispherical cap was engaged to the implant–abutment assembly, contacting the flat head of the universal testing machine. Compressive load increasing at a speed of 1mm/min was applied to the implant–abutment assembly until fracture or deformation occurred. Three implants from each group were tested and their maximum fracture load values were recorded. The average value of maximum fracture load of the tested implants served as the nominal peak value for the fatigue test.

For the fatigue testing under cyclic loading, we applied a sinusoidal oscillation with 15 Hz frequency between a nominal peak level (maximum) and a 10% value of the nominal peak level (minimum) to the implant–abutment assembly. The cyclic loading was conducted until the fracture occurred. If fracture did not occur, the cyclic loading was conducted up to a maximum number of 5×10^6 cycles. The nominal peak levels of 80%, 60%, 50%, and 40% of the maximum fracture load from the previous single-load-to-failure test were selected. Three samples for each nominal peak level group were tested and the number of cycles in which fracture occurred was recorded. If the implant–abutment assembly survived the entire loading cycle, 5×10^6 cycles were recorded. The results were then plotted on an S/N curve, which is a plot of the magnitude of an alternating stress versus the number of cycles to failure for a given material. The S/N curves were estimated by a logarithmic linear regression model utilizing the least squares method. The fatigue limit of the tested dental implant was defined as the maximum fracture load value which can withstand 5×10^6 cycles [13].

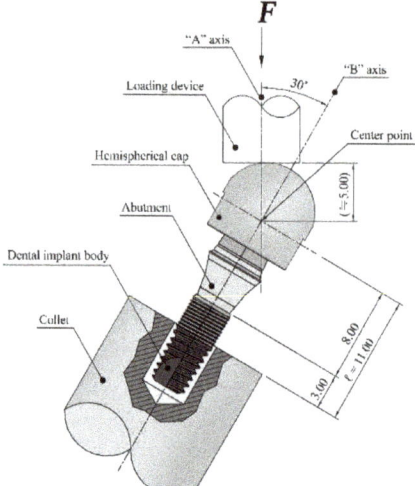

Figure 2. Schematic diagram of the loading test device according to ISO 14801:2013.

2.4. Failure Modes and Microscopic Observation

The fractured area of each implant–abutment assembly was microscopically observed and divided into three categories of failures: fixture-level, abutment-level, and screw-level. Two representative specimens were randomly selected before fatigue testing to examine the connection area of the intact implant–abutment assembly using a field emission scanning electron microscope (SEM) (S-4700, Hitachi, Tokyo, Japan). The specimens were inspected one more time after fatigue testing. The frontal and coronal sectional views of fractured specimens were microscopically examined with 15.0 kV accelerating voltage at ×25 and ×30. For the frontal view, specimens were aligned to show the loading direction from left to right. The abutment and fixture cross-sectional views were symmetrically aligned such that the loading direction could be observed from 12 o'clock and 6 o'clock respectively.

2.5. Statistical Analysis

Mean values and standard deviations of the maximum breaking loads and mean values of the performed cycle from the fatigue tests were calculated. Fisher's exact test was used to analyze the numbers of each type of failure to evaluate the difference of failure modes (fixture-level, abutment-level, and screw-level failure). All statistical analyses were performed using SAS® version 9.4 (SAS Institute, Cary, NC, USA).

3. Results

3.1. Micro-CT Image Observation

Frontal and coronal cross-sectional views of micro-CT showed the detailed design of TX and EV (Figure 3). The thinnest areas, excluding the most coronal portion of the fixture, were expected to be the initiation point of the crack; however, the initiation point was the first thread under the microthread, which does not coincide with the thinnest part.

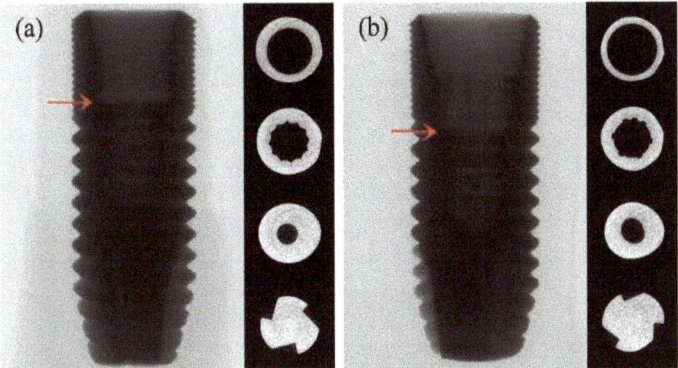

Figure 3. Frontal and cross-sectional micro-CT view: (**a**) TX, (**b**) EV; red arrow = location of the thinnest part of the implant fixture.

3.2. Maximum Breaking Load and Fatigue Limit

The TX samples that underwent single-load failure tests showed a mean maximum breaking load of 711 ± 36 N (95% CI; 670–752), and the EV samples showed an average value of 791 ± 58 N (95% CI; 725–857) (Table 2). The trend of the load and fracture of the specimens was plotted on a time–load diagram, the peak being the point when deformation occurs on the implant–abutment complex (Figure 4). Fatigue testing results are shown in Table 3 and were plotted on an S/N curve with the logarithmic values of the cycles endured on the X-axis and nominal peak level on the Y-axis (Figure 5). All three TX samples of 40% nominal peak level of 285 N endured 5×10^6 cycles, whereas the other nine specimens failed to resist breaking. The fatigue limit was 285 N to withstand 5×10^6 cycles. However, all three EV samples of 40% nominal peak level of 316 N endured 5×10^6 cycles, while the other nine samples failed. The fatigue limit was 316 N to withstand 5×10^6 cycles.

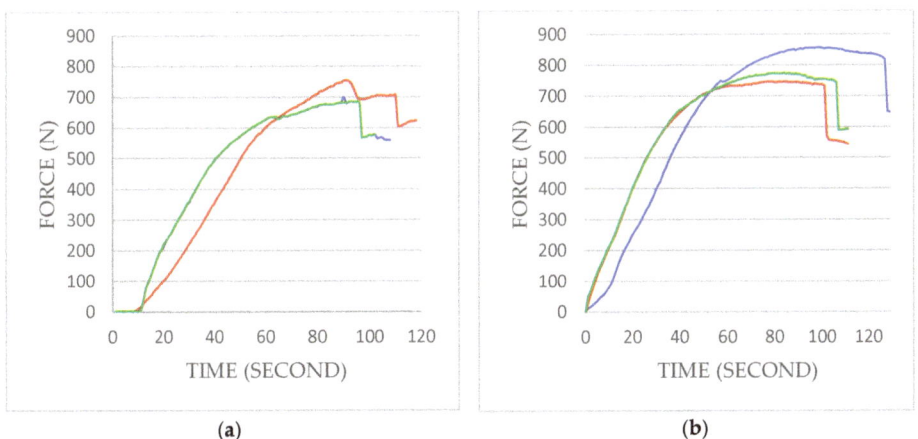

Figure 4. Single-load-to-failure test results with two different implant fixtures: (**a**) TX, (**b**) EV. Compressive load increasing at a speed of 1mm/min was applied. The peak indicates when deformation starts to occur on the implant–abutment assembly, which is the maximum breaking load. The average maximum breaking load of TX = 711 ± 36 N; EV = 791 ± 58 N.

Table 2. Values of the maximum breaking loads in single-load failure tests on three specimens each.

TX Ø4.0	Load at Break (N)	EV Ø4.2	Load at Break (N)
	698 N		856 N
	684 N		772 N
	752 N		745 N
Mean ± SD	711 ± 36 N	Mean ± SD	791 ± 58 N

Table 3. Values of the Fatigue Tests.

TX Ø4.0			
Loading Level (%)	Sinusoidal Loading (N)	Number of Performed Cycles	Mean
80	57–569	3209; 4369; 3851	3810
60	43–426	25,884; 14,353; 13,742	17,993
50	36–355	19,549; 66,014; 61,825	49,129
40	29–285	5,000,000; 5,000,000; 5,000,000	5,000,000
EV Ø4.2			
Loading Level (%)	Sinusoidal Loading (N)	Number of Performed Cycles	Mean
80	63–632	6696; 8567; 9333	8199
60	47–474	16,118; 39,423; 11,219	22,253
50	40–395	75,210; 23,584; 47,651	48,815
40	32–316	5,000,000; 5,000,000; 5,000,000	5,000,000

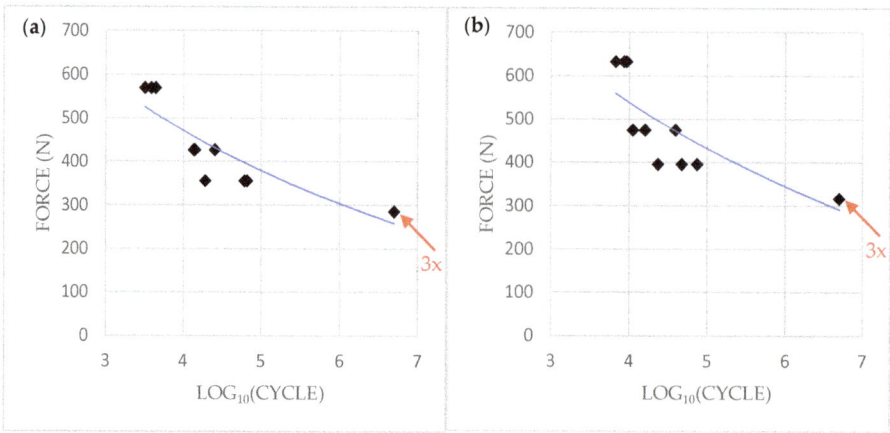

Figure 5. Plotted S/N curves from cyclic loading tests results: (**a**) TX, (**b**) EV. The x-axis represents the logarithmic value of the number of cycles performed. The loading level represents the maximum of the sinusoidal loading level; red arrow = 3 dots overlapped.

3.3. Failure Modes

Failure modes were observed to speculate the fracture mechanism of the TX and EV samples and are shown in Table 4. For the TX groups, every tested assembly except one, which showed abutment-level failure at the 80% loading level, exhibited failure at the fixture level. For the EV groups, two tested assemblies appeared to have torn-out fixtures at the 80% loading level, which were designated as fixture-level failures. The other assemblies exhibited fixture-level fractures occurring between the first and second threads. All fractures of the specimens were accompanied by screw fractures. There was no statistical difference between fractured areas between the TX and EV groups ($P > 0.99$).

Table 4. Fisher's exact test showed no difference between fractured areas ($P > 0.99$).

Fractured Area	Failure Aspect (TX Ø4.0)		Failure Aspect (EV Ø4.2)	
	Static Load	Cyclic Load	Static Load	Cyclic Load
Abutment Fracture	0	1	0	0
Fixture Fracture	3	8	3	9

3.4. Microscopic Observation

Based on the SEM examination, all the samples, except one TX sample which had an abutment-level fracture, showed a tendency of fixture-level fracture around the first and second threads apical to the microthread area. The thinnest part at the implant–abutment interface and the fractured area did not correspond for TX specimens (Figure 6). For the EV specimens, the 50% loading-level group was characterized with a clean-cut fracture tendency at the first thread level. The other groups showed a tendency to be torn out in a wavy pattern apical and coronal to the first thread. The fractured area was almost at the same level as the thinnest part of the fixture itself (Figure 7). Therefore, from the results, the null hypothesis was accepted.

Figure 6. Frontal view of TX samples (×30). Fixtures are aligned to represent a load subjected from left to right.

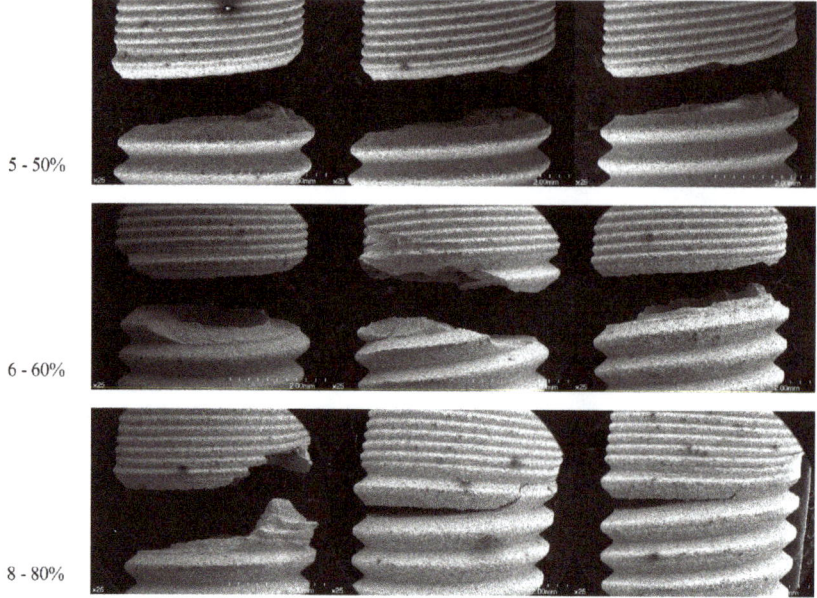

Figure 7. Frontal view of EV samples (×25). Fixtures are aligned to represent a load subjected from left to right.

4. Discussion

During masticatory function, the dental implant fixture and abutment complex should withstand high axial and lateral force of the jaw [17]. An average value of the axial direction force on a single molar implant restoration was previously reported as 120 N [18]. The reported values of maximum loads ranged from 108 to 299 N in the incisor region and from 216 to 847 N in the molar region [18–21]. In previous research, Park et al. have reported fracture strength under static loading between 799 and 1255 N in the grade 4 titanium implant–abutment assemblies with a diameter close to 4.0 mm [22]. Marchetti et al. have reported fracture strength of 430 N and a fatigue limit of 172 N (i.e., 40% of the maximum breaking load) in a grade 4 titanium implant fixture with a diameter of 3.8 mm [23]. Although a direct comparison between current findings and previous results was impossible due to the difference in the loading conditions between the studies, a similar tendency could be observed. Both TX and EV systems used in this study could overcome the normative requirements, and could be characterized by stable mechanical properties. Furthermore, the calculated fatigue strength proportion between TX and EV in our study was approximately 11%. A study conducted by Johansson and Hellqvist has previously reported that the EV system had 11–20% superior fatigue resistance compared to the TX system, which was consistent with the current findings [24]. The increased strength of EV may be the result of a more apically-located implant–abutment joint area, leading to a better stress distribution, which can be speculated from the micro-CT images. Even with similar chemical compositions, the geometry of the implant–abutment connection can affect the mechanical performance in dental implants. Therefore, the clinician should consider the mechanical properties of implant systems in the treatment planning phase, especially in locations where intraoral conditions may be harsh.

Although ISO 14801 provides a standardized protocol for cyclic loading, it does not provide a loading regimen other than starting at a nominal peak value of 80%, and so one must design the interval between the loading values. This leaves the researcher to guess a nominal peak value that can withstand 5×10^6 cycles, which in this case was 40%. However, a 50% nominal peak value seems to be too high a value to accurately estimate the fatigue limit. The 40% groups of TX and EV in this study that endured 5×10^6 cycles are equivalent to 20 years of service time in the mouth. Previous studies have shown that humans have an average of 250,000 mastication cycles per year [25,26]. Therefore, it can be assumed that 5×10^6 cycles are equivalent to 20 years of service time in the mouth. As the worst-case scenario simulates the harshest environment, it can also be assumed that both specimens can successfully survive intraoral clinical conditions. In contrast, the 50% groups fractured before an average of 50,000 cycles, which is equivalent to less than three months of service time. This "extreme" loading could have affected the failure modes as well as the estimated fatigue limit. Therefore, a loading regimen that includes a nominal peak value between 40% and 50% is recommended for future implant cyclic loading tests. In addition, we speculate that extrapolation to a clinical situation of extreme loading is less applicable for the interpretation of the 50% peak value.

The observation of fractured areas is also important in understanding the fracture mechanism in dental implants. With the advance of technology, micro-CT can be used in observing possible deformations of the dental implant [27]. In this study, the micro-CT images taken before the loading test revealed design differences between the two fixtures. The thinnest part, excluding the coronal portion of the fixture, of TX was located at the microthread area, which was approximately 0.5–1 mm coronal to the first thread of the implant fixture. The thinnest part of EV was located at the first thread of the implant fixture. The thinnest areas of each fixture are shown in Figure 4 and were expected to be the mechanically weakest parts, eventually being the fracture-prone area. However, the fracture lines initiated around the first thread of the fixture in this study. The first thread area was at the same level as the thinnest part in the EV fixture, while they were not at the same level in the TX. This suggested that the weakest part, not necessarily the thinnest part, of the TX and EV fixtures was located around the first thread area. These findings are consistent with a previous study by Shemtov-Yona et al. who tested a conical 13 mm dental implant made of titanium alloy. Three different diameters (3.3, 3.75, and 5 mm) at the implant neck were tested for fatigue performance under cyclic loading. All 5 mm implants fractured at the abutment neck and screw, while all 3.3 and 3.75 mm implants fractured at the implant body. As the implants became thinner, they showed a tendency to fracture more apically than thicker samples [10]. While the results of this study showed no statistical difference between the fracture modes of two groups in the present study ($p > 0.99$), the one sample that fractured at the abutment level may have been affected by the diameter of the implant, and not solely from the design of the connection.

A limitation of this study is the relatively small size of the samples, three specimens for each group. While ISO 14801:2013 states that at least three specimens for each group is required, the small size of samples may not be enough to extract a general conclusion. However, the tendency of the results may provide a surmise on how different implant–abutment complexes react to fatigue. Also, another limitation of this study is that loading conditions such as the number of cycles, loading force, and loading angle were not similar to intraoral masticatory conditions. However, to the best of our knowledge, no testing apparatus or protocol currently can perfectly mimic the function of physiologic mastication. Additional research with large sample size and long-term cyclic loading program is required in the future. Also, a standardized testing protocol with further detail may be a prerequisite to the research.

5. Conclusions

Within the limitation of this study, we conclude the following:

1. While both implant–abutment complexes are suitable for intraoral use, the EV fixtures in this study performed better than the TX fixtures, which indicates possible differentiation between the two implant–abutment complex designs.

2. Since all specimens with a 40% nominal peak value survived 5×10^6 cycles and 50% specimens failed before 10^5 cycles, a loading regimen with nominal peak value between 40% and 50% may be recommended for future testing of cyclic loading for the dental implant fixture.

3. The weakest parts of the tested fixtures were located at the first thread area, which happens to be the area directly coronal to the fixation simulating a 3 mm bone loss, and not necessarily the thinnest part.

From these conclusions, future researchers and implant manufacturers may benefit from starting cycling loading at the loading regimen and by considering the weakest part presented when designing an implant. On the other hand, clinicians should consider the mechanical properties of the implants they plan to use.

Author Contributions: Conceptualization, H.-I.Y., T.-H.K., E.-J.P.; methodology, H.-I.Y., T.-H.K., E.-J.P.; validation, H.-I.Y., E.-J.P.; formal analysis, N.-H.C., H.-I.Y., T.-H.K., E.-J.P.; investigation, N.-H.C.; resources, N.-H.C., H.-I.Y.; data curation, N.-H.C.; writing—original draft preparation, N.-H.C.; writing—review and editing, H.-I.Y., T.-H.K., E.-J.P.; visualization, N.-H.C.; supervision, H.-I.Y., E.-J.P.; project administration, E.-J.P.; funding acquisition, H.-I.Y., T.-H.K., E.-J.P.

Funding: This study was funded in part by Dentsply Sirona Implants (Mölndal, Sweden) and Yuhan Co. (Seoul, S. Korea) and also supported by the National Research Foundation of Korea (NRF) grant funded by the Korea government (MSIT) (NRF-2017R1C1B2007369).

Acknowledgments: This study was funded in part by Dentsply Sirona Implants (Mölndal, Sweden) and Yuhan Co. (Seoul, S. Korea). The authors would like to thank Hong-Seok Lim of Dongguk University who helped with the collection of data.

Conflicts of Interest: The authors declare no conflict of interest. The funders had no role in the design of the study; in the collection, analyses, or interpretation of data; in the writing of the manuscript, or in the decision to publish the results.

References

1. Buser, D.; Janner, S.F.M.; Wittneben, J.G.; Brägger, U.; Ramseier, C.A.; Salvi, G.E. 10-Year Survival and Success Rates of 511 Titanium Implants with a Sandblasted and Acid-Etched Surface: A Retrospective Study in 303 Partially Edentulous Patients. *Clin. Implant Dent. Relat. Res.* **2012**, *14*, 839–851. [CrossRef] [PubMed]
2. Pjetursson, B.E.; Thoma, D.; Jung, R.; Zwahlen, M.; Zembic, A. A Systematic Review of the Survival and Complication Rates of Implant-Supported Fixed Dental Prostheses(FDPs) after a Mean Observation Period of at Least 5 Years. *Clin. Oral Implants Res.* **2012**, *23*, 22–38. [CrossRef] [PubMed]
3. Jung, R.E.; Zembic, A.; Pjetursson, B.E.; Zwahlen, M.; Thoma, D.S. Systematic Review of the Survival Rate and the Incidence of Biological, Technical, and Aesthetic Complications of Single Crowns on Implants Reported in Longitudinal Studies with a Mean Follow-up of 5 Years. *Clin. Oral Implants Res.* **2012**, *23*, 2–21. [CrossRef] [PubMed]
4. Jung, R.E.; Pjetursson, B.E.; Glauser, R.; Zembic, A.; Zwahlen, M.; Lang, N.P. A Systematic Review of the 5-Year Survival and Complication Rates of Implant-Supported Single Crowns. *Clin. Oral Implants Res.* **2008**, *19*, 119–130. [CrossRef] [PubMed]
5. Åstrand, P.; Ahlqvist, J.; Gunne, J.; Nilson, H. Implant Treatment of Patients with Edentulous Jaws: A 20-Year Follow-Up. *Clin. Implant Dent. Relat. Res.* **2008**, *10*, 207–217. [CrossRef] [PubMed]
6. Blanes, R.J.; Bernard, J.P.; Blanes, Z.M.; Belser, U.C. A 10-Year Prospective Study of ITI Dental Implants Placed in the Posterior Region. I: Clinical and Radiographic Results. *Clin. Oral Implants Res.* **2007**, *18*, 699–706. [CrossRef] [PubMed]
7. Lekholm, U.; Gröndahl, K.; Jemt, T. Outcome of Oral Implant Treatment in Partially Edentulous Jaws Followed 20 Years in Clinical Function. *Clin. Implant Dent. Relat. Res.* **2006**, *8*, 178–186. [CrossRef] [PubMed]
8. Tabrizi, R.; Behnia, H.; Taherian, S.; Hesami, N. What Are the Incidence and Factors Associated With Implant Fracture? *J. Oral Maxillofac. Surg.* **2017**, *75*, 1866–1872. [CrossRef]
9. Balik, A.; Karatas, M.O.; Keskin, H. Effects of Different Abutment Connection Designs on the Stress Distribution Around Five Different Implants: A 3-Dimensional Finite Element Analysis. *J. Oral Implantol.* **2012**, *38*, 491–496. [CrossRef]
10. Shemtov-Yona, K.; Rittel, D.; Levin, L.; Machtei, E.E. Effect of Dental Implant Diameter on Fatigue Performance. Part I: Mechanical Behavior. *Clin. Implant Dent. Relat. Res.* **2014**, *16*, 172–177. [CrossRef]
11. ASTM. *ASTM E1823-10-Standard Terminology Relating to Fatigue and Fracture Testing*; ASTM International: West Conshohocken, PA, USA, 2010.

12. Imakita, C.; Shiota, M.; Yamaguchi, Y.; Kasugai, S.; Wakabayashi, N. Failure Analysis of an Abutment Fracture on Single Implant Restoration. *Implant Dent.* **2013**, *22*, 326–331. [CrossRef] [PubMed]
13. ISO14801. *Fatigue Test for Endosseous Dental Implants*; International Organization for Standardization: Geneva, Switzerland, 2013.
14. Kelly, J.R.; Benetti, P.; Rungruanganunt, P.; Bona, A.D. The Slippery Slope-Critical Perspectives on in Vitro Research Methodologies. *Dent. Mater.* **2012**, *28*, 41–51. [CrossRef] [PubMed]
15. Chieruzzi, M.; Pagano, S.; Cianetti, S.; Lombardo, G.; Kenny, J.M.; Torre, L. Effect of Fibre Posts, Bone Losses and Fibre Content on the Biomechanical Behaviour of Endodontically Treated Teeth: 3D-Finite Element Analysis. *Mater. Sci. Eng. C* **2017**, *74*, 334–346. [CrossRef] [PubMed]
16. Geng, J.P.A.; Tan, K.B.C.; Liu, G.R. Application of Finite Element Analysis in Implant Dentistry: A Review of the Literature. *J. Prosthet. Dent.* **2001**, *85*, 585–598. [CrossRef] [PubMed]
17. van der Bilt, A. Assessment of Mastication with Implications for Oral Rehabilitation: A Review. *J. Oral Rehabil.* **2011**, *38*, 754–780. [CrossRef] [PubMed]
18. Richter, E.-J. In Vivo Vertical Forces on Implants. *Int. J. Oral Maxillofac. Implants* **1995**, *10*, 99–108. [PubMed]
19. Gibbs, C.H.; Mahan, P.E.; Mauderli, A.; Lundeen, H.C.; Walsh, E.K. Limits of Human Bite Strength. *J. Prosthet. Dent.* **1986**, *56*, 226–229. [CrossRef]
20. Helkimo, E.; Carlsson, G.E.; Helkimo, M. Bite Forces Used during Chewing of Food. *J. Dent. Res.* **1959**, *29*, 133–136.
21. Waltimo, A.; Könönen, M. A Novel Bite Force Recorder and Maximal Isometric Bite Force Values for Healthy Young Adults. *Eur. J. Oral Sci.* **1993**, *101*, 171–175. [CrossRef]
22. Park, S.-J.; Lee, S.-W.; Leesungbok, R.; Ahn, S.-J. Influence of the Connection Design and Titanium Grades of the Implant Complex on Resistance under Static Loading. *J. Adv. Prosthodont.* **2016**, *8*, 388–395. [CrossRef]
23. Marchetti, E.; Ratta, S.; Mummolo, S.; Tecco, S.; Pecci, R.; Bedini, R.; Marzo, G. Mechanical Reliability Evaluation of an Oral Implant-Abutment System According to UNI En ISO 14801 Fatigue Test Protocol. *Implant Dent.* **2016**, *25*, 613–618. [CrossRef] [PubMed]
24. Johansson, H.; Hellqvist, J. Functionality of a Further Developed Implant System: Mechanical Integrity. *Clin. Oral Implants Res.* **2013**, *24*, 166.
25. Sakaguchi, R.L.; Douglas, W.H.; DeLong, R.; Pintado, M.R. The Wear of a Posterior Composite in an Artificial Mouth: A Clinical Correlation. *Dent. Mater.* **1986**, *2*, 235–240. [CrossRef]
26. DeLong, R.; Douglas, W.H. Development of an Artificial Oral Environment for the Testing of Dental Restoratives: Bi-Axial Force and Movement Control. *J. Dent. Res.* **1983**, *62*, 32–36. [CrossRef] [PubMed]
27. Stimmelmayr, M.; Edelhoff, D.; Güth, J.-F.; Erdelt, K.; Happe, A.; Beuer, F. Wear at the Titanium–Titanium and the Titanium–Zirconia Implant–Abutment Interface: A Comparative in Vitro Study. *Dent. Mater.* **2012**, *28*, 1215–1220. [CrossRef]

© 2019 by the authors. Licensee MDPI, Basel, Switzerland. This article is an open access article distributed under the terms and conditions of the Creative Commons Attribution (CC BY) license (http://creativecommons.org/licenses/by/4.0/).

Article

Axial Displacements and Removal Torque Changes of Five Different Implant-Abutment Connections under Static Vertical Loading

Ki-Seong Kim and Young-Jun Lim *

Department of Prosthodontics and Dental Research Institute, School of Dentistry, Seoul National University, Seoul 03080, Korea; namsang0249@nate.com
* Correspondence: limdds@snu.ac.kr; Tel.: +82-2-2072-2940

Received: 3 January 2020; Accepted: 31 January 2020; Published: 4 February 2020

Abstract: The aim of this study was to examine the settling of abutments into implants and the removal torque value under static loading. Five different implant-abutment connections were selected (Ext: external butt joint + two-piece abutment; Int-H2: internal hexagon + two-piece abutment; Int-H1: internal hexagon + one-piece abutment; Int-O2: internal octagon + two-piece abutment; Int-O1: internal octagon + one-piece abutment). Ten implant-abutment assemblies were loaded vertically downward with a 700 N load cell at a displacement rate of 1 mm/min in a universal testing machine. The settling of the abutment was obtained from the change in the total length of the entire implant-abutment unit before and after loading using an electronic digital micrometer. The post-loading removal torque value was compared to the initial torque value with a digital torque gauge. The settling values and removal torque values after 700 N static loading were in the following order, respectively: Ext < Int-H1, Int-H2 < Int-O2 < Int-O1 and Int-O2 < Int-H2 < Ext < Int-H1, Int-O1 ($\alpha = 0.05$). After 700 N vertical static loading, the removal torque values were statistically different from the initial values, and the post-loading values increased in the Int-O1 group and Int-H1 group ($\alpha = 0.05$) and decreased in the Ext group, Int-H2 group, and Int-O2 group ($\alpha = 0.05$). On the basis of the results of this study, it should be taken into consideration that a loss of the preload due to the settling effect can lead to screw loosening during a clinical procedure in the molar region where masticatory force is relatively greater.

Keywords: dental implants; implant-abutment connection; settling effect; static loading; removal torque

1. Introduction

Attempts have been made to understand the factors that could compromise the settling effect of different implant abutment connections [1,2]. Various implant elements including the implant-abutment interface, the types of abutments, the screw characteristics, and the cyclic loading condition have all been shown to influence settling into implants and a loss of preload [3,4].

Loosening of the abutment screws and fixture failure in implant-supported restorations reportedly occur more frequently in the premolar and molar areas than in the incisor region [5,6]. This may result from differences in masticatory force and prosthetic design. Occlusion can be critical for implant longevity due to the nature of the potential load created by tooth contacts. The mechanism and vector of force transferred by posterior teeth differ from those of anterior teeth because posterior teeth have a stronger biting force in the vertical direction. Furthermore, these forces are produced by the action of the masticatory muscles [7].

The various forces that are exerted upon dental implants during function differ in magnitude and direction. In natural dentition, the periodontal ligament has the capacity to absorb stress and allow for tooth movement, but the bone-implant interface has little capacity to allow for the movement of an implant [8,9]. The force is distributed primarily along the crest of the ridge due to the lack of micromovement of implants [10].

Cyclic loading, which simulates functional loading, can significantly influence the overall intimacy of the settling of abutments into implants and their mechanical interlocking at the bone–implant interface [2]. However, cyclic loading is not the only factor that could influence the settling phenomenon in posterior teeth. Cyclic loading and the static loading are two independent conditions, and both can affect the settling of abutments into implants after occlusal loading.

In particular, vertical forces generated on implants in the posterior region are greatest at the implant-abutment interface. This means that vertical masticatory forces can affect settling into implants and a loss of preload after occlusal static loading.

Bruxism or clenching can create destructive lateral stresses and overloading when it transfers force to the supporting bone [11]. Parafunctional movements exert a greater maximum occlusal force than natural mastication. Van Eijden measured the mean magnitudes of a maximal vertical bite force in normal dentition without implants as follows: 469 ± 85 N at the canine region, 583 ± 99 N at the second premolar region, and 723 ± 138 N at the second molar region [12]. These results were comparable to the mean maximum bite force of 738 ± 209 N measured by Braun et al. [13]. In addition, Morneburg and Pröschel investigated vertical masticatory forces in vivo on implant-supported fixed partial dentures and found a mean total masticatory force of 220 N with a maximum of 450 N [14]. On the basis of these findings, the present study evaluated the degree of settling and compared preload loss using the removal torque values before and after 700 N static vertical loading.

The aim of this study was to evaluate the settling of abutments into implants and removal torque values of five different implant-abutment connections that differ significantly in macroscopic geometry after static vertical loading at 700 N.

2. Materials and Methods

2.1. Implant-Abutment Systems Selection and Study Protocol

One external and two internal connection implant systems from the Osstem Implant (Osstem Co., Seoul, Korea) were selected for the study. The abutment–implant assemblies were divided into five groups according to the implant connection designs and abutment types (Table 1, Figure 1).

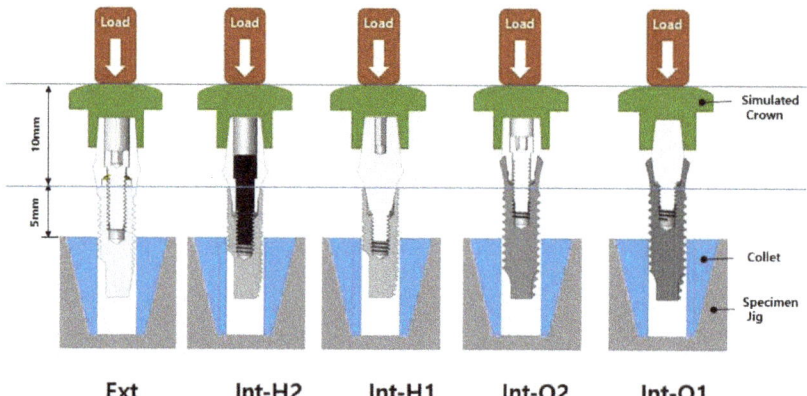

Figure 1. Schematic drawing of the test setup. Ext: external hexagon fixture + Cemented abutment; Int-H2: internal hexagon fixture + two-piece abutment; Int-H1: internal hexagon fixture + one-piece abutment; Int-O2: internal octagon fixture + two-piece abutment; Int-O1: internal octagon fixture + one-piece abutment.

Table 1. Characteristics of experimental implant-abutment systems.

Group	Ext	Int-H2	Int-H1	Int-O2	Int-O1
Implant system	US II	GS II		SS II	
Implant/abutment interface	External butt joint	11° taper internal hexagon		8° morse taper internal octagon	
Abutment type	Cemented (two-piece)	Transfer (two-piece)	Rigid (one-piece)	Comocta (two-piece)	Solid (one-piece)
Abutment material	Ti CP-Gr 3	Ti CP-Gr 3	Ti-6Al-4V	Ti CP-Gr 3	Ti-6Al-4V
Abutment diameter	Ø5.0	Ø5.0	Ø5.0	Ø4.3	Ø3.5
Abutment gingival height	2 mm	2 mm	2 mm	-	-
Abutment height (H_A)	5.5 mm	5.5 mm	5.5 mm	4 mm	4 mm
Abutment screw	Ta	WC/C Ta	-	Ta	-
Fixture material	Ti CP-Gr 4	Ti CP-Gr 4		Ti CP-Gr 4	
Fixture diameter	Ø4.0	Ø4.0	Ø4.0	Ø4.1	Ø4.1
Fixture height (H_F)	11.4 mm	11.5 mm	11.5 mm	11.5 mm	11.5 mm
Feature					

Ext: external hexagon fixture + Cemented abutment; Int-H2: internal hexagon fixture + two-piece abutment; Int-H1: internal hexagon fixture + one-piece abutment; Int-O2: internal octagon fixture + two-piece abutment; Int-O1: internal octagon fixture + one-piece abutment; Ta: titanium alloy; WC/C Ta: tungsten carbide/carbon-coated titanium alloy; H_A: Abutment height; H_F: fixture height.

Ext: External butt joint + Cemented abutment (two-piece)
Int-H2: Internal hexagon + Transfer abutment (two-piece)
Int-H1: Internal hexagon + Rigid abutment (one-piece)
Int-O2: Internal octagon + Comocta abutment (two-piece)
Int-O1: Internal octagon + Solid abutment (one-piece)

Ten implant-abutment assemblies were constructed for each group (total $n = 50$). Each assembly was held in a vise during the torque tightening procedure. The desired torque was applied to the abutment screw with a digital torque gauge (MGT12, MARK-10 Co., Hicksville, NY, USA).

The schematic diagram of experimental design based on protocol sequence is presented in Figure 2. Each abutment was tightened into the corresponding implant at 30 Ncm torque twice at 10 minute intervals. Ten minutes after the second tightening, the initial removal torque was measured with a digital torque gauge (MGT12E, Mark-10 corp, Hicksville, NY, USA). Each assembly was secured again at 30 Ncm torque, and the total length of the implant-abutment assembly was measured with an electronic digital micrometer (no. 293-561-30, Mitutoyo, Japan). After the initial measurement of the total length, a metal cap fabricated to reproduce the crown was mounted on the abutment of the assembly and the entire unit was fixed in a loading jig (Figure 3). The loading jig was designed to withstand a 700 N vertical static force applied to the implant-abutment assembly. All the specimens were tested in a universal testing machine (Instron 8841, Instron Corp., Mass, Norwood MA, USA) under 700 N vertical static loading, corresponding to the maximum biting force in posterior teeth [12,13].

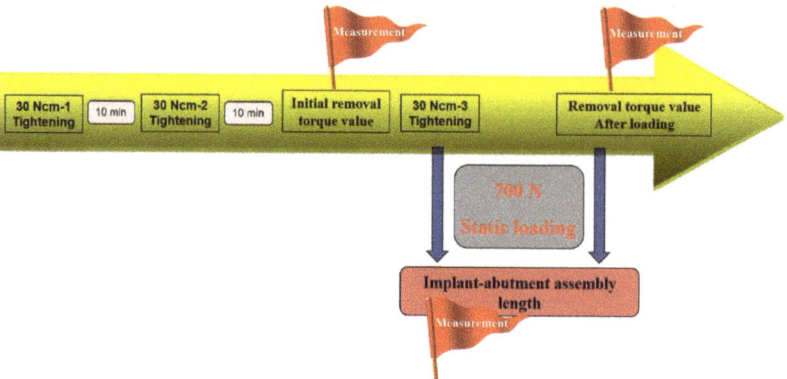

Figure 2. Schematic diagram of experimental design based on protocol sequence.

Figure 3. Loading machine and customized jig (Instron 8841, Instron Corp., Mass, Norwood MA, USA).

At the completion of static loading, the total length and removal torque of each implant-abutment specimen were measured in the same manner. The settling value of the abutment was calculated from the changes in the total lengths of the implant-abutment assembly before and after loading. The measurements were accurate up to 0.001 mm (1 µm) and the same operator performed all of the specimen preparations and testing in random order. The details of the experimental protocol and the overall outcomes between the magnitude of applied torque and the axil displacement of abutments into implants in external and internal implant-abutment connections were reported in previous studies [1,2].

2.2. Statistical Analysis

One-way ANOVA and Tukey's honestly significant difference (HSD) tests were used to analyze settling lengths and removal torque of the five implant-abutment systems before and after 700 N vertical static loading. A paired t-test was performed to compare the initial and post-loading removal torques for each implant connection system. $p < 0.05$ was considered to represent a statistically significant difference.

3. Results

The mean lengths and settling values of the specimen groups after vertical static loading are presented in Tables 2 and 3 and Figure 4. After 700 N static loading, there were statistically significant differences in the settling values in the Ext group (0.8 ± 0.45 µm), Int-H1 group (10.2 ± 0.84 µm), Int-H2 group (11.2 ± 0.84 µm), Int-O2 group (19.2 ± 4.21 µm), and Int-O1 group (25.6 ± 2.97 µm) ($\alpha = 0.05$). In the internal octagon groups with an 8° Morse taper interface, there were greater increases compared with those seen in the other groups. A multiple comparison test by Tukey's HSD exhibited differences in the settling values in each group after 700 N static loading in the following order: Ext < Int-H1, Int-H2 < Int-O2 < Int-O1 (see Tables 2 and 3).

Table 2. Mean total lengths and standard deviations of the implant-abutment specimens before and after 700 N static loading.

Group	Ext (mm)	Int-H2 (mm)	Int-H1 (mm)	Int-O2 (mm)	Int-O1 (mm)
Tightening torque 30 Ncm-③ *	18.6096 ±0.0054	18.9624 ±0.0153	19.0456 ±0.0261	18.9564 ±0.0222	18.9992 ±0.0041
Load 700 N Static **	18.6088 ±0.0054	18.9512 ±0.0151	19.0354 ±0.0266	18.9372 ±0.0222	18.9736 ±0.0035

* Additional tightening at 30 Ncm after measuring the initial removal torque after the second 30 Ncm tightening.
** After 700 N vertical static loading.

Table 3. Mean settling values after 700 N static loading in each group and multiple comparisons using Tukey's honestly significant difference (HSD).

Group	Settling Values Mean ± SD (µm)	Group Comparisons †
Ext	0.8 ± 0.45	Ext < Int-H1, Int-H2 < Int-O2 < Int-O1
Int-H2	11.2 ± 0.84	
Int-H1	10.2 ± 0.84	
Int-O2	19.2 ± 4.21	
Int-O1	25.6 ± 2.97	Settling value = (total lengths of the implant-abutment assemblies at 30 Ncm-③) minus (total lengths of the implant-abutment assemblies after 700 N static loading)

† Tukey's HSD method was performed for between group comparisons ($p < 0.05$).

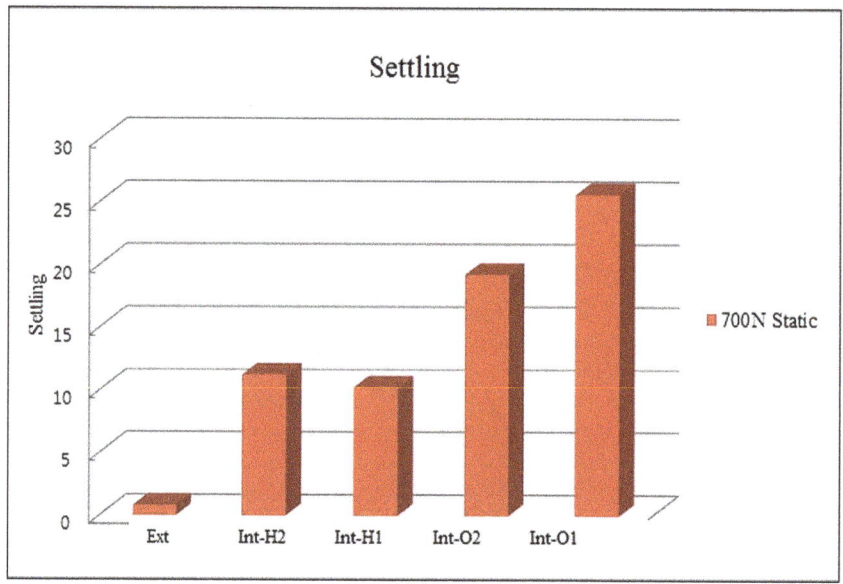

Figure 4. Settling of abutments into the implants after static loading (μm).

The mean values of removal torque after loading are presented in Tables 4–6 and Figure 5. After 700 N static loading, the Int-O1 group exhibited the highest removal torque of 39.64 ± 4.28 Ncm. The other groups are shown in the following decreasing order: Int-H1 (36.38 ± 6.25 Ncm), Ext (22.78 ± 0.40 Ncm), Int-H2 (11.62 ± 0.56 Ncm), and Int-O2 (1.14 ± 0.40 Ncm). Using Tukey's HSD, the specific group-wise comparisons in the post-loading removal torque values were as follows: Int-O2 < Int-H2 < Ext < Int-H1, Int-O1.

Table 4. Multiple comparisons of mean values of initial removal torque and removal torque after 700 N static loading.

Test	Group (n = 5)	Tightening Torque (Ncm)	Removal Torque (Ncm)	Significance †
Initial removal torque	Ext	30	24.22 ± 0.81	Int-H2 < Ext, Int-O2 < Int-H1 < Int-O1
	Int-H2	30	21.22 ± 1.04	
	Int-H1	30	27.44 ± 0.92	
	Int-O2	30	25.38 ± 1.86	
	Int-O1	30	30.54 ± 0.56	
Removal torque after 700N static loading	Ext	30	22.78 ± 0.40	Int-O2 < Int-H2 < Ext < Int-H1, Int-O1
	Int-H2	30	11.62 ± 0.56	
	Int-H1	30	36.38 ± 6.25	
	Int-O2	30	1.14 ± 0.40	
	Int-O1	30	39.64 ± 4.28	

† Tukey's HSD method was performed for between group comparisons ($p < 0.05$).

Table 5. Comparison of the mean values of initial and post-loading removal torque in each group.

Group	Initial R/T [a]	R/T after Static Load [b]	Significance †
Ext	24.22 ± 0.81	22.78 ± 0.40	*
Int-H2	21.22 ± 1.04	11.62 ± 0.56	**
Int-H1	27.44 ± 0.92	36.38 ± 6.25	NS
Int-O2	25.38 ± 1.86	1.14 ± 0.40	*
Int-O1	30.54 ± 0.56	39.64 ± 4.28	*

[a] Removal torque values before loading; [b] Removal torque values 700 N static loading; † Paired *t*-test was performed to compare the removal torque values before and after loading: NS, not significant; * $p < 0.01$; ** $p < 0.001$.

Table 6. Comparisons of the mean values of initial removal torque and removal torque after static loading in each group.

Group	Removal Torque	t/P Value
Ext	Initial	6.279
	after 700 N static loading	/0.003 *
Int-H2	Initial	16.204
	after 700 N static loading	/0.000 *
Int-H1	Initial	−3.313
	after 700 N static loading	/0.030 *
Int-O2	Initial	6.413
	after 700 N static loading	/0.003 *
Int-O1olid	Initial	−4.768
	after 700 N static loading	/0.009 *

* indicates values that were statistically different ($p < 0.05$).

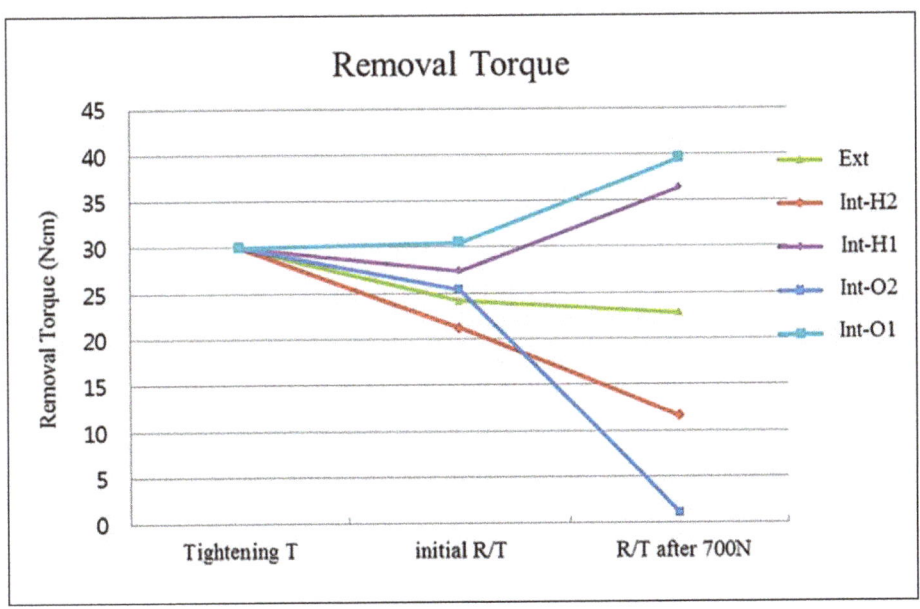

Figure 5. Removal torque (Ncm) after 700 N static loading.

In cases in which one-piece abutments were used for the internal connection system (Int-H1 group and Int-O1 group), the removal torque was increased compared to the initial removal torque. In cases where two-piece abutments were used for the internal connection system (Int-H2 group and Int-O2 group), after 700 N vertical static loading, the removal torque was decreased compared to the initial removal torque to a greater extent. In the Int-O2 group in particular, the abutment screw nearly came loose from the abutment. After 700 N loading, the removal torque value also exhibited a small but significant decrease in the Ext group (Table 6).

4. Discussion

Along with the expanded indications for implants and the changing clinical protocols, the relationship between implant design and load distribution at the implant–bone interface has become an important issue. The inadequate interaction between these two factors may result in both mechanical and biologic complications such as screw loosening and peri-implant bone loss. Whether an implant prosthesis is placed in function after an undisturbed healing period or immediately after placement, the biomechanical environment is, thereafter, a critical factor that influences implant duration and bone preservation. Loads applied to teeth and implants during physiologic oral functions including chewing, clenching, swallowing, or grinding may vary because the anchorage of natural and artificial abutments in the jaw is not of the same type and quality [15].

Most of the studies related to axial displacement [1–3] are on the magnitude of tightening torque and the duration of cyclic loading, and few studies have applied with static loading. Ko et al. [4] reported that axial displacement and reverse torque loss occurred at significantly low levels after the cyclic and static loading in the case of wide-type implants of 5.0 mm diameter. In addition, the CAD/CAM (Computer Aided Design/Computer Aided Manufacturing) customized abutments, which are currently in the spotlight, may show differences in the fabricating process from the stock abutments produced by manufacturers. Therefore, using implant fixtures and abutments made by the same manufacturer, we wanted to prove that axial displacement could occur even at static loading of 700 N, and the difference comes from different connection types.

For osseointegrated dental implants, previous studies have revealed that occlusal interferences and parafunctional activities may lead to mechanical and biologic complications [16]. Many investigators have attempted to evaluate maximum bite forces. Typical maximum bite force magnitudes exhibited by adults are affected by age, sex, degree of edentulism, bite location, and especially parafunction. In centric occlusion involving swallowing and clenching, forces are transmitted bilaterally, predominantly by molars and premolars. For a single tooth or implant in the molar region, the greatest forces occur along the axial direction [17]. Therefore, the results of this study showed the settling effect in relation to a loss of removal torque after 700 N vertical static loading, corresponding to the maximum masticatory force.

The settling effect after 700 N loading showed a clinical association between screw loosening with a loss of preload and an increase in friction. The results followed a similar pattern with cyclic loading in our previous study [2]. The Ext group showed the lowest settling due to its flat platform interface. Likewise, the internal hexagon and octagon groups had statistically greater settling due to their tapered interface. In particular, the internal octagon group with an 8° Morse taper showed the highest settling value compared to the internal hexagon group with an 11° taper.

The removal torque values after 700 N vertical static loading may be influenced by the amount of settling and the type and configuration characteristics of the abutment used. When a two-piece abutment, as seen in the Int-H2 and Int-O2 groups, is used, the screw joint connection is based on the tension mechanism, where a screw may become loose due to a loss of preload by settling. Therefore, the settling effect of the Int-H2 and Int-O2 groups produced a significant decrease in the removal torque even to the extent of the loss of the abutment screw in the Int-O2 group. On the other hand, when a one-piece abutment is used, the main retention mechanism is friction. As a result, the settling effect of the one-piece abutment in the Int-H1 and Int-O1 groups created a greater compressive force at the implant-abutment interface, which resulted in the increased post-loading values of removal torque.

The metal cap used in this experimental protocol was inserted into the abutment by friction only, and without dental cement. The simulated crown had a gap between the abutment and the metal cap in order to prevent any forces from being transferred to the abutment during the removal of the crown. However, because the margin of the crown was seated on the fixture in the original internal octagon design, there was no such space. Consequently, this discrepancy may have led to greater settling values than the actual value due to the lack of a vertical stop. In addition, this study could not use the direct method as described by Haack et al., where the change in the preload was evaluated by measuring the length of an elongated screw [18]. Therefore, further studies are warranted to evaluate the actual measurement of an elongated screw as a value of tightening torque.

5. Conclusions

The current study strived to gain a better understanding of the nature of the implant-abutment screw joint on the basis of the settling effect and removal torque. On the basis of the findings of this study, in the molar region where masticatory force is relatively greater, a loss of preload due to the axial displacement and the possibility of screw loosening should be taken into account in clinical procedures.

The clinical implication of this study is that when the implant fixture of a regular platform with a diameter of 4.0 mm is placed in the posterior molar region, the settling of abutments into implants caused by the vertical force may cause a problem of lowering the occlusion after the prosthesis is mounted.

Author Contributions: Conceptualization, writing—Original draft preparation, data curation, K.-S.K.; supervision, visualization, validation, writing—Review and editing, Y.-J.L. All authors have read and agreed to the published version of the manuscript.

Funding: The APC was funded by the Dental Research Institute, School of Dentistry, Seoul National University, Seoul, Republic of Korea. This work was supported by grant no. 02-2015-0004 from the Seoul National University Dental Hospital Research Fund, Seoul National University Dental Hospital, Seoul, Korea.

Conflicts of Interest: The authors declare no conflict of interest.

References

1. Kim, K.S.; Lim, Y.J.; Kim, M.J.; Kwon, H.B.; Yang, J.H.; Lee, J.B.; Yim, S.H. Variation in the total lengths of abutment/implant assemblies generated with a function of applied tightening torque in external and internal implant-abutment connection. *Clin. Oral Implants Res.* **2011**, *22*, 834–839. [CrossRef] [PubMed]
2. Kim, K.S.; Han, J.S.; Lim, Y.J. Settling of abutments into implants and changes in removal torque in five different implant-abutment connections. Part 1: Cyclic loading. *Int. J. Oral Maxillofac. Implants* **2014**, *29*, 1079–1084. [CrossRef] [PubMed]
3. Lee, J.H.; Lee, W.; Huh, Y.H.; Park, C.J.; Cho, L.R. Impact of Intentional Overload on Joint Stability of Internal Implant-Abutment Connection System with Different Diameter. *J. Prosthodont.* **2019**, *28*, e649–e656. [CrossRef] [PubMed]
4. Ko, K.H.; Huh, Y.H.; Park, C.J.; Cho, L.R. Axial displacement in cement-retained prostheses with different implant-abutment connections. *Int. J. Oral Maxillofac. Implants* **2019**, *34*, 1098–1104. [CrossRef] [PubMed]
5. Bianco, G.; Di Raimondo, R.; Luongo, G.; Paoleschi, C.; Piccoli, P.; Piccoli, C.; Rangert, B. Osseointegrated implant for single-tooth replacement: A retrospective multicenter study on routine use in private practice. *Clin. Implant Dent. Relat. Res.* **2000**, *2*, 152–158. [CrossRef] [PubMed]
6. Palmer, R.M.; Palmer, P.J.; Smith, B.J. A 5-year prospective study of Astra single tooth implants. *Clin. Oral Implants Res.* **2000**, *11*, 179–182. [CrossRef] [PubMed]
7. Dean, J.S.; Throckmorton, G.S.; Ellis, E., III; Sinn, D.P. A preliminary study of maximum voluntary bite force and jaw muscle efficiency in pre-orthognathic surgery patients. *J. Oral Maxillofac. Surg.* **1992**, *50*, 1284–1288. [CrossRef]
8. Koyano, K.; Esaki, D. Occlusion on oral implants: Current clinical guidelines. *J. Oral Rehabil.* **2015**, *42*, 153–161. [CrossRef] [PubMed]
9. Kim, Y.; Oh, T.; Misch, C.E.; Wang, H.L. Occlusal considerations in implant therapy: Clinical guidelines with biomechanical rationale. *Clin. Oral Implants Res.* **2005**, *1*, 26–35. [CrossRef] [PubMed]

10. Rieger, M.R.; Mayberry, M.; Brose, M.O. Finite element analysis of six endosseous implants. *J. Prosthet. Dent.* **1990**, *63*, 671–676. [CrossRef]
11. Misch, C.E. The effect of bruxism on treatment planning for dental implants. *Dent. Today* **2002**, *2*, 76–81.
12. van Eijden, T.M. Three-dimensional analyses of human bite-force magnitude and moment. *Arch. Oral Biol.* **1991**, *36*, 535–539. [CrossRef]
13. Braun, S.; Hnat, W.P.; Freudenthaler, J.W.; Marcotte, M.R.; Hönigle, K.; Johnson, B.E. A study of maximum bite force during growth and development. *Angle Orthod.* **1996**, *66*, 261–264. [PubMed]
14. Morneburg, T.R.; Pröschel, P.A. Measurement of masticatory forces and implant loads: A methodologic clinical study. *Int. J. Prosthodont.* **2002**, *15*, 20–27. [PubMed]
15. Richter, E.J. In vivo vertical forces on implants. *Int. J. Oral Maxillofac. Implants* **1995**, *10*, 99–108. [PubMed]
16. Chambrone, L.; Chambrone, L.A.; Lima, L.A. Effects of occlusal overload on peri-implant tissue health: A systematic review of animal-model studies. *J. Periodontol.* **2010**, *81*, 1367–1378. [CrossRef] [PubMed]
17. Rigsby, D.F.; Bidez, M.W.; Misch, C.E. Bone Response to Mechanical Loads. In *Contemporary Implant Dentistry*, 2nd ed.; Misch, C.E., Ed.; Mosby: St. Louis, Missouri, USA, 1998; pp. 317–328.
18. Haack, J.E.; Sakaguchi, R.L.; Sun, T.; Coffey, J.P. Elongation and preload stress in dental implant abutment screws. *Int. J. Oral Maxillofac. Implants* **1995**, *10*, 529–536. [PubMed]

© 2020 by the authors. Licensee MDPI, Basel, Switzerland. This article is an open access article distributed under the terms and conditions of the Creative Commons Attribution (CC BY) license (http://creativecommons.org/licenses/by/4.0/).

Article

Dental Implants with Different Neck Design: A Prospective Clinical Comparative Study with 2-Year Follow-Up

Pietro Montemezzi [1,2,*], Francesco Ferrini [1,2], Giuseppe Pantaleo [3], Enrico Gherlone [1,2] and Paolo Capparè [1,2]

1. Dental School, Vita-Salute San Raffaele University, 20132 Milan, Italy; ferrini.f@gmail.com (F.F.); gherlone.enrico@hsr.it (E.G.); cappare.paolo@hsr.it (P.C.)
2. Department of Dentistry, IRCCS San Raffaele Hospital, 20132 Milan, Italy
3. UniSR-Social.Lab (Research Methods), Faculty of Psychology, Vita-Salute San Raffaele University, 20132 Milan, Italy; pantaleo.giuseppe@hsr.it
* Correspondence: m.montemezzi@libero.it

Received: 24 December 2019; Accepted: 20 February 2020; Published: 25 February 2020

Abstract: The present study was conducted to investigate whether a different implant neck design could affect survival rate and peri-implant tissue health in a cohort of disease-free partially edentulous patients in the molar–premolar region. The investigation was conducted on 122 dental implants inserted in 97 patients divided into two groups: Group A (rough wide-neck implants) vs. Group B (rough reduced-neck implants). All patients were monitored through clinical and radiological checkups. Survival rate, probing depth, and marginal bone loss were assessed at 12- and 24-month follow-ups. Patients assigned to Group A received 59 implants, while patients assigned to Group B 63. Dental implants were placed by following a delayed loading protocol, and cemented metal–ceramic crowns were delivered to the patients. The survival rates for both Group A and B were acceptable and similar at the two-year follow-up (96.61% vs. 95.82%). Probing depth and marginal bone loss tended to increase over time (*follow-up*: t_1 = 12 vs. t_2 = 24 months) in both groups of patients. Probing depth (p = 0.015) and bone loss (p = 0.001) were significantly lower in Group A (3.01 vs. 3.23 mm and 0.92 vs. 1.06 mm; Group A vs. Group B). Within the limitations of the present study, patients with rough wide-neck implants showed less marginal bone loss and minor probing depth, as compared to rough reduced-neck implants placed in the molar–premolar region. These results might be further replicated through longer-term trials, as well as comparisons between more collar configurations (e.g., straight vs. reduced vs. wide collars).

Keywords: dental implants; dental implant neck design; peri-implant bone loss; peri-implant probing depth

1. Introduction

The scientific debate on dental implant macro-design is a well-known topic in the field of implant dentistry. The ideal fixture design should bring together the most suitable and distinctive characteristics for implant osseointegration, such as type of material (zirconium or titanium), body shape (cylindrical or conical), neck geometry (straight, reduced, or wide), threads depth, width, and pitch, as well as tapered or non-tapered apical portion, body length, and diameter. Although there is no perfect implant design [1,2], nor a best surface treatment [3], scientific evidence has consistently demonstrated that different dental implant macro-designs affect long-term implant success [4,5] and also accelerate the healing process, to allow implant therapy in the population of patients who are more prone to failure [6,7]. Implant collar, being the portion of the implant that connects the fixture with the oral cavity throughout a prosthetic device, is a very important feature related to the peri-implant tissue's health conditions.

Several studies about implant neck design and marginal bone loss can be found in the literature, but the results are controversial. In vivo animal studies reported a greater crestal bone height and thickness of surrounding implant tissue in dental implants with triangular neck designs [8]; smaller crestal bone loss but similar peri-implant tissue thickness in narrow ring extra-shorts implants [9]; and greater bone loss in dental implants with micro-rings on the neck, as compared to open-thread implant collars [10]. Human model studies reported improved biomechanical behavior for stress/strain distribution pattern in dental implants with divergent collar design [11]; no additional bone loss in non-submerged dental implants with a short smooth collar compared to similar but longer implant collar design [12].

Other clinical findings suggest that specific implant neck design might be suitable in anterior areas, where bone loss, even if acceptable, can lead to adverse aesthetic results [13,14].

The purpose of the present study is to compare peri-implant hard- and soft-tissue health conditions in partially edentulous patients who received the same dental implants but with two different implant neck designs, at a two-year follow-up. In this study, the null hypothesis led to the expectation of no differences in survival rate, probing depth, and marginal bone loss among patients who received dental implants with wide or reduced collar morphology.

2. Materials and Methods

2.1. Patients

Study participants were selected from patients who attended the Dental Department of IRCCS San Raffaele Hospital, Milan, Italy asking for partial fixed implant-prosthetic rehabilitation. Recruitment occurred from February 2016 to November 2017, and the investigation was conducted following all the ethical regulations related to the institution.

Patients had to meet the following inclusion criteria: (1) hopeless teeth to be extracted at least four months prior to surgery in molar/premolar region; (2) no previous dental implants already in place adjacent to surgical site; (3) natural antagonistic teeth (composite resin restorations allowed); (4) absence of diabetes, periodontitis, bruxism, and smoking; (5) absence of chemotherapy or radiation therapy of head and neck district, as well as anti-resorptive drug therapy (i.e., bisphosphonates); and (6) neither mucosal lesions (lichen planus, epulis fissuratum) nor bone lesions (i.e., simple bone cyst or odontomas). Eligible areas for surgery of edentulous maxilla or mandible were selected to receive 1 to a maximum of 3 dental implants. Participants were verbally informed about the purpose of the study but not assigned to a specific group, as they were randomly chosen either to receive a wide-neck implant (Group A) or a reduced-neck implant (Group B).

Patients were assigned to conditions according to a computer-generated random list, prescribing the use of the reduced vs. wide implant. Clinical measures (i.e., survival rate, peri-implant probing depth, and mean marginal bone loss) were taken at 12 and 24 months. Thus, the design amounted to a 2 (implant: wide vs. reduced) X 2 (time: 12 vs. 24 month follow-up) *mixed* factorial design, following the *Consolidated Standards of Reporting Trials* (CONSORT) guidelines available as supplementary material to this manuscript and on http://www.consort-statement.org/.

Written informed consent was signed before the start of the study; patients were allowed to leave the research at any time, without any consequence.

Implant macrogeometry regarding the two different collar designs used in the present study is shown in Figure 1 (CSR, Sweden & Martina, Due Carrare, Italy).

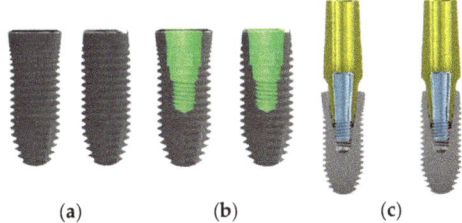

Figure 1. Image shows the CSR full-treatment ZirTi conical dental implant collar with different macro-design. (**a**) Rough wide neck compared with rough reduced neck; (**b**), wide-neck and reduced-neck designs with double conical implant–abutment connection with internal hexagon for prosthetic repositioning; and (**c**) wide-neck and reduced-neck designs with same contact length and tapered angles at the interface.

2.2. Implant Surgery

The study was based on a single blind design, with patients being unaware of which type of implant neck design (wide or reduced) was used for the therapy.

Local anesthesia was induced with local infiltration of lidocaine 20 mg/mL with 1:50.000 adrenaline (Ecocain, Molteni Dental, Firenze, Italy). A crestal horizontal incision was made, with buccal relieving incisions in the medial and distal portions of the main incision. A full-thickness flap was raised, and dental implants were placed in edentulous sites of 0.5 mm, subcrestally, with a minimum insertion torque of 35 Ncm. Cover screw was positioned, and a periosteal incision was performed in order to allow flap passivation in search for primary intention healing of the wound. Vertical mattress suturing technique was used with a 4-0 coated braided absorbable suture (Vicryl, ETHICON, Johnson & Johnson, New Brunswick, NJ, USA). Sterile dry gauze compression was performed on the wound to control post-operative bleeding. Ice packages were delivered to the patients immediately after surgery, with instruction to apply cold to the surgical area for the following 24 h. Semi-liquid cold diet was recommended for the first 48 h.

At-home pharmacological therapy prescribed was amoxicillin 1 g, every 12 hours, for six days, and non-steroid anti-inflammatory drug ibuprofen 400 mg, every 12 hours, for four days, post-operatively. All implants were loaded after a 4-month healing period, through a delayed loading protocol, with a composite resin temporary restoration, followed by metal–ceramic cemented crowns. Definitive abutments used for both Group A and B were the same and had conical connection with Double Action Tight (DAT), a system that presents a conical interface between the abutment and the implant, plus one more conical interface between the screw and the abutment.

Clinically, abutment screws were tightened at 25 Ncm by using a dental torque wrench.

2.3. Parameters

Dental implant survival rate was defined as the fixtures being osseointegrated and staying in situ; and capable to guarantee stability for prosthetic support along the 2-year observation period following the surgical placement. Peri-implant probing depth was estimated through a CP12 University of North Carolina color-coded periodontal probe (Hu Friedy, Chicago, IL, USA), in the mesial, distal, buccal, and lingual/palatal surfaces of the fixture. Distance in mm between the mucosal margin and the tip of the probe was considered as pocket depth.

Intraoral radiographs were taken, using extension cone paralleling system (XCP, Dentsply international, RINN), and mean marginal bone loss was calculated, using Digora Optime digital intraoral imaging system (Soredex, Tuusula, Finland).

A line was traced parallel to the long axis of the implant in order to measure in mm the distance between the crestal bone level at the margin of the implant neck and the top of the apical portion of the implant.

2.4. Statistical Analysis

All analyses were run at the implant level. Peri-implant probing depth and marginal bone loss were submitted to separate 2 (follow-up: $t_1 = 12$ vs. $t_2 = 24$ months) X 2 (*neck design*: reduced vs. wide) multivariate analyses of variance (MANOVA$_s$), in order to distinguish the effects of follow-up time, implant neck design, and additionally assess any interactive effect(s) of the two factors. Mean values were complemented by standard errors of the mean (*Se*) and 95% confidence intervals (CI).

3. Results

A total of 97 patients (56 men and 41 women) aged between 33 and 75 years (mean 58.2 ± 6.22 years) were selected for the present study. None of them withdrew from the research, and 122 fixtures were placed in the molar/premolar region.

Fixtures made of titanium grade 4 had a standard length (≥10 mm) and a diameter of 3.8 and 4.2 mm for wide-neck implants and 4.2 and 5.0 mm for reduced-neck ones. Dental implants received the same subtraction procedure, according to the Zir-Ti full-surface treatment (Zirconium Oxide Sand-Blasted and Acid Etched Titanium). The apical portion was tapered with 50° accentuated triangular threads and four longitudinal incisions, to increase penetration ability and anti-rotation features. Fifty patients formed Group A (rough wide-neck design) and received 59 implants. Group B (rough reduced-neck design) was composed of forty-eight patients, who received 63 implants.

The two groups were compared at one-year and two-year follow-ups. Survival rate, probing depth, and marginal bone loss were recorded through clinical and radiological checkups. Radiological records for different dental implants placed in Group A and B patients are shown in Figures 2 and 3.

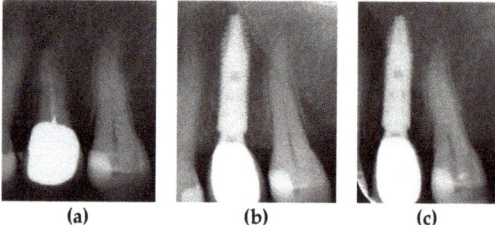

Figure 2. Periapical X-rays showing marginal bone level of CSR dental implant with a reduced neck. (**a**) Pre-operative X-ray; (**b**) post-operative follow-up at 12 months; and (**c**) post-operative follow-up at 24 months.

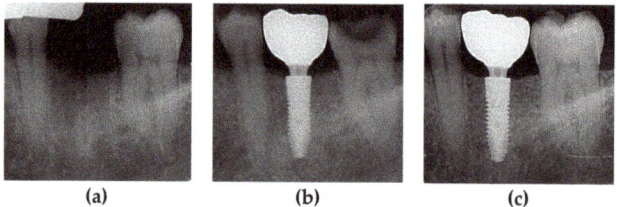

Figure 3. Periapical X-rays showing marginal bone level of CSR dental implant with a wide neck. (**a**) Pre-operative X-ray; (**b**) post-operative follow-up at 12 months; and (**c**) post-operative follow up at 24 months.

The overall survival rate of CSR dental implants at the two-year follow-up was 96.72% (four implant failures out of 122 implants placed). Both groups showed similar outcomes: At 12 months, survival rate was 98.30% for Group A and 98.41% for Group B, while it decreased at 96.61% for Group A and 96.82% for Group B at the 24-month follow-up.

Regarding peri-implant probing depth, a 2 (*follow-up*: $t_1 = 12$ vs. $t_2 = 24$ months) X 2 (*neck design*: reduced vs. wide) multivariate analysis of variance (MANOVA) affirmed a main effect of follow-up, $F(1, 116) = 10.69$, $p < 0.001$, such that probing depth was generally lower at 12 months (3.06 mm ± $Se = 0.046$; 95% CI = 2.96, 3.15) than at 24 months (3.18 mm ± $Se = 0.050$; 95% CI = 3.08, 3.28), independently of type of neck design. Furthermore, the analysis also revealed a main effect of neck design, $F(1, 116) = 6.28$, $p < 0.015$, such that probing depth was generally lower for wide (Group A: 3.01 mm ± $Se = 0.063$; 95% CI = 2.88, 3.13) than for reduced-neck implants (Group B: 3.23 mm ± $Se = 0.061$; 95% CI = 3.11, 3.35), independently of time of follow-up. More specifically, the difference between the two groups, considered at one and two years of follow-up were, respectively, as follows: Group A (one year): 2.93 mm ± $Se = 0.07$; 95% CI = 2.79, 3.07 vs. Group B (one year): 3.18 mm ± $Se = 0.05$; 95% CI = 3.07, 3.28 ($p = 0.007$); and Group A (two years): 3.09 mm ± $Se = 0.07$; 95% CI = 2.95, 3.24 vs. Group B (two years): 3.28 mm ± $Se = 0.06$; 95% CI = 3.15, 3.40 ($p = 0.061$). The interaction *follow-up* ($t_1 = 12$ vs. $t_2 = 24$ months) X *neck design* (reduced vs. wide) was not significant, $F(1, 116) = 0.58$, $p = 0.45$, n.s.

A 2 (*follow-up*: $t_1 = 12$ vs. $t_2 = 24$ months) X 2 (*neck design*: reduced vs. wide) multivariate analysis of variance (MANOVA) was also conducted for marginal bone loss and revealed a main effect of follow-up, $F(1, 116) = 198.85$, $p < 0.001$, such that marginal bone loss was generally lower at 12 months (0.89 mm ± $Se = 0.02$; 95% CI = 0.86, 0.93) than at 24 months (1.08 mm ± $Se = 0.01$; 95% CI = 1.06, 1.11), independently of type of neck design. Furthermore, the analysis also revealed a main effect of neck design, $F(1, 116) = 34.04$, $p < 0.001$, such that marginal bone loss was generally lower for wide (Group A: 0.92 mm ± $Se = 0.02$; 95% CI = 0.88, 0.95) than for reduced-neck implants (Group B: 1.06 mm ± $Se = 0.02$; 95% CI = 1.03, 1.10), independently of time of follow-up. More specifically, the difference between the two groups, considered at one and two years of follow-up, were, respectively, as follows: Group A (one year): 0.84 mm ± $Se = 0.03$; 95% CI = 0.78, 0.88 vs. Group B (one year): 0.95 mm ± $Se = 0.02$; 95% CI = 0.91, 0.99 ($p = 0.001$); and Group A (two years): 1.00 mm ± $Se = 0.02$; 95% CI = 0.97, 1.03 vs. Group B (two years): 1.17 mm ± $Se = 0.02$; 95% CI = 1.14, 1.20 ($p = 0.001$). Importantly, the two-way interaction *follow-up* ($t_1 = 12$ vs. $t_2 = 24$ months) X *neck design* (reduced vs. wide) was statistically significant, $F(1, 116) = 3.91$, $p = 0.05$, showing that the increase in bone loss for reduced-neck implants (Group B) was steeper than the increase observed for wide-neck implants.

4. Discussion

Our study focused on dental implants' macro-design, particularly on the clinical performance of the same type of fixture but with two different rough collar designs in partially edentulous patients, using a delayed loading protocol. Examined parameters were peri-implant probing depth, marginal bone loss, and survival rate at two-year follow-up. Both groups of patients showed an acceptable but almost similar implant survival rate. However, patients who received implants with a wide-neck design presented lower probing depth and minor marginal bone loss compared to reduced neck; thus, the null hypothesis of no differences between dental implants with different neck designs was partially rejected. From a clinical point of view, differences in probing depth and marginal bone loss between Group A and B were not relevant at the two-year follow-up. Since the absence of signs of soft-tissue inflammation and the absence of further additional bone loss following initial healing were found, according to peri-implant health definition by Renvert et al. [15], it can be affirmed that both groups of patients showed peri-implant tissue health conditions.

Implant therapy is a very helpful discipline when it comes to rehabilitating dental patients. Even if bone loss around oral implants is described to be an unavoidable and physiologic foreign-body reaction of bone against titanium [16–18], the key for success resides in the neutralization of risk factors at multiple levels: patient level, implant level, and prosthetic level.

Risk factors such as diabetes, periodontitis, bruxism, smoking, antidepressants intake, bone augmentation procedures, head and neck radiotherapy [19–22] play a principal role in long-term implants' outcome. These factors are found at the patient level, meaning that they are poorly controllable over time, as they can worsen along with local or systemic health conditions. Here, we must recall that patients included in the present study where disease-free individuals.

Other factors that are set at prosthesis level also interfere with the success of implant therapy and should not be underestimated. According to Vazquez-Alvarez et al. [23], the distance between the implant platform and the horizontal component of the prosthesis has a significant influence on peri-implant bone loss, and to be adequate, it should range from 3.3 to 6 mm. According to Lemos et al. [24], the retention system for implant-supported prostheses may lead to a different bone-loss pattern, as cement-retained restorations showed less marginal bone loss than screw-retained restorations, and implant survival rate was in favor of cement-retained prosthesis.

Restorations for the present study were cemented crowns where a minimum distance of 3.5 mm was kept between implant-abutment junction and horizontal prosthetic component, and where extreme attention was payed to remove any cement excess that could be found underneath them.

Accuracy of dental impression used, whether traditionally or digitally taken, may lead to differences in the fit of the definitive restoration [25]. In our case, prosthetic rehabilitations were performed by passing through light and putty consistency polyvinylsiloxane materials.

The type of prosthetic material itself is described to be capable of having an effect on the peri-implant tissues [26]. In this study, the decision for metal–ceramic crowns was supported by appropriate biomechanical properties, as it was demonstrated in the literature [27–29].

Occlusal forces were exerted against natural antagonistic teeth in the molar/premolar region, to standardize the procedure and avoid contact with previously installed dental restorations made of unknown or undefined material properties (e.g., a preexisting zirconium-based bridge in the antagonistic region).

Finally, implant therapy risk factors are also found at the implant level, being the fixture macro-design capable to affect the osseointegration process, as reported by several authors [4,5,30–33]. Fixture micro- and macro-designs can be adequately selected before treatment, and with the ideal concept design, implant success rate would be more predictable.

Starting from the type of material from which implants are manufactured, different osseointegration processes (amount of bone attachment to the surface and strength of the bone-surface interaction) may occur at the bone level.

Recently reported by Taek-Ka et al. [34], a qualitative different osseointegration was found through higher bone-surface interaction in commercially pure titanium grade 2 implants compared to grade 4. Apart from titanium, zirconia has also been proposed as an alternative material for oral fixtures. At the moment, despite its optimal biocompatibility, no definitive decision is available on the clinical performance of such implants [35,36].

Back to implant collar, the manner in which it is configured appears to be of relevant interest: The maximum loading stress distribution in bone is localized at the neck of the implants, as described by Anitua et al. [37] and Huang et al. [38]. Several studies are available in the literature, but no consensus on which collar design is more suitable for osseointegration was agreed on by the authors.

Our study would qualify rough wide-neck implants to reduce bone loss over time, being conscious that a longer follow-up period is necessary to confirm these findings. This may be related to the platform-switching concept, which has been described to be beneficial for osseointegration [39–43]. In fact, even in the case that a platform-matched abutment is used in such implants, a minimal effect of switching platform still exists, being that the neck of the implant is wider in diameter with respect to the main body. Otherwise, reduced-neck implants are less likely to benefit from the platform-switching effect because of their narrower platform.

According to Eshkol-Yogev et al. [44], round neck implants may significantly increase primary stability when compared to triangular neck design. In a paper by Mendoca et al. [45], bone remodeling showed to be of benefit around implants with rough collar design, in mandible but not in maxilla, if compared to machined collar surface implants. In a review by Koodaryan et al. [46], rough-surfaced micro-threaded neck implants appeared to lose less bone compared to polished and rough-surfaced neck implants.

CSR implants placed in this study had roughened surface collars with no microthreads at the bone cervical region. Presence or absence of microthreads, as well as the amount of surface roughness, may have an effect on bone preservation. Despite that an implant collar with a microthread can help in the maintenance of peri-implant bone against prosthetic loading, [47] this study was focused on conventional rough-surface dental implants, not to add confounding aspects related to numerous available surface topography (e.g., smooth, polished neck vs. machined surface vs. microthread design). Furthermore, CSR implants had a moderate degree of roughness, as no beneficial effect seemed to be associated with an increase in surface roughness. In fact, a 20-year follow-up clinical trial by Donati et al. [48] reported no peri-implant bone preservation related to implants with an increased surface roughness.

Another relevant issue to consider is the implant–abutment connection system. Implants in the present study were provided with DAT connection. Consisting of a double conical interface and internal hexagon for prosthetic repositioning, this type of connection follows the recent literature's outcomes. According to Caricasulo et al. [49], internal connection, particularly conical interfaces seem to better maintain crestal bone level around dental implants.

As stated by Kim et al. [50], transmission of the occlusal load from the restoration to the implant, and then from the implant to the surrounding bone, is essential to stimulate osteoblasts activity. This is to say, to avoid minimum but regular and continuous bone resorption, described to be around 1 mm for the first year and of 0.2 mm per year thereafter [51], bone deposition must be encouraged.

The concept of biocompatibility related to implant-prosthetic rehabilitation can be considered as the ultimate key for success: proper design of the fixture, together with a correct function of the implant–abutment connection, and optimal adaptation of the prosthetic restoration generates a self-defensive mechanism that guarantees long-term survival rates.

Considering multiple and confounding aspects which affect implant failure, with risk factors set at patient, implant, and prosthetic level, it is important to affirm that bone loss in not solely determined by collar morphology. Further studies should be conducted on multiple heterogeneous implant collar design in different populations (e.g., diabetic vs. nondiabetic) and with different prosthetic restorations (e.g., screwed vs. cemented). Longer follow-up periods could highlight the enhancement of the clinical performance of dental implants with specific neck configurations.

5. Conclusions

Within the limitations of the present prospective clinical comparative study, peri-implant probing depth and marginal bone level around dental implants placed in edentulous sites in molar/premolar region were affected by different neck designs. Patients who received implants with rough wide-neck design presented lower probing depth and minor marginal bone loss compared to patients with rough reduced-neck implants.

Reduced-neck implants showed a tendency to lose comparatively more bone over time if compared with wide-neck implants.

However, dental implants' survival rate was acceptable and satisfactory for both groups of patients and showed no differences at the two-year follow-up.

Supplementary Materials: The following are available online at http://www.mdpi.com/1996-1944/13/5/1029/s1, Consort Statement and 2010 Checklist.

Author Contributions: P.M.: conceptualization, writing-original draft preparation; F.F.: investigation, data curation; G.P.: study design, research methodology, statistical analysis, drafting and final approval of the manuscript; E.G.: supervision, project administration; P.C.: conceptualization, investigation. All authors have read and agreed to published version of the manuscript.

Funding: This research received no external funding.

Conflicts of Interest: The authors declare no conflicts of interest.

References

1. Steigenga, J.T.; Al-Shammari, K.F.; Nociti, F.H.; Misch, C.E.; Wang, H.L. Dental implant design and its relationship to long-term implant success. *Implant Dent.* **2003**, *12*, 306–317. [CrossRef] [PubMed]
2. Ogle, O.E. Implant surface material, design, and osseointegration. *Dent. Clin. N. Am.* **2015**, *59*, 505–520. [CrossRef] [PubMed]
3. Rupp, F.; Liang, L.; Geis-Gerstorfer, J.; Scheideler, L.; Huttig, F. Surface characteristics of dental implants: A review. *Dent. Mater.* **2018**, *34*, 40–57. [CrossRef]
4. Spies, B.C.; Bateli, M.; Ben Rahal, G.; Christmann, M.; Vach, K.; Kohal, R.J. Does oral implant design affect marginal bone loss? Results of a parallel group randomized controlled equivalence trial. *BioMed Res. Int.* 2018. [CrossRef] [PubMed]
5. Ormianer, Z.; Matalon, S.; Block, J.; Kohen, J. Dental implant thread design and the consequences on long-term marginal bone loss. *Implant Dent.* **2016**, *25*, 471–477. [CrossRef] [PubMed]
6. Lesmes, D.; Laster, Z. Innovations in dental implant design for current therapy. *Oral. Maxillofac. Surg. Clin. N. Am.* **2011**, *23*, 193–200. [CrossRef]
7. Boyan, B.D.; Cheng, A.; Olivares-Navarrete, R.; Schwartz, Z. Implant surface design regulates mesenchymal stem cell differentiation and maturation. *Adv. Dent. Res.* **2016**, *28*, 10–17. [CrossRef]
8. Perez-Albacete, M.A.; Perez-Albacete, C.; Mate-Sanchez de Val, J.E.; Ramos-Oltra, M.L.; Fernandez-Dominguez, M.; Calvo-Guirado, J.L. Evaluation of a new dental implant cervical design in comparison with a conventional design in an experimental american foxhound model. *Materials* **2018**, *11*, 462. [CrossRef]
9. Calvo-Guirado, J.L.; Morales-Melendez, H.; Perez-Martinez, C.; Morales-Schwarz, D.; Kolerman, R.; Fernandez-Dominguez, M.; Gehrke, S.A.; Mate-Sanchez de Val, J.E. Evaluation of the surrounding ring of two different extra-short implant designs in crestal bone maintenance: A histologic study in dogs. *Materials* **2018**, *11*, 1630. [CrossRef]
10. Calvo-Guirado, J.L.; Jimenez-Soto, R.; Perez-Martinez, C.; Fernandez-Dominguez, M.; Gehrke, S.A.; Mate-Sanchez de Val, J.E. Influence of implant neck design on peri-implant tissue dimensions: A comparative study in dogs. *Materials* **2018**, *11*, 2007. [CrossRef]
11. Shen, W.L.; Chen, C.S.; Hsu, M.L. Influence of implant collar design on stress and strain distribution in the crestal compact bone: A three-dimensional finite element analysis. *Int. J. Oral Maxillofac. Implant.* **2010**, *25*, 901–910. [PubMed]
12. Hanggi, M.P.; Hanggi, D.C.; Schoolfield, J.D.; Meyer, J.; Cochran, D.L.; Hermann, J.S. Crestal bone changes around titanium implants. Part I: A retrospective radiographics evaluation in humans comparing two non-submerged implant designs with different machined collar lengths. *J. Periodontol.* **2005**, *76*, 791–802. [CrossRef] [PubMed]
13. Crespi, R.; Capparè, P.; Polizzi, E.; Gherlone, E. Fresh-Socket implants of different collar length: Clinical evaluation in the aesthetic zone. *Clin. Implant. Dent. Relat. Res.* **2015**, *17*, 871–878. [CrossRef] [PubMed]
14. Chappuis, V.; Bornstein, M.M.; Buser, D.; Belser, U. Influence of implant neck design on facial bone crest dimensions in the esthetic zone analyzed by cone beam CT: A comparative study with a 5-to-9-year follow-up. *Clin. Oral Implant. Res.* **2016**, *27*, 1055–1064. [CrossRef] [PubMed]
15. Renvert, S.; Person, G.R.; Pirih, F.Q.; Camargo, P.M. Peri-implant health, peri implant mucositis, and peri-implantitis: Case definitions and diagnostic considerations. *J. Periodontol.* **2018**, *89*, s304–s312. [CrossRef]
16. Albrektsson, T.; Dahlin, C.; Jemt, T.; Sennerby, L.; Turri, A.; Wennerberg, A. Is marginal bone loss around dental implants the result of a provoked foreign body reaction? *Clin. Implant Dent. Relat. Res.* **2014**, *16*, 155–165. [CrossRef]
17. Albrektsson, T.; Canullo, L.; Cochran, D.; De Bruyn, H. Peri-implantitis: A complication of a foreign body or a man-made "Disease". Facts and fiction. *Clin. Implant Dent. Relat. Res.* **2016**, *18*, 840–849. [CrossRef]
18. Buser, D.; Sennerby, L.; De Bruyn, H. Modern implant dentistry based on osseointegration: 50 years of progress, current trends and open questions. *Periodontol 2000*, *73*, 7–21. [CrossRef]
19. Tecco, S.; Grusovin, M.G.; Sciara, S.; Bova, F.; Pantaleo, G.; Capparè, P. The association between three attitude-related indexes of oral hygiene and secondary implant failures: a retrospective longitudinal study. *Int. J. Dent. Hyg.* **2018**, *16*, 372–379. [CrossRef]

20. Kandasamy, B.; Kaur, N.; Tomar, G.K.; Bharadwaj, A.; Manual, L.; Chauhan, M. Long-term retrospective study based on implant succes rate in patients with risk factor: 15-year follow-up. *J. Contemp. Dent. Pract.* **2018**, *19*, 90–93. [CrossRef]
21. Chrcanovic, B.R.; Kisch, J.; Albrektsson, T.; Wennerberg, A. Factors influencing early dental implant failures. *J. Dent. Res.* **2016**, *95*, 995–1002. [CrossRef] [PubMed]
22. Gherlone, E.F.; Capparè, P.; Tecco, S.; Polizzi, E.; Pantaleo, G.; Gastaldi, G.; Grusovin, M.G. A prospective longitudinal study on implant prosthetic rehabilitation in controlled HIV-positive patients with 1-year follow-up: The role of CD4+ level, smoking habits, and oral hygiene. *Clin. Implant Dent. Relat. Res.* **2016**, *18*, 955–964. [CrossRef] [PubMed]
23. Vazquez-Alvarez, R.; Perez-Sayans, M.; Gayoso-Diz, P.; Garcia-Garcia, A. Factors affecting peri-implant bone loss: A post-five-year retrospective study. *Clin. Oral Implant. Res.* **2015**, *26*, 1006–1014. [CrossRef] [PubMed]
24. Lemos, C.A.; de Souza Batista, V.E.; Almeida, D.A.; Santiago Junior, J.F.; Verri, F.R.; Pellizzer, E.P. Evaluation of cement-retained versus screw-retained implant-supported restorations for marginal bone loss: A systematic review and meta-analysis. *J. Prosthet. Dent.* **2016**, *115*, 419–427. [CrossRef] [PubMed]
25. Cappare, P.; Sannino, G.; Minoli, M.; Montemezzi, P.; Ferrini, F. Conventional versus digital impression for full arch screw-retained maxillary rehabilitations: A randomized clinical trial. *Int. J. Environ. Res. Public Health* **2019**, *16*, 829. [CrossRef] [PubMed]
26. Maminskas, J.; Puisys, A.; Kuoppala, R.; Raustia, A.; Juodzbalys, G. The prosthetic influence and biomechanics on peri-implant strain: A systematic literature review of finite element studies. *J. Oral Maxillofac. Res.* **2016**, *7*, e4. [CrossRef] [PubMed]
27. Augustin-Panadero, R.; Soriano-Valero, S.; Labaig-Rueds, C.; Fernandez-Estevan, L.; Sola-Ruiz, M.F. Implant-supported metal-ceramic and resin.modified ceramic crowns: A 5-year prospective clinical study. *J. Prosthet. Dent.* **2019**. [CrossRef]
28. Schwartz, S.; Schroder, C.; Hassel, A.; Bomicke, W.; Rammelsber, P. Survival and chipping of zirconia-based and metal-ceramic implant-supported single crowns. *Clin. Oral Implant. Res.* **2012**, *14*, e119–e125. [CrossRef]
29. Pjetursson, B.E.; Valente, N.A.; Strasding, M.; Zwahlen, M.; Liu, S.; Sailer, I. A systematic review of the survival and complication rates of zirconia-ceramic and metal-ceramic single crowns. *Clin. Oral Implant. Res.* **2018**, *16*, 199–214. [CrossRef]
30. Vivan-Cardoso, M.; Vandamme, K.; Chaudhari, A.; De Rycker, J.; Van Meerbeek, B.; Naert, I.; Duyck, J. Dental implant macro-design features can impact the dynamics of osseointegration. *Clin. Implant Dent. Relat. Res.* **2015**, *17*, 639–645. [CrossRef]
31. Lima de Andrade, C.; Carvalho, M.A.; Bordin, D.; da Silva, W.J.; Del Bel Cury, A.A.; Sotto-Maior, B.S. Biomechanical behavior of the dental implant macrodesign. *Int. J. Oral Maxillofac. Implant.* **2017**, *32*, 264–270. [CrossRef] [PubMed]
32. Jimbo, R.; Tovar, N.; Marin, C.; Teixera, H.S.; Anchieta, R.B.; Silveira, L.M.; Janal, M.N.; SHibli, J.A.; Coelho, P.G. The impact of a modified cutting flute implant design on osseointegration. *Int. J. Oral Maxillofac. Surg.* **2014**, *43*, 883–888. [CrossRef] [PubMed]
33. Triplett, R.G.; Frohberg, U.; Sykaras, N.; Woody, R.D. Implant materials, design, and surface topographies: The influence on osseointegration of dental implants. *J. Long Term Eff. Med. Implant.* **2003**, *13*, 485–501. [CrossRef]
34. Taek-Ka, K.; Jung-Yoo, C.; Jae-Il, P.; In-Sung, L.Y. A clue to the existence of bonding between bone and implant surface: An in vivo study. *Materials* **2019**, *12*, 1187. [CrossRef]
35. Hashim, D.; Cionca, N.; Courvoisier, D.S.; Mombelli, A. A systematic review of the clinical survival of zirconia implants. *Clin. Oral Investig.* **2016**, *20*, 1403–1417. [CrossRef] [PubMed]
36. Cionca, N.; Hashim, D.; Mombelli, A. Zirconia dental implants: Where are we now, and where are we heading? *Periodontol 2000* **2017**, *73*, 241–258. [CrossRef]
37. Anitua, E.; Tapia, R.; Luzuriaga, F.; Orive, G. Influence of implant length, diameter, and geometry on stress distribution: A finite element analysis. *Int. J. Periodontics Restor. Dent.* **2010**, *30*, 89–95.
38. Huang, Y.M.; Chou, I.C.; Jiang, C.P.; Wu, Y.S.; Lee, S.Y. Finite element analysis of dental implant neck effects on primary stability and osseointegration in a type IV bone mandible. *Bio Med. Mater. Eng.* **2014**, *24*, 1407–1415. [CrossRef]

39. Salamanca, E.; Lin, J.C.; Tsai, C.Y.; Hsu, Y.S.; Huang, H.M.; Teng, N.C.; Wang, P.D.; Feng, S.W.; Chen, M.S.; Chang, W.J. Dental implant surrounding marginal bone level evaluation: Platform switching versus platform matching-one-year retrospective study. *BioMed Res. Int.* **2017**. [CrossRef]
40. Santiago, J.J.F.; Batista, V.E.; Verri, F.R.; Honorio, H.M.; de Mello, C.C.; Almeida, D.A.; Pellizzer, E.P. Platform-switching implants and bone preservation: A systematic review and meta-analysis. *Int. J. Oral Maxillofac. Surg.* **2016**, *45*, 332–345. [CrossRef]
41. Rocha, S.; Wagner, W.; Wiltfang, J.; Nicolau, P.; Moergel, M.; Messias, A.; Behrens, E.; Guerra, F. Effect of platform switching on crestal bone levels around implants in the posterior mandible: 3 years results from a multicentre randomized clinical trial. *J. Clin. Periodontol.* **2016**, *43*, 374–382. [CrossRef] [PubMed]
42. Strietzel, F.P.; Neumann, K.; Hertel, M. Impact of platform switching on marginal peri-implant bone-level changes. Asystematic review and meta-analysis. *Clin. Oral Implant. Res.* **2015**, *26*, 342–358. [CrossRef] [PubMed]
43. Bouazza-Juanes, K.; Martinez-Gonzalez, A.; Peiro, G.; Rodenas, J.J.; Lopez-Molla, M.V. Effect of platform switching on the peri-implant bone: A finite element study. *J. Clin. Exp. Dent.* **2015**, *7*, e483–e488. [CrossRef] [PubMed]
44. Eshkol-Yogev, I.; Tandlich, M.; Shapira, L. Effect of implant neck design on primary and secondary implant stability in the posterior maxilla: A prospective randomized controlled study. *Clin. Oral Implant. Res.* **2019**. [CrossRef] [PubMed]
45. Mendoca, J.A.; Senna, P.M.; Francischone, C.E.; Francischone Junior, C.E.; de Souza Picorelli Assis, N.M.; Sotto-Maior, B. Retrospective evaluation of the influence of the collar surface topography on peri-implant bone preservation. *Int. J. Oral Maxillofac. Implant.* **2017**, *32*, 858–863. [CrossRef] [PubMed]
46. Koodaryan, R.; Hafezeqoran, A. Evaluation of implant collar surfaces for marginal bone loss: A systematic review and meta-analysis. *BioMed Res. Int.* **2016**. [CrossRef]
47. Lee, D.W.; Choi, Y.S.; Park, K.H.; Kim, C.S.; Moon, I.S. Effect of microthread on the maintenance of marginal bone level: A 3-year prospective study. *Clin. Oral Implant. Res.* **2007**, *18*, 465–470. [CrossRef]
48. Donati, M.; Ekestubbe, A.; Lindhe, J.; Wennstrom, J.L. Marginal bone loss at implants with different surface characteristics: A 20-year follow-up of a randomized controlled clinical trial. *Clin. Oral Implant. Res.* **2018**, *29*, 480–487. [CrossRef]
49. Caricasulo, R.; Malchiodi, L.; Ghensi, P.; Fantozzi, G.; Cucchi, A. The influence of implant-abutment connection to peri-implant bone loss: A systematic review and meta-analysis. *Clin. Implant Dent. Relat. Res.* **2018**, *20*, 653–664. [CrossRef]
50. Kim, J.J.; Lee, J.H.; Kim, J.C.; Lee, J.B.; Yeo, I.L. Biological responses to the transitional area of dental implants: Material and structure dependent responses of peri-implant tissue to abutments. *Materials* **2019**, *13*, 72. [CrossRef]
51. Albrektsson, T.; Zarb, G.; Worthington, P.; Eriksson, A.R. The long-term efficacy of currently used dental implants: A review and proposed criteria of success. *Int. J. Oral Maxillofac. Implant.* **1986**, *1*, 11–25.

© 2020 by the authors. Licensee MDPI, Basel, Switzerland. This article is an open access article distributed under the terms and conditions of the Creative Commons Attribution (CC BY) license (http://creativecommons.org/licenses/by/4.0/).

Article

A Hybrid Model for Predicting Bone Healing around Dental Implants

Pei-Ching Kung, Shih-Shun Chien and Nien-Ti Tsou *

Department of Materials Science and Engineering, National Chiao Tung University, Ta Hsueh Road, Hsinchu 300, Taiwan; gong1014.mse06g@nctu.edu.tw (P.-C.K.); play_58032.05g@g2.nctu.edu.tw (S.-S.C.)
* Correspondence: tsounienti@nctu.edu.tw; Tel.: +886-3-5712121 (ext. 55308)

Received: 24 May 2020; Accepted: 23 June 2020; Published: 25 June 2020

Abstract: Background: The effect of the short-term bone healing process is typically neglected in numerical models of bone remodeling for dental implants. In this study, a hybrid two-step algorithm was proposed to enable a more accurate prediction for the performance of dental implants. Methods: A mechano-regulation algorithm was firstly used to simulate the tissue differentiation around a dental implant during the short-term bone healing. Then, the result was used as the initial state of the bone remodeling model to simulate the long-term healing of the bones. The algorithm was implemented by a 3D finite element model. Results: The current hybrid model reproduced several features which were discovered in the experiments, such as stress shielding effect, high strength bone connective tissue bands, and marginal bone loss. A reasonable location of bone resorptions and the stability of the dental implant is predicted, compared with those predicted by the conventional bone remodeling model. Conclusions: The hybrid model developed here predicted bone healing processes around dental implants more accurately. It can be used to study bone healing before implantation surgery and assist in the customization of dental implants.

Keywords: dental implant; tissue differentiation; bone remodeling; mechano-regulation theory; short-term healing; long-term healing

1. Introduction

Implant stability is one of the important indexes to determine dental implant survival rates in the clinic [1]. It is dominated by the bone healing around the surgery site. Bone healing is a series of complex physiological processes that involved the regulation of several tissue phenotypes. At the beginning of bone healing, micro-vessels, and new connective tissue form on the surface of the wound, which is collectively referred to as granulation tissue [2–5]. After the formation of granulation tissue, further tissue differentiation initiates. Cells then transfer into fibrous connective tissues, cartilages, and new bones according to biophysical stimulus [6–8]. The final stage of bone healing is referred to as bone remodeling, which is a lifelong process, where the skeletal system maintains a dynamic equilibrium, related to the regulation of osteoclasts and osteoblasts [9–11]. When the balance is disrupted by external forces, a new equilibrium state can be achieved spontaneously. Thus, according to the healing process mentioned above, tissue differentiation and bone remodeling stages have a great impact on the short-term and long-term stability of implants, respectively [12,13].

To efficiently predict the short-term stability of implants in advance, we aim to simulate the tissue differentiation process by using the mechano-regulation algorithm. The origin of the method is based on Pauwels [14] who first specified that distortional stress and hydrostatic compression dominate tissue differentiation. Carter et al. [15] implemented the theory into a finite element model (FEM), revealing the evolution of connective tissues. Prendergast et al. [16] modified the methods by adopting octahedral shear strain and fluid flow as the solid and fluid stimuli. Lacroix further improved the model by using poroelastic finite elements, which can describe the biological mechanism in bones more accurately [5]. In recent years, many studies have revealed the effect of the mechanical environment and the geometric design of implants on the performance of tissue differentiation [17,18].

Bone adapts itself based on its mechanical environment and loading conditions, greatly affecting the morphology of bone and long-term stability of implants [11,13,19]. Many scientists have developed numerical methods to describe the behavior of bone remodeling [20–24]. Carter et al. [15,25,26] proposed that bone apparent density is dominated by strain energy density and studied the energy transfer in hip stems. The internal changes in bone morphology and the aging of connective tissues affected by the external loads were also predicted by FEM. Huiskes et al. [27] adopted a similar approach and simulated the femoral cortex around intramedullary prostheses to reveal the relationship between stress shielding effect and bone resorption. The bone remodeling algorithm was then extended to predict the variation of bone apparent density after implantation treatment [28–31]. The algorithm was verified by computed tomographic (CT) images, showing a high degree of similarity [32]. Most of the studies assumed a simple initial state of the models, where uniform material properties were assigned around the implants [29,31,33,34], i.e., the short-term bone healing has no effect on bone remodeling results. However, short-term healing is crucial since bone remodeling is an iterative process, where different initial conditions may lead to different bone density distribution around dental implants.

In order to test the null hypothesis of the bone healing process in the conventional model, we proposed a hybrid algorithm that regards the procedure of bone healing as two stages: (1) the short-term stage which was simulated by a tissue differentiation model and (2) the long-term stage which simulated by a bone remodeling model. At the beginning of the tissue differentiation model, it was assumed that the wound was filled with granulation tissue. The mechano-regulation algorithm was then applied to determine the tissue phenotypes for the following time steps. Once a stable tissue differentiation has been reached, the current tissue distribution with the material properties, such as Young's modulus, apparent bone density, and Poisson's ratio, in callus around the implant then served as the initial condition for the bone remodeling model. Then, the resulting long-term distribution of Young's modulus and the remodeling stimulus will be discussed. and compared with the results which were similar to those done by Chou et al. [29], where the effect of short-term tissue differentiation was not considered.

The objective of this study is to develop a hybrid model that can predict the stability of dental implants and the strength of the surrounding bones with consideration of both the short-term and long-term bone healing process. The results of the current work can reveal the effect of bone with different material properties on bone healing, providing useful information for dental clinics. Furthermore, the proposed model can be used to rapidly examine the morphology design of dental implants (such as implant radius, length, thread geometry) and the placement protocol (such as insertion angle and depth) to improve the osseointegration between implants and bones.

2. Materials and Methods

Figure 1 shows the flowchart of the current hybrid algorithm, including short-term and long-term bone healing models. The distribution of strain, fluid velocity, and stem cell diffusion in the initial model ($t = 0$) was firstly calculated by FEM. Granulation tissues then differentiated into various tissue phenotypes based on the mechano-regulation algorithm. Next, the rule of mixture and smoothing procedure [35] was applied to determine the updated material properties and the detail will be discussed in Section 2.2. After the short-term healing process finished, the distribution of tissue phenotypes and the corresponding material properties around the implants was obtained and assigned to the bone remodeling model at the initial state for simulating the long-term healing process. Where bone remodeling algorithm adjusted the bone apparent density of each element iteratively until the equilibrium state of the remodeling stimulus under the given loading condition was achieved. The procedures will be discussed in more detail in Sections 2.2 and 2.3.

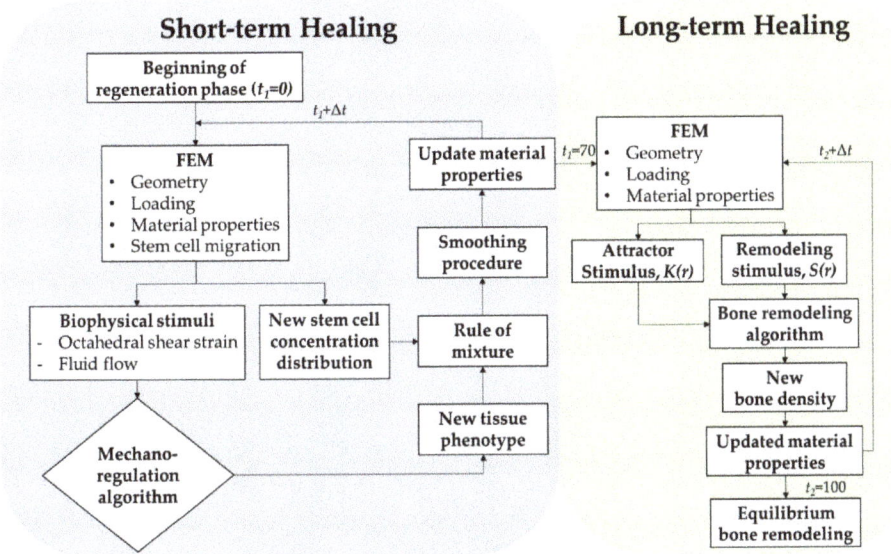

Figure 1. Flow chart of bone healing preoperative evaluation.

2.1. Three-Dimensional FEM Model

The current hybrid algorithm was applied to study the bone healing of the mandibular second molar (back teeth in the upper jaw), where the bone geometry, density, and other material properties were adopted based on Chou et al. [29]. The geometry of the bone structure was obtained by extruding a planar CT image with a thickness of 80 mm, as shown in Figure 2a. It was referred to as the bone-tooth system, consisting of a layer of cortical bone overlying on cancellous bone, and a natural tooth. The physiological stimulus of the healthy state, i.e., bone-tooth system, was served as the objective function (i.e., attractor stimulus) for the calculation of bone remodeling in the bone-implant-prosthesis system, details of the calculation will be introduced in the next section. Next, the bone-implant-prosthesis system replaced the tooth in the bone-tooth system by a prosthesis and a short implant with a size of 5.0 × 5.1 mm; the remaining region, i.e., the extraction socket, was filled with callus, as shown in Figure 2b.

The models of the two systems were built by a commercial finite element package ANSYS 18.0 (ANSYS, Inc., Canonsburg, PA, USA). The implant, tooth, and prosthesis were meshed by the built-in element type, SOLID185; the remaining parts of the tissues were meshed by CPT215, which allows the calculation of poroelastic material properties, such as the fluid velocity and pressure in the pores of the bones. There were approximately 138,000 elements and 94,500 nodes for both systems. To maintain the balance between computation time and accuracy, finer meshes were applied around the interfaces between bone and the tooth/implant, as shown in Figure 2. The interfaces were set to allow sliding with a friction coefficient of 0.3. A symmetry boundary condition was applied to the mesial side of the model. All of the nodes in the distal side were constrained in all degrees of freedom. A displacement of 10.5 μm was applied at nodes on the top of the tooth/prosthesis. This value was equivalent to a biting force of 100 N [36–38]; the angle of the displacement was set according to it used in Chou et al. [29]. Note that loading, setting, and properties of all materials, including bones, prosthesis, tooth, implant, and bone graft, used in the current work were based on Chou et al. [29] for comparison, as shown in Table 1. It is worth mentioning that, in the tissue differentiation process, the material properties of elements in the callus region transformed with iterations, i.e., they evolved according to the corresponding tissue phenotypes during the iteration process. Details of the mechanism of the mechano-regulation algorithm will be explained in the next section.

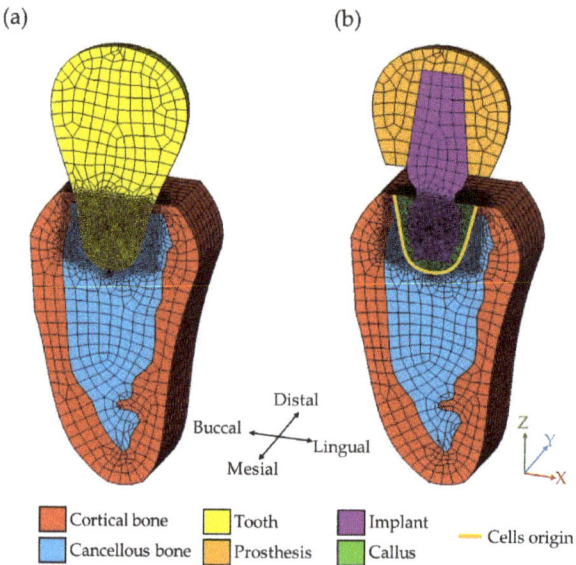

Figure 2. FEM model of (a) bone-tooth and (b) bone-implant-prosthesis systems.

Table 1. The material properties used in the current work [29].

Materials	Young's Modulus (GPa)	Poisson's Ratio	Permeability (m⁴/Ns)
Ti6Al4V	113.8	0.34	N/A
Tooth	20	0.3	N/A
Prosthesis	80	0.3	N/A
Bone graft	2	0.3	N/A
Cortical bone	13.7	0.3	10^{-17}
Cancellous bone	2	0.3	3.7×10^{-13}
Granulation tissue	0.001	0.17	10^{-14}
Fibrous tissue	0.002	0.17	10^{-14}
Cartilage	0.01	0.17	5×10^{-15}
Immature bone	1	0.3	10^{-13}
Mature bone	6	0.3	3.7×10^{-13}

N/A: not applicable.

2.2. Mechano-Regulation Algorithm

The mechano-regulation algorithm proposed by Lacroix and Prendergast [5,35] was adopted in the current work to predict the distribution of tissue phenotypes. The procedures of the algorithm, including the calculation in each iteration, updates of material properties, and post-processing to the results, were implemented by MATLAB 2019 (The MathWorks, Inc., Natick, MA, USA). At the beginning of the calculation, elements in the callus region were set with the material properties of granulation tissues. According to the theory, tissue differentiation (TD) is induced by the combination of octahedral shear strain (γ) and fluid flow (v) caused by external loads. It is referred to as biophysical stimulus S_{TD} such that:

$$S_{TD} = \frac{\gamma}{a} + \frac{v}{b} \qquad (1)$$

where a = 0.0375 and b = 3 µm/s are empirical constants. Then, the tissue phenotype for the next iteration can be determined based on the value of S_{TD} as shown in Table 2. The material properties of the tissue phenotype were updated accordingly.

Table 2. The ranges of biophysical stimulus for different tissue phenotypes.

S_{TD}			Tissue Phenotypes
3<	S_{TD}		Fibrous tissue
1<	S_{TD}	≤3	Cartilage
0.266<	S_{TD}	≤1	Immature bone
0.010<	S_{TD}	≤0.266	Mature bone
	S_{TD}	≤0.010	Initial resorption

The concentration of mesenchymal stem cells determined the level of the transition from granulation tissue to the other tissue phenotypes. The migration [39] and proliferate [40] of mesenchymal stem cell can be simplified as a classical isotropic diffusion, such that:

$$\frac{dn}{dt} = D\nabla^2 n \qquad (2)$$

where t is time; D is the diffusion coefficient; n is the current concentration of mesenchymal stem cell. The migration of stem cells started from the boundary of the extraction socket, which is called cells origin and marked by yellow lines in Figure 2b. At the last iteration, the concentration of the stem cell reached the maximal value. In the current work, D is set as 8.85×10^{-14} m^2/s. Then, the effective material properties of tissues for the next iteration, including Young's modulus, Poisson's ratio, and permeability, can be obtained by a linear combination between the granulation tissue (X_g) and the differentiated tissue (X_d) as follows:

$$X_{mix} = \frac{n_{max} - n}{n_{max}} X_g + \frac{n}{n_{max}} X_d \qquad (3)$$

where n^{max} is the maximum concentration; n is the current concentration of stem cells determined by Equation (2); X_d is material properties of the differentiated tissue phenotype shown in Table 1.

To avoid the instability and dramatic change of material properties between iterations, Lacroix and Prendergast [35] suggested a smooth procedure to average the material properties from the previous nine iterations, which can be written as:

$$X_i = \frac{1}{N}\left(X_{mix} + X_{i-1} + X_{i-2} + \ldots + X_{i-(N-1)}\right) \qquad (4)$$

where $N = 10$; i is the current iteration number; X_{mix} is the effective material properties calculated from Equation (3). Note that when the iteration number is $i < 9$, smoothing operation is applied to the iteration i to the first [35]. The short-term healing time was set as 70 days, which is the average healing period for implantation surgeries [41–43]. The short-term healing process and the differentiated tissue phenotypes can then be predicted.

2.3. Bone Remodeling Algorithm

After the numerical calculation for the short-term healing, long-term healing, i.e., bone remodeling (BR), occurred to alter the internal structure of bones and reach a new equilibrium state based on the mechanical environment. Huskies et al. [27] proposed a bone remodeling theory assuming the driving force of self-adaptive activity is determined by remodeling stimulus (S_{BR}, unit: J/kg), such that:

$$S_{BR}(\vec{r},t) = \frac{u(\vec{r},t)}{\rho(r,t)} \quad (5)$$

where u is strain energy density (unit: J/m³); ρ is the apparent bone density (kg/m³), t is time; and $\vec{r}$ is the position vector [27]. When the value of remodeling stimulus S_{BR} is greater than the given threshold, bone formation occurred and Young's modulus and bone density increase accordingly. On the contrary, when the remodeling stimulus S_{BR} is less than the threshold, bone resorption occurred, and Young's modulus and bone density decrease. In addition, Carter [25] stated that bones maintain a state of homeostasis when the remodeling stimulus is in a certain range, which is referred to as a "lazy zone." Thus, the bone remodeling process can be expressed by nonlinear functions of the remodeling stimulus [27]:

$$\frac{d\rho}{dt} = \begin{cases} A_f[S_{BR} - (1+s)K(\vec{r})]^2, & S_{BR} \geq (1+s)K(\vec{r}) \quad \text{(Formation)} \\ 0, & (1-s)K(\vec{r}) < S_{BR} \leq (1+s)K(\vec{r}) \quad \text{(Lazy zone)} \\ A_r[S_{BR} - (1-s)K(\vec{r})]^3, & S_{BR} \leq (1-s)K(\vec{r}) \quad \text{(Resorption)} \end{cases} \quad (6)$$

where A_f and A_r are formation and resorption coefficients; s is the threshold of the lazy zone, which is set as 0.75 [44]; $K(\vec{r})$ is the attractor stimulus induced by the biting force in the bone-tooth system, which is determined by Equation (5). The value of $K(\vec{r})$ in the region of callus was set to 5, which is the average of the overall remodeling stimulus in bone element. It is worth noting that the rate of bone resorption is greater than that of bone formation based on clinical observations, resulting in a greater exponential term of resorption in Equation (6). The apparent density of a bone element m at the jth iteration can be derived by integrating Equation (6) with a forward Euler method [27,45–47], such that:

$$\rho_m^j = \begin{cases} \rho_m^{j-1} + A_f \Delta t [S_{BR}^{j-1} - (1+s)K(\vec{r})]^2, & S_{BR}^{j-1} \geq (1+s)K(r) \quad \text{(Formation)} \\ 0, & (1-s)K(r) < S_{BR}^{j-1} \leq (1+s)K(r) \quad \text{(Lazy zone)} \\ \rho_m^{j-1} + A_r \Delta t [S_{BR}^{j-1} - (1-s)K(\vec{r})]^3, & S_{BR}^{j-1} \leq (1-s)K(r) \quad \text{(Resorption)} \end{cases} \quad (7)$$

where Δt is the time increment [21,48]. Young's modulus (E, unit: GPa) of bone elements was associated with the corresponding apparent density based on the finding of Carter and Hayes [26], such that:

$$E = C\rho^3 \quad (8)$$

where C is constant. Note that the integration coefficients $A_f \Delta t$ and $A_r \Delta t$ in Equation (7) were set as 1×10^{-11} and the constant was set as 3.79, based on the setting used in Chou et al. [29]. The value of Young's modulus indicated the strength of the internal structure of bone, which affected the calculation at the next iteration. The average remodeling stimulus (S_{ave}) in each iteration was recorded and served as a measure of convergence of the model, such that:

$$S_{ave} = \frac{1}{N_{total}} \sum_{k=1}^{N_{total}} S_k \quad (9)$$

where N_{total} is the total number of bone elements; S_k is remodeling stimulus of the local bone element k. It is worth noting that long-term bone healing, i.e., bone remodeling, is a lifelong process and the bone system evolves to reach an equilibrium state according to its current mechanical environment. Thus, the time step used here is, in fact, a computational increment and is not associated with a real-life time scale. The end of the calculation depends on the convergence of S_{ave} as mentioned above.

3. Results

3.1. Short-Term Healing and Tissue Differentiation

Tissue differentiation and the evolution of bone ingrowth around the implant were evaluated by the mechano-regulation model. Figure 3 shows the percentage of tissue phenotypes in each day during the short-term healing process. In the early stage of tissue differentiation, granulation tissues still existed in the callus region. This is because tissue differentiation was initiated when the concentration of stem cells above certain levels. At this stage, the inner callus region has relatively low concentration as the stem cells diffused from the boundary of the callus region. Moreover, it was found that the soft tissue with a higher biophysical stimulus ($S_{TD} > 1$) such as cartilage and fibrous tissue decrease with time while the bone tissue increased continuously until the middle of the differentiation process (i.e., around the 35th day). After the 35th day, bones possessed a certain degree of strength, i.e., higher Young's modulus. This resulted in the decreasing values of bone stimuli, and thus, maturate and immature bones gradually became the dominant tissue phenotype in the entire callus region. Then, most of the immature bones transformed into maturate bones as the healing process was closed to the 70th day.

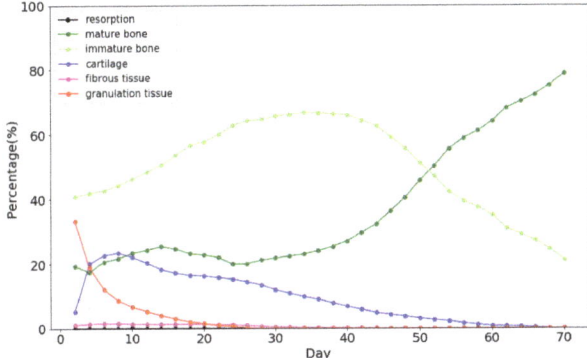

Figure 3. Percentage of various tissue phenotype in each day during the short-term bone healing process.

Details of the tissue phenotype at the specific days are shown in Figure 4. On the 4th day, more than half of the callus region remained as granulation tissue, as shown in Figure 4. It is observed that cartilages and fibrous tissues occurred around the threads in the middle part and the bottom of the implant, respectively. It is because the applied load caused the stress concentration around the threads and the bottom, giving higher biophysical stimulus to the elements in that region. Note that, although there were mature and immature bones, the effective material properties, such as Young's modulus, were still close to that of granulation tissue based on Equation (3), due to the low concentration of stem cell on the 4th day. Then, it can be observed that, on the 10th day, granulation tissues gradually transformed into mature and immature bones as the concentration of stem cells increased with time. It is worth noting that cartilages accumulated at tips of the threads since stress concentration was partially released with the increasing maturity of the surrounding bones. On the 30th day, granulation tissues were disappeared entirely. Cartilages were mainly located at the lingual side because of the oblique biting force. Then, maturate and immature bones gradually dominated the entire system around the implant on the 50th to 70th days. It is worth noting that there were very few elements of bone resorption on the 10th day and were not shown in the current cross-section in Figure 4. In this way, the short-term healing pattern around the implant was obtained.

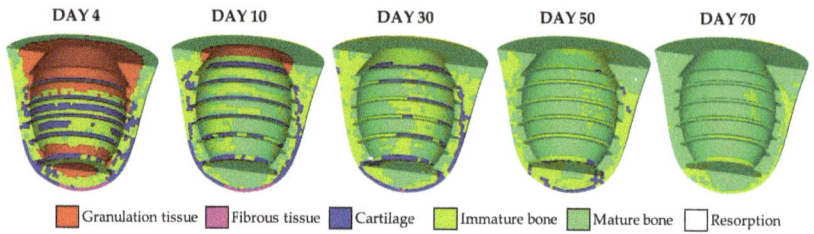

Figure 4. The tissue differentiation history predicted by the mechano-regulation algorithm.

3.2. Bone Remodeling

Now consider two bone remodeling models. The first, i.e., the current model, adopted the result at the end of short-term healing predicted by the mechano-regulation algorithm as the initial state; the second was the conventional model based on bone remodeling algorithm with the assumption that the extraction socket was filled with bone graft and regarded as the initial state where the material properties were set as uniform. The setting of the second model followed the work done by the literature [29]. Both models shared the same distribution of the objective attractor stimulus $K(\vec{r})$ based on the natural tooth, as shown in Figures 5a and 6a. Then, the bone remodeling models altered the bone apparent density of all the bone elements in the following iterations to achieve that objective distribution.

In the current model, the distribution of Young's modulus of the initial state is shown in Figure 5d. It can be observed that certain regions in the callus in Figure 5d were in dark and light blue colors giving low and high Young's modulus, ranging from 0.001 to 6 GPa based on Table 1, due to the non-uniform stem cell concentration and the presence of different tissue phenotypes. It is worth noting that the average value of Young's modulus in that region was around 2 GPa, having a good agreement with the value used in the conventional model where uniform Young's modulus was assumed, as shown in Figure 6d. The corresponding bone apparent density of each element in the callus region can then be obtained by Equation (8). The resulting model with the updated bone apparent density was then subjected to the biting force, giving the distribution of the remodeling stimulus (S_{BR}) for the 1st iteration, as shown in Figure 5b. The values are shown in Figure 5a,b were substituted into Equation (7) to obtain the updated bone apparent density for the next iteration. The corresponding distribution of Young's modulus in the 1st iteration was shown in Figure 5e. It can be observed that bone regions with high Young's modulus (above 12.18 GPa, colored in red) fully covered around the implant; most of the bones attached to the surface of the implant were also with non-uniform Young's modulus (ranging from 3.04 to 6.08 GPa). The calculation continued until the 100th iteration was achieved. It is worth noting that the average remodeling stimulus (S_{ave}) of the current model quickly converged around the 10th iteration, showing great stability. The converged values S_{ave} for both the current and conventional models were identical. The distribution of S_{BR} of the 100th iteration is shown in Figure 5c. It can be observed that most of the regions had the value of S_{BR} closed to those of the target, i.e., the objective attractor stimulus $K(\vec{r})$. The final state of bone remodeling gave the distribution of Young's modulus, shown in Figure 5f.

The average Young's modulus of the entire system predicted by the current model was around 4.77 GPa. It is worth noting that bone resorptions (colored in gray) occurred around the threads toward the lingual side and around the neck of the implant. The total volume fraction of bone resorption was 0.042%. Similar to the 1st iteration, bone regions with high Young's modulus were still fully covered around the implant. Two additional high strength bone tissue bands connected to cortical bones were formed in the bottom right (the lingual side) and top left (the buccal side) of the implant, which provided extra supports and enhanced the stability of the implant.

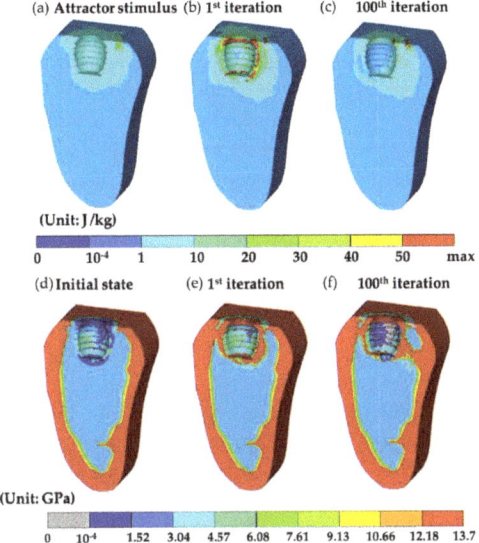

Figure 5. The results generated by the current model. (**a**) The target distribution of the attractor stimulus $K(\vec{r})$ based on the natural tooth. (**b**,**c**) are bone remodeling stimulus (S_{BR}) at the 1st and 100th iterations. (**d**–**f**) are the corresponding distribution of Young's modulus during the bone remodeling process at the initial state, 1st, and 100th iteration.

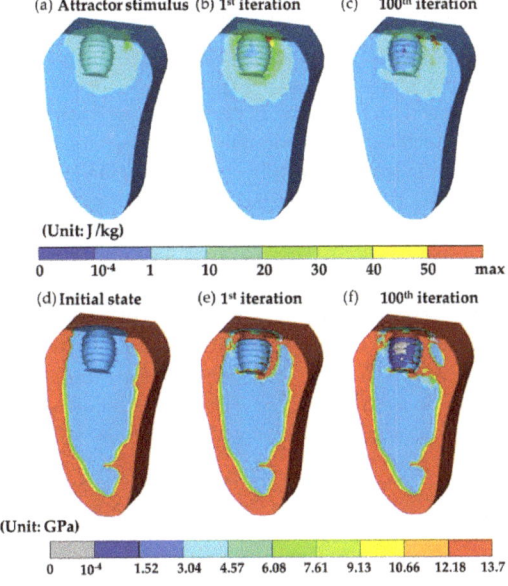

Figure 6. The results generated by the conventional model. (**a**) The target distribution of the attractor stimulus $K(\vec{r})$ based on the natural tooth. (**b**,**c**) are bone remodeling stimulus (S_{BR}) at the 1st and 100th iterations. (**d**–**f**) are the corresponding distribution of Young's modulus during the bone remodeling process at the initial state, 1st, and 100th iteration.

Next, Figure 6d shows the distribution of Young's modulus of the initial state adopted in the conventional model. It can be observed that constant Young's modulus of 2 GPa in the callus in Figure 6d was assumed without considering the result of differentiating tissue phenotypes during the short-term healing. Then, the distribution of the remodeling stimulus (S_{BR}) for the 1st iteration can be determined as shown in Figure 6b. In the 1st iteration, the distribution of S_{BR} was similar to that generated by the current model, apart from there was no high stimulus bands around the implant. However, such a small difference resulted in a very different distribution of Young's modulus shown in Figure 6e compared with that in Figure 5e, wherein, the high strength bones partially covered around the implant; most of the bones attached to the surface of the implant remained a constant Young's modulus of 2 GPa. Then, similar to the current model, the average remodeling stimulus (S_{ave}) quickly converged. Finally, the 100th iteration was achieved, giving the distribution of S_{BR} as shown in Figure 6c. The resulting Young's modulus distribution is shown in Figure 6f. The volume fraction of bone resorption was at the value of 0.044%, which was very similar to the current model. A significant difference between the results of the two models was the location of bone resorption. It was found that bone resorption (colored in grey) occurred in the buccal side in the conventional model while it appeared on the lingual side in the current model. This will be discussed in more detail in the next section. Another obvious difference between the two models was the distribution of high Young's modulus bands. The high Young's modulus band was absent in the buccal side, and thus, no supporting connection between the surface of the implant and cortical bone. The average Young's modulus at the 100th iteration was around 3.65 GPa which was relatively lower than that generated by the current model.

4. Discussion

The results mentioned above show that the initial state of bone remodeling can greatly affect the distribution of Young's modulus at the final state. In the current model, the initial state of bone remodeling was the result of the mechano-regulation model (short-term healing process), giving the top and bottom regions in the callus with lower Young's modulus. This leads to a higher strain energy density and a low corresponding bone apparent density base on Equation (8). Then, these regions had higher remodeling stimulus based on Equation (5), promoting the formation of bones, i.e., Young's modulus and apparent density increased. This non-uniform Young's modulus at the initial state leads to a dramatic change of Young's modulus in the entire callus region in the 1st iteration. On the contrary, in the conventional model, Young's modulus was assumed uniform, resulting in the change of Young's modulus around the implant only in the 1st iteration.

In the 100th iteration, it can be observed that the stability of the implants was greatly influenced by the initial states. In the current model, there were two high strength bone tissue bands connected to cortical bones, while there was only one connected bone tissue band in the result generated in the conventional model. In addition, the average Young's modulus in the current model was higher than it predicted by the current model. Thus, the stability of the implant was underestimated in the conventional model where uniform Young's modulus in the callus region at the initial state was assumed. This indicated that short-term healing can greatly affect the results of bone remodeling and cannot be neglected.

In the results of short-term healing generated by the mechano-regulation model, soft tissues occurred in the early stage and then replaced by bone tissues due to the decrease of biophysical stimulus in the later stage. This was in accordance with both the experimental [14] and computational [5] works in the literature. Where the literature reported that bone tissue forms after the formation of soft tissues (i.e., fibrous tissue and cartilage), and then bone began to differentiate, giving the increase of fluid flow since woven bone is more permeable. Furthermore, three additional features can be found in the remodeling result in the 100th iteration predicted by both the current and conventional models, which were in accordance with the clinical observations. Firstly, both models predicted a region of bone resorption at the top surface in the lingual (right) side. This phenomenon was known as marginal bone loss [49,50], which was an important factor for implant stability. Secondly, both models generated a bone resorption region in the middle of the bone-implant interface. This is the so-called stress shielding effect [27,47], which was the result of the occurrence of high strength bone tissue (colored in red) at the region around the first thread. However, the locations of the bone resorption region predicted by the two models were different, where the region predicted by the current model was in the lingual (right) side and that predicted by the conventional model was in the buccal (left) side. According to the literature [51], bone resorption occurred on the right-hand side when the loading direction was similar to that in the current work, i.e., the load was applied from the top right to the bottom left. The current model successfully captured this feature, reproduce the bone resorption region at the right-hand side at the bone-implant interface, while the conventional model predicted the location of bone resorption at the opposite side. This shows that the current model can give a more accurate result of bone remodeling procedure in the bone-implant-prosthesis system. Third, the result in the 100th iteration predicted by the current model shows that around 70% of the surface of the implant was covered by bone tissues (i.e., the elements with Young's modulus greater than that of immature bone, 2 GPa). This value was similar to that reported by Lian et al. [31], where they suggested around 60% contact between bone and implant when an equilibrium of bone remodeling is reached. Based on the features mentioned above, we conclude the rejection of the null hypothesis that short-term bone healing has no effect on bone remodeling results.

Since the current hybrid model was implemented by the finite element method, which can perform virtual tests on a wide variety of people characteristics and dental materials by simply changing the geometry of the model, boundary conditions, and material properties of bones. Thus, the current model can potentially provide estimated information and may even provide optimized dental implants for dentist clinics. Next, to demonstrate the applicability of the current model and reveal the effect of individual differences of the patients, we adopted the case of middle-aged male adults with higher strength material property of bone, referred to as to higher bone strength case. Where Young's modulus of cortical, cancellous, immature, and mature bone was 1.4 times [52] than it in Table 1, while the properties of the remaining tissues, such as fibrous tissue and cartilage, stayed unchanged. Then, the short- and long-term bone healing processes were evaluated by the current model. The corresponding results are shown in Figures 7 and 8, respectively. It was found that the trend of tissue differentiation in the short-term healing process was almost identical to it the standard case as shown in Figure 4. The only difference between the two cases was the averaged Young's modulus in the high strength case was about two times higher than it was in the standard case, which can be seen in Figure 8a, where the material properties in the callus region were assigned according to the tissue differentiation result. In the 100th iteration, there were several significant differences between the two cases. Firstly, there was merely no bone resorption in the higher bone strength case. The volume fraction of bone resorption was at a value of 0.0068%. Secondly, the connective tissue bands in the higher bone strength case were thicker than in the standard case. The two features indicate that the bone system in the higher bone strength case can provide excellent supports and enhance the stability of the implant. This result also has good agreement with the clinical observation, where the dental implant failure rate for the middle-aged male adults was relatively low [53].

Although the current model considered both the short-term and long-term healing process and reproduced many features that were discovered in the experiments, the model can be further improved by considering the physiological mechanism listed as follows. For example, a more complex diffusion mechanism of stem cell migration in the short-term healing process, such as the growth of vessels which can be implemented by the random-walk model [54]; periodontal ligaments (PDL), which play a crucial role in bone remodeling, can be simulated by taken the anisotropic and nonlinear elastic stress-strain behavior into account [55–57]. It is expected a more accurate prediction can be achieved if these factors were considered.

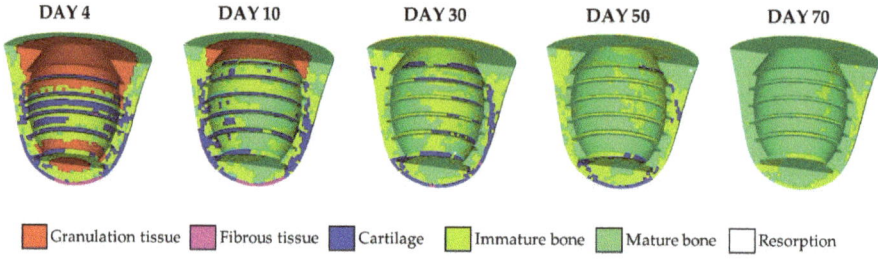

Figure 7. The tissue differentiation history predicted by the mechano-regulation algorithm in higher bone strength case.

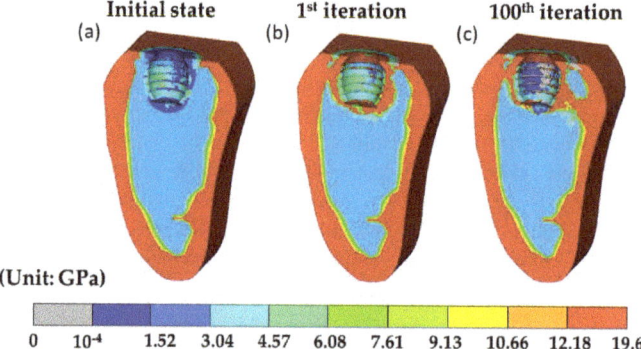

Figure 8. The results generated by the current model in higher bone strength case. (**a**–**c**) are the corresponding distribution of Young's modulus during the bone remodeling process at the initial state, 1st, and 100th iteration.

5. Conclusions

In this study, a hybrid numerical bone healing algorithm was developed to predict the morphology of bone around dental implants with the consideration of both short-term and long-term bone healing. The results showed that the effect of short-term bone healing should not be ignored, and the assumption of uniform material properties for the initial state in the bone remodeling model is inappropriate. The current hybrid model can reveal many bone healing features having a very good agreement with the literature. It can be extended to simulate different implant geometries, applied loads, and bone properties of patients, enabling an early prediction of the performance of clinical treatments.

Author Contributions: Conceptualization, S.-S.C. and N.-T.T.; methodology, P.-C.K. and S.-S.C.; software, P.-C.K. and S.-S.C.; validation, P.-C.K. and N.-T.T.; writing—original draft preparation, P.-C.K.; writing—review and editing, N.-T.T.; visualization, P.-C.K.; supervision, N.-T.T.; project administration, N.-T.T. All authors have read and agreed to the published version of the manuscript.

Funding: This research was supported by Ministry of Science and Technology Taiwan, Grant No. MOST 106-2218-E-009-003 and Grant No. MOST 107-2218-E-080-001.

Acknowledgments: The authors would like to thank Ming-Jun Li for his comments and assistance to this work. The authors wish to acknowledge the National Center for High-Performance Computing, Taiwan, for providing the computational platform.

Conflicts of Interest: The authors declare no conflict of interest.

References

1. Meredith, N. Assessment of implant stability as a prognostic determinant. *Int. J. Prosthodont.* **1998**, *11*, 491–501. [PubMed]
2. Ammon, C.; Kreutz, M.; Rehli, M.; Krause, S.W.; Andreesen, R. Platelets induce monocyte differentiation in serum-free coculture. *J. Leukoc. Biol.* **1998**, *63*, 469–476. [CrossRef] [PubMed]
3. Spisani, S.; Giuliani, A.L.; Cavalletti, T.; Zaccarini, M.; Milani, L.; Gavioli, R.; Traniello, S. Modulation of neutrophil functions by activated platelet release factors. *Inflammation* **1992**, *16*, 147–158. [CrossRef] [PubMed]
4. Tang, Y.; Wu, X.; Lei, W.; Pang, L.; Wan, C.; Shi, Z.; Zhao, L.; Nagy, T.R.; Peng, X.; Hu, J.; et al. TGF-β1-induced migration of bone mesenchymal stem cells couples bone resorption with formation. *Nat. Med.* **2009**, *15*, 757. [CrossRef] [PubMed]
5. Lacroix, D.; Prendergast, P.J. A mechano-regulation model for tissue differentiation during fracture healing: Analysis of gap size and loading. *J. Biomech.* **2002**, *35*, 1163–1171. [CrossRef]
6. Crisan, M.; Yap, S.; Casteilla, L.; Chen, C.W.; Corselli, M.; Park, T.S.; Andriolo, G.; Sun, B.; Zheng, B.; Zhang, L.; et al. A perivascular origin for mesenchymal stem cells in multiple human organs. *Cell Stem Cell* **2008**, *3*, 301–313. [CrossRef] [PubMed]
7. Wollert, K.C.; Meyer, G.P.; Lotz, J.; Lichtenberg, S.R.; Lippolt, P.; Breidenbach, C.; Fichtner, S.; Korte, T.; Hornig, B.; Messinger, D.; et al. Intracoronary autologous bone-marrow cell transfer after myocardial infarction: The BOOST randomised controlled clinical trial. *Lancet* **2004**, *364*, 141–148. [CrossRef]
8. Ashton, B.A.; Allen, T.D.; Howlett, C.R.; Eaglesom, C.C.; Hattori, A.; Owen, M. Formation of bone and cartilage by marrow stromal cells in diffusion chambers in vivo. *Clin. Orthop. Relat. Res.* **1980**, *151*, 294–307. [CrossRef]
9. Hadjidakis, D.J.; Androulakis, I.I. Bone remodeling. *Ann. N. Y. Acad. Sci.* **2006**, *1092*, 385–396. [CrossRef]
10. Frost, H.M. Tetracycline-based histological analysis of bone remodeling. *Calcif. Tissue Int.* **1969**, *3*, 211–237. [CrossRef]
11. Wolff, J. *The Law of Bone Remodelling*, 1st ed.; Springer: Berlin/Heidelberg, Germany, 1986; ISBN 978-3-642-71031-5.
12. Ghiasi, M.S.; Chen, J.; Vaziri, A.; Rodriguez, E.K.; Nazarian, A. Bone fracture healing in mechanobiological modeling: A review of principles and methods. *Bone Rep.* **2017**, *6*, 87–100. [CrossRef] [PubMed]
13. Mohaghegh, K.; Pérez, M.A.; García-Aznar, J.M. Accelerating numerical simulations of strain-adaptive bone remodeling predictions. *Comput. Methods Appl. Mech. Eng.* **2014**, *273*, 255–272. [CrossRef]
14. Pauwels, F. A new theory on the influence of mechanical stimuli on the differentiation of supporting tissue. The tenth contribution to the functional anatomy and causal morphology of the supporting structure. *Z. Anat. Entwicklungsgesch.* **1960**, *121*, 478–515. [CrossRef] [PubMed]
15. Carter, D.R.; Fyhrie, D.P.; Whalen, R.T. Trabecular bone density and loading history: Regulation of connective tissue biology by mechanical energy. *J. Biomech.* **1987**, *20*, 785–794. [CrossRef]
16. Prendergast, P.J.; Huiskes, R.; Søballe, K. Biophysical stimuli on cells during tissue differentiation at implant interfaces. *J. Biomech.* **1997**, *30*, 539–548. [CrossRef]
17. Chou, H.Y.; Müftü, S. Simulation of peri-implant bone healing due to immediate loading in dental implant treatments. *J. Biomech.* **2013**, *46*, 871–878. [CrossRef]
18. Sotto-maior, B.S.; Graciliano, E.; Mercuri, F.; Mendes, P.; Maria, N.; Picorelli, S.; Francischone, C.E.; Maria, N.; Picorelli, S.; Francischone, C.E.; et al. Computer Methods in Biomechanics and Biomedical Engineering Evaluation of bone remodeling around single dental implants of different lengths: A mechanobiological numerical simulation and validation using clinical data. *Comput. Methods Biomech. Biomed. Engin.* **2016**, *19*, 699–706. [CrossRef]
19. Fernández, J.R.; García-Aznar, J.M.; Martínez, R.; Viaño, J.M. Numerical analysis of a strain-adaptive bone remodelling problem. *Comput. Methods Appl. Mech. Eng.* **2010**, *199*, 1549–1557. [CrossRef]
20. Cowin, S.C.; Hegedus, D.H. Bone remodeling I: Theory of adaptive elasticity. *J. Elast.* **1976**, *6*, 313–326. [CrossRef]
21. Frost, H.M. Bone "mass" and the "mechanostat": A proposal. *Anat. Rec.* **1987**, *219*, 1–9. [CrossRef] [PubMed]
22. Kim, Y.; Park, S.; Shin, K. Effect of cycloid movement on plate-to-roll gravure offset printing. *Microsyst. Technol.* **2016**, *22*, 357–365. [CrossRef]

23. Hegedus, D.H.; Cowin, S.C. Bone remodeling II: Small strain adaptive elasticity. *J. Elast.* **1976**, *6*, 337–352. [CrossRef]
24. Hart, R.T.; Davy, D.T.; Heiple, K.G. A computational method for stress analysis of adaptive elastic materials with a view toward applications in strain-induced bone remodeling. *J. Biomech. Eng.* **1984**, *106*, 342–350. [CrossRef] [PubMed]
25. Carter, D.R. Mechanical loading histories and cortical bone remodeling. *Calcif. Tissue Int.* **1984**, *36*, S19–S24. [CrossRef] [PubMed]
26. Carter, D.R.; Hayes, W.C. The compressive behavior of bone as a two-phase porous structure. *J. Bone Jt. Surg. Am.* **1977**, *59*, 954–962. [CrossRef]
27. Huiskes, R.; Weinans, H.; Grootenboer, H.J.; Dalstra, M.; Fudala, B.; Slooff, T.J. Adaptive bone-remodeling theory applied to prosthetic-design analysis. *J. Biomech.* **1987**, *20*, 1135–1150. [CrossRef]
28. Lin, D.; Li, Q.; Li, W.; Zhou, S.; Swain, M.V. Design optimization of functionally graded dental implant for bone remodeling. *Compos. Part B Eng.* **2009**, *40*, 668–675. [CrossRef]
29. Chou, H.-Y.; Romanos, G.; Müftü, A.; Müftü, S. Peri-implant bone remodeling around an extraction socket: Predictions of bone maintenance by finite element method. *Int. J. Oral Maxillofac. Implants* **2012**, *27*, 39–48.
30. Crupi, V.; Guglielmino, E.; LaRosa, G.; VanderSloten, J.; VanOosterwyck, H. Numerical analysis of bone adaptation around an oral implant due to overload stress. *Proc. Inst. Mech. Eng. Part H J. Eng. Med.* **2004**, *218*, 407–415. [CrossRef]
31. Lian, Z.; Guan, H.; Ivanovski, S.; Loo, Y.-C.; Johnson, N.W.; Zhang, H. Effect of bone to implant contact percentage on bone remodelling surrounding a dental implant. *Int. J. Oral Maxillofac. Surg.* **2010**, *39*, 690–698. [CrossRef]
32. Reina, J.M.; García-Aznar, J.M.; Domínguez, J.; Doblaré, M. Numerical estimation of bone density and elastic constants distribution in a human mandible. *J. Biomech.* **2007**, *40*, 828–836. [CrossRef] [PubMed]
33. Lin, D.; Li, Q.; Li, W.; Swain, M. Bone remodeling induced by dental implants of functionally graded materials. *J. Biomed. Mater. Res. Part B Appl. Biomater.* **2010**, *92*, 430–438. [CrossRef] [PubMed]
34. Li, J.; Li, H.; Shi, L.; Fok, A.S.L.; Ucer, C.; Devlin, H.; Horner, K.; Silikas, N. A mathematical model for simulating the bone remodeling process under mechanical stimulus. *Dent. Mater.* **2007**, *23*, 1073–1078. [CrossRef]
35. Lacroix, D.; Prendergast, P.J. A homogenization procedure to prevent numerical instabilities in poroelastic tissue differentiation models. In Proceedings of the Eighth Annual Symposium: Computational Methods in Orthopaedic Biomechanics, Orlando, FL, USA, 11 March 2000.
36. Barbier, L.; VanderSloten, J.; Krzesinski, G.; Schepers, E.; Van DerPerre, G. Finite element analysis of non-axial versus axial loading of oral implants in the mandible of the dog. *J. Oral Rehabil.* **1998**, *25*, 847–858. [CrossRef] [PubMed]
37. Jansen, V.K.; Conrads, G.; Richter, E.J. Microbial leakage and marginal fit of the implant-abutment interface. *Int. J. Oral Maxillofac. Implants* **1997**, *12*, 23.
38. Streckbein, P.; Streckbein, R.G.; Wilbrand, J.F.; Malik, C.Y.; Schaaf, H.; Howaldt, H.P.; Flach, M. Non-linear 3D evaluation of different oral implant-abutment connections. *J. Dent. Res.* **2012**, *91*, 1184–1189. [CrossRef]
39. Zohar, R.; Cheifetz, S.; McCulloch, C.A.G.; Sodek, J. Analysis of intracellular osteopontin as a marker of osteoblastic cell differentiation and mesenchymal cell migration. *Eur. J. Oral Sci.* **1998**, *106*, 401–407. [CrossRef]
40. Iwaki, A.; Jingushi, S.; Oda, Y.; Izumi, T.; Shida, J.I.; Tsuneyoshi, M.; Sugioka, Y. Localization and quantification of proliferating cells during rat fracture repair: Detection of proliferating cell nuclear antigen by immunohistochemistry. *J. Bone Miner. Res.* **1997**, *12*, 96–102. [CrossRef]
41. Monjo, M.; Lamolle, S.F.; Lyngstadaas, S.P.; Rønold, H.J.; Ellingsen, J.E. In vivo expression of osteogenic markers and bone mineral density at the surface of fluoride-modified titanium implants. *Biomaterials* **2008**, *29*, 3771–3780. [CrossRef]
42. Papalexiou, V.; Novaes, A.B., Jr.; Grisi, M.F.M.; Souza, S.S.L.S.; Taba, M., Jr.; Kajiwara, J.K. Influence of implant microstructure on the dynamics of bone healing around immediate implants placed into periodontally infected sites: A confocal laser scanning microscopic study. *Clin. Oral Implants Res.* **2004**, *15*, 44–53. [CrossRef]
43. JR, S.A.; Allegrini, M.R.F.; Yoshimoto, M.; JR, B.K.; Mai, R.; Fanghanel, J.; Gedrange, T. Soft tissue integration in the neck area of titanium implants–an animal trial. *J. Physiol. Pharmacol.* **2008**, *59*, 117–132.

44. Huiskes, R.; Weinans, H.; VanRietbergen, B. The relationship between stress shielding and bone resorption around total hip stems and the effects of flexible materials. *Clin. Orthop. Relat. Res.* **1992**, 124–134. [CrossRef]
45. Turner, C.H.; Anne, V.; Pidaparti, R.M. VA uniform strain criterion for trabecular bone adaptation: Do continuum-level strain gradients drive adaptation? *J. Biomech.* **1997**, *30*, 555–563. [CrossRef]
46. Turner, C.H. Three rules for bone adaptation to mechanical stimuli. *Bone* **1998**, *23*, 399–407. [CrossRef]
47. Weinans, H.; Huiskes, R.; Grootenboer, H.J. The behavior of adaptive bone-remodeling simulation models. *J. Biomech.* **1992**, *25*, 1425–1441. [CrossRef]
48. Frost, H.M. Bone's mechanostat: A 2003 update. *Anat. Rec. Part A Discov. Mol. Cell. Evol. Biol.* **2003**, *275*, 1081–1101. [CrossRef]
49. Akca, K.; Cehreli, M.C. Biomechanical consequences of progressive marginal bone loss around oral implants: A finite element stress analysis. *Med. Biol. Eng. Comput.* **2006**, *44*, 527–535. [CrossRef]
50. Esposito, M.; Ekestubbe, A.; Gröndahl, K. Radiological evaluation of marginal bone loss at tooth surfaces facing single Brånemark implants. *Clin. Oral Implants Res.* **1993**, *4*, 151–157. [CrossRef]
51. Arifin, A.; Sulong, A.B.; Muhamad, N.; Syarif, J.; Ramli, M.I. Material processing of hydroxyapatite and titanium alloy (HA/Ti) composite as implant materials using powder metallurgy: A review. *Mater. Des.* **2014**, *55*, 165–175. [CrossRef]
52. Miyaura, K.; Matsuka, Y.; Morita, M.; Yamashita, A.; Watanabe, T. Comparison of biting forces in different age and sex groups: A study of biting efficiency with mobile and non-mobile teeth. *J. Oral Rehabil.* **1999**, *26*, 223–227. [CrossRef]
53. Moy, P.K.; Medina, D.; Shetty, V.; Aghaloo, T.L. Dental implant failure rates and associated risk factors. *Int. J. Oral. Maxillofac. Implant.* **2005**, *20*, 569–577.
54. Pérez, M.A.; Prendergast, P.J. Random-walk models of cell dispersal included in mechanobiological simulations of tissue differentiation. *J. Biomech.* **2007**, *40*, 2244–2253. [CrossRef]
55. Toms, S.R.; Lemons, J.E.; Bartolucci, A.A.; Eberhardt, A.W. Nonlinear stress-strain behavior of periodontal ligament under orthodontic loading. *Am. J. Orthod. Dentofac. Orthop.* **2002**, *122*, 174–179. [CrossRef]
56. Vollmer, D.; Bourauel, C.; Maier, K.; Jäger, A. Determination of the centre of resistance in an upper human canine and idealized tooth model. *Eur. J. Orthod.* **1999**, *21*, 633–648. [CrossRef]
57. Provatidis, C.G. A comparative FEM-study of tooth mobility using isotropic and anisotropic models of the periodontal ligament. *Med. Eng. Phys.* **2000**, *22*, 359–370. [CrossRef]

© 2020 by the authors. Licensee MDPI, Basel, Switzerland. This article is an open access article distributed under the terms and conditions of the Creative Commons Attribution (CC BY) license (http://creativecommons.org/licenses/by/4.0/).

Article

Load-Bearing Capacity of Zirconia Crowns Screwed to Multi-Unit Abutments with and without a Titanium Base: An In Vitro Pilot Study

Hadas Heller [1], Adi Arieli [1], Ilan Beitlitum [2], Raphael Pilo [1] and Shifra Levartovsky [1,*]

1. Department of Oral Rehabilitation, the Maurice and Gabriela Goldschleger School of Dental Medicine, Tel Aviv University, Tel Aviv, Israel; heller.hadas@gmail.com (H.H.); dr.arieli@gmail.com (A.A.); rafipilo@gmail.com (R.P.)
2. Department of Periodontology and Dental Implantology, Maurice and Gabriela Goldschleger School of Dental Medicine, Tel Aviv University, Tel Aviv, Israel; beilan1@bezeqint.net
* Correspondence: shifralevartov@gmail.com; Tel.: +972-52-3515403; Fax: +972-3-6481148

Received: 23 August 2019; Accepted: 17 September 2019; Published: 20 September 2019

Abstract: The static and dynamic load-bearing capacities and failure modes of zirconia crowns screwed to multi-unit abutments (MUAs) with and without a titanium base (T-base) were determined. Thirty-six monolithic zirconia crowns screwed to straight MUAs torqued to laboratory analogs (30 Ncm) were assigned to two groups (n = 18). In group A, the zirconia crowns were screwed directly to the MUAs; in group B, the zirconia crowns were cemented to the T-base and screwed to the MUAs. All specimens were aged in 100% humidity (37 °C) for one month and subjected to thermocycling (20,000 cycles). Afterwards, the specimens underwent static and dynamic loading tests following ISO 14801. The failure modes were evaluated by stereomicroscopy (20×). There was an unequivocally similar trend in the S-N plots of both specimen groups. The load at which the specimens survived 5,000,000 cycles was 250 N for both groups. Group A failed mainly within the metal, and zirconia failure occurred only at a high loading force. Group B exhibited failure within the metal mostly in conjunction with adhesive failure between the zirconia and T-base. Zirconia restoration screwed directly to an MUA is a viable option, but further studies with larger sample sizes are warranted.

Keywords: monolithic zirconia; multi-unit abutment; titanium base

1. Introduction

As esthetic demands in dentistry increase, porcelain fused to metal has become more frequently substituted with zirconia-based ceramic for teeth and implant-supported fixed dental prostheses (FDPs) [1]. With the introduction of computer-aided design/computer-aided manufacturing (CAD/CAM) technologies and the development of yttrium oxide partially stabilized tetragonal zirconia polycrystalline (Y-TZP), bilaminar or preferably monolithic zirconia structures have become the material of choice. Monolithic zirconia restorations demonstrate a high flexural strength (900–1200 MPa) and fracture toughness (9–10 Mpa m$^{0.5}$) compared to those of other ceramic materials and exhibit minimal wear of the antagonist teeth [2,3].

Implant-supported prostheses are utilized via two methods of retention; they are either screw-retained or cement-retained. Both modalities have benefits and shortcomings in clinical application [4]. In recent years, with the increased risk of implant loss due to biological complications such as peri-implantitis, which is associated mainly with cemented implant restorations, the advantages of retrievability as well as accessibility for maintenance and replacement have accelerated the use of screw-retained implant restorations. Due to these reasons and their apparently higher biological compatibility, the screw-retained modalities are currently preferable [5–8]. When connecting several implants with a screw-retained implant restoration, there is a need for an interim part, a multi-unit abutment (MUA), to correct the differences in implant angles and to create a common path of insertion. The first MUA was introduced for the Branemark implant system and was configured as a two-piece titanium abutment cylinder [9]. Currently, a one-piece abutment, which can be straight or angled, is commonly used. These definitive MUAs enable better hemidesmosomal adherence between the soft tissue and titanium and therefore might reduce bone resorption around the implants [10–12]. This "one abutment at one time" concept is especially advantageous in immediately restored implants for partial and full edentulous cases, whereas non removal of the multi-unit abutments placed at the time of surgery results in a statistically significant reduction in crestal bone resorption around the implants [11].

The high demand for aesthetic outcomes, combined with new material technologies and the need for retrievability in implant-supported FDPs, has led to an increased use of screwed monolithic zirconia implant-supported restorations. Usually, this type of zirconia is cemented to a milled titanium sleeve (T-base) and then screwed to the MUA (screwed-cemented type of restoration). This design was suggested by McGlumphy et al. [13], Rajan and Gunaseelan [14], and Uludag and Celik [15] and is known as "the combination implant crown". Hussien et al. [16] showed that this type of combination (screwed-cemented) does not affect the fatigue failure load of monolithic zirconia, monolithic lithium disilicate, or veneered zirconia ceramic crowns when compared to cemented crowns. Moreover, significantly higher fatigue failure loads have been recorded for monolithic zirconia crowns than for the other two types of crowns [16]. The retention of monolithic zirconia copings to the T-base might be the weakest link of this type of screwed-cemented restoration; therefore, a reliable bond between the zirconia, cement, and T-base is essential for the longevity of the restoration [17,18].

In contrast to the screwed-cemented restoration, there is another mode of retention of monolithic zirconia crowns to MUAs: Screwing the zirconia directly to the MUA (screwed restoration) without the use of a T-base, thus avoiding the dependence on the cemented joint between the zirconia and the titanium. Although restorations with zirconia crowns attached to MUAs with and without titanium bases are widely used by clinicians, there is no evidence-based data to favor either of these approaches; moreover, no research study has compared the mechanical failure between the two methods. Some believe that the titanium base is vital for the survival of the zirconia restoration, while others are concerned with the detrimental impact of the cement that is placed between the zirconia and the titanium base.

The aim of the current study was to compare the static and dynamic load-bearing capacities and determine the failure mode of 3Y-TZP (zirconia) crowns screwed to MUAs with and without a T-base. The null hypothesis was that there would be no difference in either the failure load or the failure mode between the two groups.

2. Results

Table 1 presents the results of the static load compressive test for three specimens of each group. From the mean maximal failure load, the 80% level load was set for the start of the dynamic test. The 80% level set for group A was slightly higher than that for group B.

Table 1. Mean maximal load-bearing capacity (N) and 80% level for each experimental group.

Specimen	Group A(N)	Group B(N)
1	872	718
2	811	680
3	675	707
Mean	786	702
80% Level	629	561

Group A—screwed restoration; Group B—screwed-cemented restoration; 80% Level—80% of the average static load compressive test result.

The stress-number of cycles (S-N) curve results, as well as the failure modes, are presented in Tables 2 and 3 for group A (screwed restoration) and group B (screwed-cemented restoration), respectively. In both groups, when the applied load was equal to 250 N, the three specimens tested (samples 13–15) remained intact after five million cycles (Tables 2 and 3). The S-N curves for both groups are presented in Figure 1. However, despite the small number of samples, an unequivocally similar trend of both specimen groups on the S-N plot, could be noted.

Table 2. Fatigue test results for group A (screwed restoration).

Specimen	Load (N)	No. of Cycles	Average No of Cycles	Failure Mode
1	450	35,670	-	Zirconia + metal (screw)
2	450	209,543	80,545	Metal (MUA + screw)
3	450	176,422	-	Zirconia + metal (screw)
4	400	1,254,436	-	Metal (MUA + screw)
5	400	125,878	533,457	Metal (MUA + screw)
6	400	220,056	-	Metal (MUA + screw)
7	350	160,749	-	Metal (MUA + screw)
8	350	546,871	266,048	Metal (MUA + screw)
9	350	905,205	-	Metal (MUA + screw)
10	300	3,401,784	-	Metal (MUA + screw)
11	300	437,068	2,946,284	Metal (MUA + screw)
12	300	5,000,000	-	Metal (MUA + screw)
13	250	5,000,000	-	No Failure
14	250	5,000,000	5,000,000	No Failure
15	250	5,000,000	-	No Failure

With a maximum load of 250 N, three specimens completed five million cycles without fracturing; Load—maximum load applied in each cycle; Number of cycles—number of cycles before the test was interrupted; Average number of cycles- for particular load values; MUA—multi-unit abutment.

Table 3. Fatigue test results for group B (screwed-cemented restoration).

Specimen	Load (N)	No. of Cycles	Average No of Cycles	Failure Mode
1	550	6607	-	Adhesive + metal (screw)
2	550	30,341	22,450	Metal (screw)
3	550	30,341	-	Adhesive + metal (screw)
4	450	99,751	-	Adhesive + metal (MUA)
5	450	178,367	158,818	Metal (screw)
6	450	198,337	-	Adhesive + metal (MUA)
7	350	2,950,685	-	Adhesive + metal (screw)
8	350	200,751	1,764,689	Adhesive + metal (MUA)
9	350	2,142,632	-	Adhesive + metal (MUA)
10	300	553,418	-	Adhesive + metal (MUA)
11	300	2,264,147	2,472,765	Metal (screw)
12	300	4,600,732	-	Adhesive + metal (MUA
13	250	5,000,000	-	No Failure
14	250	5,000,000	5,000,000	No Failure
15	250	5,000,000	-	No Failure

With a maximum load of 250 N, three specimens completed five million cycles without fracturing; Load—maximum load applied in each cycle; Number of cycles—number of cycles before the test was interrupted; Average number of cycles- for particular load values; MUA—multi-unit abutment.

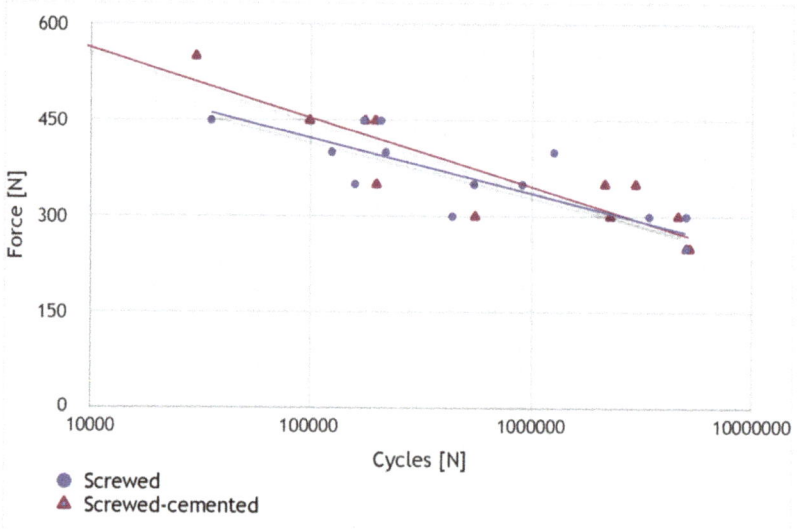

Figure 1. S-N curve. Chart showing the applied load as a function of the number of cycles. After the samples underwent five million cycles, the test was stopped. The horizontal and vertical lines are used as a guide for the eye.

Stereomicroscopic examination of the failure mode of all the specimens revealed differences between the groups (Tables 2 and 3). Group A exhibited mainly failure within the metal (MUA/screw), and zirconia failure occurred only at a high loading force. The representative failure photographs are presented in Figures 2 and 3. Figure 2 shows deformation and fracturing within the cone of the MUA, and Figure 3 shows fracturing of the zirconia concomitantly with bending of the restoration screw head. Group B exhibited failure within the metal (the cone of the MUA/the restoration screw head), mostly in conjunction with adhesive failure between the zirconia and T-base. Figure 4 shows adhesive failure between the zirconia and T-base in addition to fracturing within the MUA cone, which was the predominant failure mode in group B.

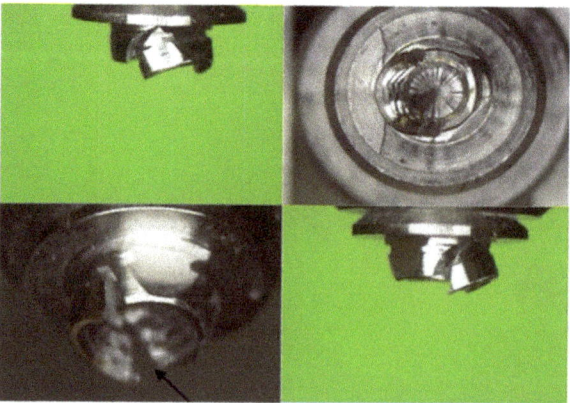

Figure 2. Specimen number seven in group A shows deformation and fracturing within the cone of the multi-unit abutment (MUA) under an applied load of 350 N. This mode of failure was predominant in group A.

Figure 3. Specimen number one in group A shows fracturing of the zirconia concomitantly with bending of the restoration screw head under an applied load of 450 N. This failure was observed in group A at only high loading forces.

Figure 4. Specimen number 10 in group B exhibits adhesive failure between the zirconia and T-base in addition to deformation and fracturing within the cone of the MUA under an applied load of 300 N. This mode of failure was predominant in group B.

3. Discussion

Numerous studies have evaluated the fracture strength of different abutments for cemented implant restorations. No available data exists on the fracture strength of screwed implant restorations connected to MUAs, even though these types of restorations have gained popularity with the increased use of immediate loading of multiple implants.

The current study compared the static and dynamic load-bearing capacities and determined the failure mode of zirconia crowns screwed to MUAs with and without a T-base, an issue that has not yet been addressed in the literature. Our null hypothesis was partially proven because both groups demonstrated that three specimens survived 5,000,000 cycles, according to International Standard (ISO) 14801, at 250 N. This result may indicate that the two modes of retention (with and without a T-base) have no effect on the mechanical performance of the prosthesis under cyclic loading.

The range of human biting forces is 50–300 N in normal chewing as opposed to 175–800 N for maximum voluntary biting forces [19–21]. These values are highly dependent on the measurement tools, mainly strain gauges, piezo-electric sensors, and pressure sheets. In the current study, the occlusal forces range from 250 N to 550 N, but the specimens survived 5,000,000 cycles only at 250 N. Indeed, the 250 N value fits well within the range of normal chewing forces but should be interpreted with caution because this value is valid for the specific configuration of the current study design. The conclusion that zirconia screw-type prostheses, with and without a titanium base, are equally adequate for implant-supported prostheses with MUAs must be based on the unequivocally similar trend in the S-N plots of both specimen groups and not on the 250 N value.

On the other hand, the failure modes differ between the groups. Metal failure (MUA/screw) was demonstrated in both groups; however, in group A (screwed), zirconia failure was also observed, mainly at high loading forces. This pattern was not observed in group B (screwed-cemented). With the latter failure pattern, which included metal failure, most specimens exhibited an adhesive mode of failure between the cement and zirconia. In this study, we used 3Y-TZP, although a newer form of zirconia material with better translucency (5Y-TZP) has been introduced. Research has shown that 3Y-TZP has significantly higher flexural strength than that of 5Y-TZP [22]. Our purpose in the current study was to compare both restorations for posterior clinical use; in that case, 3Y-TZP was preferred.

This research was performed according to ISO standard 14801 because the international standard requires that stress is exerted on the implant-abutment-restoration complex in the "worst-case" scenario loading situation [23]. In addition, the ISO standard established a methodology that can be replicated by other researchers studying the same subject. Studies applying dynamic loading tests do not always comply with ISO 14801. Koyama et al. [24] and Ding et al. [25] used the "staircase method" to determine the mean cyclic fatigue force in accordance with the current study. However, Duan et al. [26] used variable loading amplitude, and each specimen was subjected to increasing load amplitude over time and thus experienced multiple load amplitudes. Shemtov-Yona and Rittel [27] suggested a new method for the dynamic loading test based on random spectrum loading, noting that the ISO 14801 protocol is questionable for statistical analysis.

In the screwed-cemented implant-supported restoration, the impact of oral fluids on the solubility of the luting cement and the retention of the restoration is always of concern. To simulate the oral cavity environment, artificial aging, such as water storage and thermocycling, is performed. In the current study, an aging protocol of storage at 37 °C under 100% humidity for one month, followed by thermal cycling, was conducted. Naumova et al. [28] and Güngör and Nemli [29] investigated the effect of various resin cements on the retentive strength of zirconia abutments bonded to titanium inserts (T-bases). Both studies showed that exposure to the liquid environment had a negative effect on the retention force of the resin cements, regardless of whether thermocycling was performed. The specimens were sandblasted, after which a primer was applied to some of the specimens. Irrespective of the surface treatment, the typical mode of failure was adhesive failure. In contrast, Zenthofer et al. [30] demonstrated that artificial aging had no effect on the retentive force of the resin cement, which might be influenced by the type of luting cement. In the current study, we used a self-curing resin-based cement (Multilink Hybrid Abutment cement, Ivoclar Vivadent). Güngör and Nemli [29] investigated the effect of three types of resin cement on the retentive strength of zirconia abutments bonded to titanium inserts and found that the retentive force of the multilink hybrid abutment cement was the same as that of Panavia F 2.0 but was lower than that of the Zirconite cement.

The failure mode in group B (screwed-cemented restoration, with a T-base) was expected to be mainly adhesive failure. However, our results demonstrated that at all force levels, metal failure also occurred (screw/MUA). This result is in agreement with the results of other reports [31–33], which differ from our study because they evaluated the strength of the zirconia abutments with and without a T-base, while we evaluated the zirconia prostheses with and without a T-base. Metal failure was also observed at all force levels in group A (without a T-base). One would expect that because there was no luting cement, zirconia failure would occur in all the specimens, as this was the rationale for using the T-base. Our results showed that zirconia failure occurred at only high loading forces. This phenomenon is probably related to the high flexural strength (900–1200 Mpa) and fracture toughness (9–10 Mpa m$^{0.5}$) of the Y-TZP monolithic structure [3]. Under continuous cyclic loading, a catastrophic failure will occur within the metal (titanium) at lower force levels than those affecting the zirconia. At higher forces, because zirconia is a brittle material that is not capable of deforming, catastrophic failure will occur within the zirconia concomitantly with metal failure. When there is a cement mediator, as in group B, the weakest link is the cement, and thus an adhesive failure will occur and cause the zirconia to separate from the titanium sleeve, even at high force levels. Zirconia failure was thus not observed.

From a clinical point of view, separation of the zirconia crown from the T-base, as was found in the screwed-cemented restoration, is easy to repair by re-cementation. Even, metal failure, either of the MUA or of the screw head, is reparable with some cost, by simply replacing the damaged elements. However, when a zirconia fracture occurs, as was seen in the screwed restoration (without a T-base) at high loading forces, the zirconia crown should be replaced, with a higher cost.

The limitation of this study was the number of experiments, which was too small for full-scale statistical analysis. However, this study highlighted an importance issue that has not yet been addressed in the literature. Because this is a pilot study, further studies with larger sample sizes are warranted. In addition, similar to other in vitro studies, a laboratory study design cannot fully reproduce clinical conditions, even if the samples are artificially aged by chewing simulation and thermocycling. One might argue that the geometry of the hemisphere with the zirconia coping is different from the geometry of the in vivo zirconia crown. We followed the ISO protocol that simulates the functional loading of an endosseous dental implant and its prosthesis under "worst-case" conditions; however, this method cannot fully predict the in vivo behavior of the dental implant and its prosthesis. Therefore, the results of the present study should be interpreted with caution.

4. Materials and Methods

Thirty-six straight MUAs (Alpha-Bio Tec., Petah Tikva, Israel) were torqued to laboratory analogs (6 mm width) (Alpha-Bio Tec.) with a digital torque ratchet (30 Ncm) (CEDAR Model DIW 15) and were assigned to two groups (24 specimens per group):

Group A (screwed restoration) (n = 18): Spherically shaped monolithic zirconia copings (Ceramill® material, Amann Girrbach AG, Koblach, Austria) with a height of 10 mm and a wall thickness of 1.5 mm were designed and milled with an internal ledge for the screw head and then torqued to the MUAs with a digital torque ratchet (25 Ncm) (CEDAR Model DIW 15) (Figure 5a).

The screwed zirconia cap has an internal zirconia ledge that supports the screw head and is in intimate contact with the MUAs (Figure 5c).

Group B (screwed-cemented restoration) (n = 18): Spherically shaped monolithic zirconia copings (Ceramill® material, Amann Girrbach AG, Koblach, Austria) with a height of 10 mm and a wall thickness of 1.5 mm were designed and milled to fit a titanium sleeve (a T-base) (Alpha-Bio Tec.) (Figure 5b). The cementation of the monolithic zirconia crowns to the T-base was performed according to commonly accepted protocols [34]. The intaglio surface of the crowns and the outer surface of the T-base were airborne particle-abraded using 50 µ alumina oxide particles at pressures of 1 and 2 bar, respectively; then, the surfaces were cleaned with alcohol. The pretreated T-base and the intaglio surface of the crown were coated with a single component primer to promote adhesion (Monobond Plus, Ivoclar Vivadent). The primer agent was allowed to react for 60 min, after which it was dispersed by a stream of air. The specimens were then cemented to the T-base with a self-curing resin-based cement (Multilink Hybrid Abutment cement, Ivoclar Vivadent) according to the manufacturer's instructions. The copings were seated with a device that allowed a predefined pressure of 50 N to be applied along the longitudinal axis of the abutment for 7 min [35,36]. Excess resin was removed from the bonded margins before it set completely. After cementation, the specimens were torqued to the MUAs with a digital torque ratchet (25 Ncm) (CEDAR Model DIW 15).

All specimens in both groups were stored at 37 °C under 100% humidity for one month, followed by thermal cycling between water temperatures of 5 °C and 55 °C for 20,000 cycles with a 10 s dwell time (Y. Manes, Tel-Aviv, Israel).

After aging, a fatigue test was performed according to the ISO standard 14801 [23]. To comply with this standard, every specimen was embedded in a designated metal holder inclined 30° to the longitudinal force axis direction. Additionally, to allow a uniform stress distribution and to avoid stress concentration, the load was applied to a hemisphere modeled exactly on the occlusal surface of each coping without cementation (Figure 6a,b).

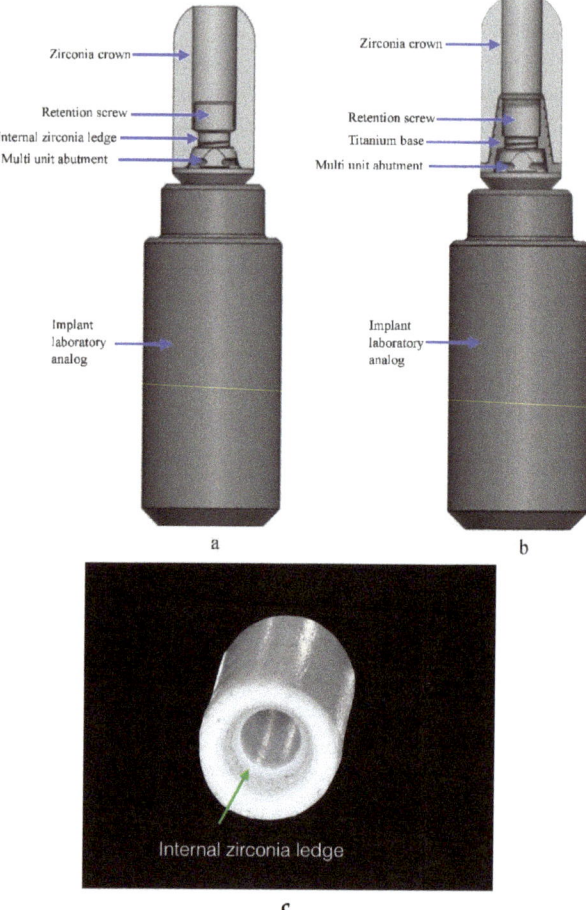

Figure 5. Components of the zirconia-abutment restorations. (**a**) Screwed restoration; (**b**) screwed-cemented restoration; (**c**) apical view of the screwed zirconia cap. Note the internal zirconia ledge for the support of the screw head.

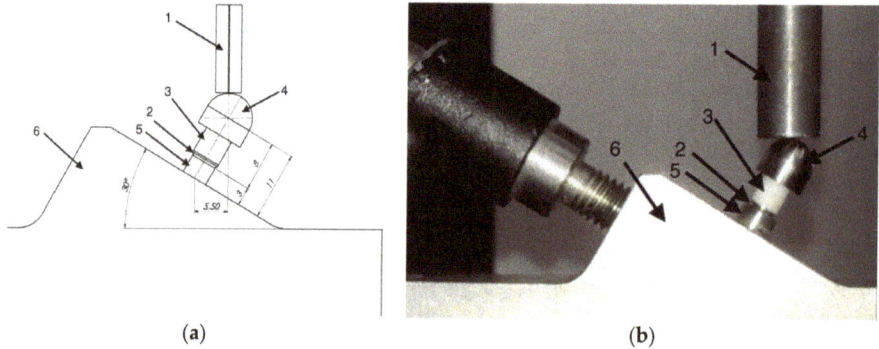

Figure 6. Test setup following ISO 14801:2016. (**a**) Schematic illustration of the test design: (1) loading device; (2) nominal bone level; (3) zirconia coping; (4) hemispherical loading member; (5) dental implant analog; and (6) metal specimen holder; (**b**) setup for the study.

To determine the static maximum load that the assembly could resist, which corresponds to one cycle on the S-N curve (plotted as force versus log number of cycles), three samples of each group were subjected to an increased static compressive load at a crosshead speed of 1 mm/min until catastrophic failure using a uniaxial universal testing machine (Instron, ElectroPuls E3000, Instron Corp., Buckinghamshire, high Wycombe, UK). The average failure rate from this test was calculated and defined as the maximal failure load.

For the dynamic loading, the remaining specimens were tested in progressively lower loads, starting at 80% of the maximal failure load, using an MTS Systems Corporation machine (Instron, ElectroPuls E3000, Instron Corp., Buckinghamshire, high Wycombe, UK) (Tables 2 and 3). The compressive fatigue limits were determined by testing according to the staircase or up-and-down method [37] in a 14 Hz frequency cycle. In this method, the tests were conducted sequentially, with the maximum applied stress in each successive test increasing or decreasing by a fixed amount according to whether the previous stress resulted in failure, until a load was reached at which three specimens survived 5,000,000 cycles, which is equivalent to five years of in vivo mastication [38]. The failure mode of all the specimens was evaluated by stereomicroscopy at 20× magnification (StarLite 150, OGP, Rochester, NY, USA).

Due to the small number of samples, no statistical analysis could be performed with sufficient statistical power.

5. Conclusions

Within the limitations of this pilot study, zirconia restoration screwed directly to an MUA without a T-base is shown to be a viable procedure, and the survival odds of this restoration are not inferior to those of the screwed-cemented restoration. Failure will probably occur within the metal (MUA/screw) before zirconia failure. Further studies with larger sample sizes are warranted.

Author Contributions: Conceptualization, H.H. and S.L.; Formal analysis, I.B., R.P. and S.L.; Investigation, H.H. and A.A.; Methodology, R.P. and S.L.; Writing—original draft, H.H.; Writing—review & editing, S.L.

Funding: Alpha-Bio Tec., Petach Tikva, Israel.

Acknowledgments: Special thanks to H Gryner and O Havkin for their help during the experiments and with the creation of the figures. Y Peer from the Peer Dental Laboratory helped with the CAD-CAM zirconia preparation. The authors thank Alpha-Bio Tec., Petach Tikva, Israel, for providing the materials and financial support for this study.

Conflicts of Interest: The authors declare that they have no conflict of interest.

References

1. Guncu, M.B.; Cakan, U.; Muhtarogullari, M.; Canay, S. Zirconia-based crowns up to 5 years in function: A retrospective clinical study and evaluation of prosthetic restorations and failures. *Int. J. Prosthodont.* **2015**, *28*, 152–157. [CrossRef] [PubMed]
2. Beuer, F.; Stimmelmayr, M.; Gueth, J.F.; Edelhoff, D.; Naumann, M. In vitro performance of full-contour zirconia single crowns. *Dent. Mater.* **2012**, *28*, 449–456. [CrossRef] [PubMed]
3. Guess, P.C.; Att, W.; Strub, J.R. Zirconia in fixed implant prosthodontics. *Clin. Implant Dent. Relat. Res.* **2012**, *14*, 633–645. [CrossRef] [PubMed]
4. Michalakis, K.X.; Hirayama, H.; Garefis, P.D. Cement-retained versus screw-retained implant restorations: A critical review. *Int. J. Oral Maxillofac. Implant.* **2003**, *18*, 719–728.
5. Wilson, T.G., Jr. The positive relationship between excess cement and peri-implant disease: A prospective clinical endoscopic study. *J. Periodontol.* **2009**, *80*, 1388–1392. [CrossRef] [PubMed]
6. Sailer, I.; Muhlemann, S.; Zwahlen, M.; Hammerle, C.H.; Schneider, D. Cemented and screw-retained implant reconstructions: A systematic review of the survival and complication rates. *Clin Oral Implants Res* **2012**, *23*, 163–201. [CrossRef] [PubMed]

7. Linkevicius, T.; Puisys, A.; Vindasiute, E.; Linkeviciene, L.; Apse, P. Does residual cement around implant-supported restorations cause peri-implant disease? A retrospective case analysis. *Clin. Oral Implant. Res.* **2013**, *24*, 1179–1184. [CrossRef]
8. Wittneben, J.G.; Millen, C.; Bragger, U. Clinical performance of screw-versus cement-retained fixed implant-supported reconstructions-A systematic review. *Int. J. Oral Maxillofac. Implant.* **2014**, *29* (Suppl. 2014), 84–98. [CrossRef] [PubMed]
9. Lewis, S. An esthetic titanium abutment: Report of a technique. *Int. J. Oral Maxillofac. Implant.* **1991**, *6*, 195–201.
10. Canullo, L.; Bignozzi, I.; Cocchetto, R.; Cristalli, M.P.; Iannello, G. Immediate positioning of a definitive abutment versus repeated abutment replacements in post-extractive implants: 3-Year follow-up of a randomised multicentre clinical trial. *Eur. J. Oral Implant.* **2010**, *3*, 285–296.
11. Degidi, M.; Nardi, D.; Piattelli, A. One abutment at one time: Non-removal of an immediate abutment and its effect on bone healing around subcrestal tapered implants. *Clin. Oral Implant. Res.* **2011**, *22*, 1303–1307. [CrossRef] [PubMed]
12. Grandi, T.; Guazzi, P.; Samarani, R.; Garuti, G. Immediate positioning of definitive abutments versus repeated abutment replacements in immediately loaded implants: Effects on bone healing at the 1-year follow-up of a multicentre randomised controlled trial. *Eur. J. Oral Implant.* **2012**, *5*, 9–16.
13. McGlumphy, E.A.; Papazoglou, E.; Riley, R.L. The combination implant crown: A cement- and screw-retained restoration. *Compendium* **1992**, *13*, 34–41. [PubMed]
14. Rajan, M.; Gunaseelan, R. Fabrication of a cement- and screw-retained implant prosthesis. *J. Prosthet. Dent.* **2004**, *92*, 578–580. [CrossRef] [PubMed]
15. Uludag, B.; Celik, G. Fabrication of a cement- and screw-retained multiunit implant restoration. *J. Oral Implant.* **2006**, *32*, 248–250. [CrossRef] [PubMed]
16. Hussien, A.N.; Rayyan, M.M.; Sayed, N.M.; Segaan, L.G.; Goodacre, C.J.; Kattadiyil, M.T. Effect of screw-access channels on the fracture resistance of 3 types of ceramic implant-supported crowns. *J. Prosthet. Dent.* **2016**, *116*, 214–220. [CrossRef] [PubMed]
17. Silva, N.R.; Teixeira, H.S.; Silveira, L.M.; Bonfante, E.A.; Coelho, P.G.; Thompson, V.P. Reliability and failure modes of a hybrid ceramic abutment prototype. *J. Prosthodont.* **2018**, *27*, 83–87. [CrossRef]
18. Linkevicius, T.; Caplikas, A.; Dumbryte, I.; Linkeviciene, L.; Svediene, O. Retention of zirconia copings over smooth and airborne-particle-abraded titanium bases with different resin cements. *J. Prosthet. Dent.* **2019**, *121*, 949–954. [CrossRef] [PubMed]
19. Röhrle, O.; Saini, H.; Ackland, D.C. Occlusal loading during biting from an experimental and simulation point of view. *Dent. Mater.* **2018**, *34*, 58–68. [CrossRef]
20. Umesh, S.; Padma, S.; Asokan, S.; Srinivas, T. Fiber Bragg Grating based bite force measurement. *J. Biomech.* **2016**, *49*, 2877–2881. [CrossRef]
21. Fernandes, C.P.; Glantz, P.O.; Svensson, S.A.; Bergmark, A. A novel sensor for bite force determinations. *Dent Mater.* **2003**, *19*, 118–126. [CrossRef]
22. Kwon, S.J.; Lawson, N.C.; McLaren, E.E.; Nejat, A.H.; Burgess, J.O. Comparison of the mechanical properties of translucent zirconia and lithium disilicate. *J. Prosthet. Dent.* **2018**, *120*, 132–137. [CrossRef] [PubMed]
23. ISO. *International Organization for Standardization Endosseous Dental Implants: Dentistry Implants-Dynamic Fatigue Test, ISO 14801*; ISO: Geneva, Switzerland, 2016.
24. Koyama, T.; Sato, T.; Yoshinari, M. Cyclic fatigue resistance of yttria-stabilized tetragonal zirconia polycrystals with hot isostatic press processing. *Dent. Mater. J.* **2012**, *31*, 1103–1110. [CrossRef] [PubMed]
25. Ding, Q.; Zhang, L.; Bao, R.; Zheng, G.; Sun, Y.; Xie, Q. Effects of different surface treatments on the cyclic fatigue strength of one-piece CAD/CAM zirconia implants. *J. Mech. Behav. Biomed. Mater.* **2018**, *84*, 249–257. [CrossRef] [PubMed]
26. Duan, Y.; Gonzalez, J.A.; Kulkarni, P.A.; Nagy, W.W.; Griggs, J.A. Fatigue lifetime prediction of a reduced-diameter dental implant system: Numerical and experimental study. *Dent. Mater.* **2018**, *34*, 1299–1309. [CrossRef] [PubMed]
27. Shemtov-Yona, K.; Rittel, D. Random spectrum loading of dental implants: An alternative approach to functional performance assessment. *J. Mech. Behav. Biomed. Mater.* **2016**, *62*, 1–9. [CrossRef] [PubMed]

28. Naumova, E.A.; Roth, F.; Geis, B.; Baulig, C.; Arnold, W.H.; Piwowarczyk, A. Influence of luting materials on the retention of cemented implant-supported crowns: An in vitro study. *Materials (Basel)* **2018**, *11*, 1853. [CrossRef]
29. Gungor, M.B.; Nemli, S.K. The effect of resin cement type and thermomechanical aging on the retentive strength of custom zirconia abutments bonded to titanium inserts. *Int. J. Oral Maxillofac. Implant.* **2018**, *33*, 523–529. [CrossRef]
30. Zenthofer, A.; Rues, S.; Krisam, J.; Rustemeyer, R.; Rammelsberg, P.; Schmitter, M. Debonding forces for two-piece zirconia abutments with implant platforms of different diameter and use of different luting strategies. *Int. J. Oral Maxillofac. Implant.* **2018**, *33*, 1041–1046. [CrossRef]
31. Chun, H.J.; Yeo, I.S.; Lee, J.H.; Kim, S.K.; Heo, S.J.; Koak, J.Y.; Han, J.S.; Lee, S.J. Fracture strength study of internally connected zirconia abutments reinforced with titanium inserts. *Int. J. Oral Maxillofac. Implant.* **2015**, *30*, 346–350. [CrossRef]
32. Kelly, J.R.; Rungruanganunt, P. Fatigue behavior of computer-aided design/computer-assisted manufacture ceramic abutments as a function of design and ceramics processing. *Int. J. Oral Maxillofac. Implant.* **2016**, *31*, 601–609. [CrossRef] [PubMed]
33. Yilmaz, B.; Salaita, L.G.; Seidt, J.D.; McGlumphy, E.A.; Clelland, N.L. Load to failure of different zirconia abutments for an internal hexagon implant. *J. Prosthet. Dent.* **2015**, *114*, 373–377. [CrossRef] [PubMed]
34. Gehrke, P.; Alius, J.; Fischer, C.; Erdelt, K.J.; Beuer, F. Retentive strength of two-piece CAD/CAM zirconia implant abutments. *Clin. Implant Dent. Relat. Res.* **2014**, *16*, 920–925. [CrossRef] [PubMed]
35. Pilo, R.; Agar-Zoizner, S.; Gelbard, S.; Levartovsky, S. The retentive strength of laser-sintered cobalt-chromium-based crowns after pretreatment with a desensitizing paste containing 8% arginine and calcium carbonate. *Int. J. Mol. Sci.* **2018**, *19*, 4082. [CrossRef] [PubMed]
36. Pilo, R.; Folkman, M.; Arieli, A.; Levartovsky, S. Marginal fit and retention strength of zirconia crowns cemented by self-adhesive resin cements. *Oper. Dent.* **2018**, *43*, 151–161. [CrossRef]
37. Garoushi, S.; Lassila, L.V.; Tezvergil, A.; Vallittu, P.K. Static and fatigue compression test for particulate filler composite resin with fiber-reinforced composite substructure. *Dent. Mater.* **2007**, *23*, 17–23. [CrossRef]
38. Ongthiemsak, C.; Mekayarajjananonth, T.; Winkler, S.; Boberick, K.G. The effect of compressive cyclic loading on retention of a temporary cement used with implants. *J. Oral Implant.* **2005**, *31*, 115–120. [CrossRef]

© 2019 by the authors. Licensee MDPI, Basel, Switzerland. This article is an open access article distributed under the terms and conditions of the Creative Commons Attribution (CC BY) license (http://creativecommons.org/licenses/by/4.0/).

Article

Effects of Liner-Bonding of Implant-Supported Glass–Ceramic Crown to Zirconia Abutment on Bond Strength and Fracture Resistance

Yong-Seok Jang [1], Sang-Hoon Oh [2], Won-Suck Oh [3], Min-Ho Lee [1], Jung-Jin Lee [4] and Tae-Sung Bae [1],*

1. Department of Dental Biomaterials, Institute of Biodegradable Materials, BK21 plus Program, School of Dentistry, Chonbuk National University, Jeonju 54896, Korea
2. Haruan Dental Clinic, Department of Dental Biomaterials, Institute of Biodegradable Materials, BK21 plus Program, School of Dentistry, Chonbuk National University, Jeonju 54896, Korea
3. Department of Biologic and Materials Sciences Division of Prosthodontics, School of Dentistry, University of Michigan, Ann Arbor, MI 48109, USA
4. Department of Prosthodontics, Institute of Oral Bio-Science, School of Dentistry, Chonbuk National University and Research Institute of Clinical Medicine of Chonbuk National University-Biomedical Research Institute of Chonbuk National University Hospital, Jeonju 54907, Korea
* Correspondence: bts@jbnu.ac.kr; Tel.: +82-63-270-4041

Received: 12 August 2019; Accepted: 28 August 2019; Published: 30 August 2019

Abstract: This study was conducted to test the hypothesis that heat-bonding with a liner positively affects the bond strength and fracture resistance of an implant-supported glass–ceramic crown bonded to a zirconia abutment produced by a computer-aided design/computer-aided milling (CAD/CAM) procedure. Lithium disilicate-reinforced Amber Mill-Q glass ceramic blocks were bonded to 3 mol% yttria stabilized tetragonal zirconia polycrystal (3Y-TZP) blocks by heat-bonding with a liner or cementation with a dual-cure self-adhesive resin cement for a microtensile bond strength test. CAD/CAM implant-supported glass ceramic crowns were produced using Amber Mill-Q blocks and bonded to a milled 3Y-TZP zirconia abutments by heat-bonding or cementation for a fracture test. A statistical analysis was conducted to investigate the significant differences between the experimental results. The mode of failure was analyzed using high-resolution field emission scanning electron microscopy. Chemical bonding was identified at the interface between the zirconia ceramic and liner. The mean tensile bond strength of the liner-bonded group was significantly higher than that of the cement-bonded group. The initial chipping strength of the liner-bonded group was significantly higher than that of the cement-bonded group, although no statistically significant difference was found for the fracture strength. The mode of failure was mixed with cohesive fracture through the liner, whereas the cement-bonded group demonstrated adhesive failure at the interface of bonding.

Keywords: CAD/CAM all-ceramic restoration; fracture strength; liner treatment; resin cement; tensile bond strength; zirconia abutment

1. Introduction

Implant-supported crowns restore oral functions and aesthetics without affecting the integrity of adjacent teeth. These crowns are usually connected to the implants by means of titanium abutments. However, the use of a metallic abutment may compromise the gingival aesthetics of implant crowns, particularly in patients presenting with a thin biotype gingival architecture. The grayish metallic color can be pronounced with a light reflected from the metallic surface of the abutment [1,2]. In addition, the submucosal placement of a crown margin can restrict the cement removal procedure and lead to develop peri-implantitis [3].

Ceramic abutments are biocompatible and aesthetic by mimicking the color of natural teeth [4,5]. In addition, the use of ceramic abutments is versatile with the development of CAD (computer-aided design)/CAM (computer-aided milling) technology, which allows for the design and milling of zirconia ceramics. Though concerns regarding the mechanical properties of zirconia abutments (i.e., brittleness, stress corrosion cracking, and low temperature degradation) have been addressed, their clinical application has been expanded for the anterior and premolar regions [6–9]. Furthermore, all abutments made of zirconia or titanium can be customized to produce a desired size, shape, angle, and location of a crown margin for better aesthetics [10–12].

Zirconia abutments are commonly designed as one-piece system to support implant crowns and establish the implant–abutment interface with a zirconia ceramic. This system eliminates the negative effects of titanium abutments related to aesthetics. However, the implant–abutment interface may experience an excessive wear under occlusal loading due to dissimilarities between the zirconia ceramic and the titanium implant [13,14]. Complications may include a discoloration of the peri-implant mucosa associated with the embedment of titanium particles dislodged from implants, abutment screw loosening, and the fracture of ceramic abutments and crowns [14,15]. This type of abutments have also shown very low fracture resistance when compared to zirconia abutments using a titanium base [16,17].

The two-piece system of zirconia abutments was designed to avoid the possible consequences related to the use of the one-piece system. This system consists of a ceramic core and a titanium link to establish the implant–abutment interface with titanium-to-titanium without compromising the aesthetics of the ceramic abutment. The ceramic core is customized using CAD/CAM technology and is bonded to the titanium base with adhesive resin cement [18]. This two-piece system was found to demonstrate a higher flexural resistance than one-piece system [1,19] without compromising the emergence profile, crown orientation, and coronal contour matching the prosthesis and anatomical shape of the mucosa [20].

However, problems may occur when bonding a final prosthesis to a zirconia ceramic abutment [18,21,22]. The resin cement may not adhere to the zirconia ceramic abutment because of the lack of undercut features commonly created by blasting with air-borne particle abrasion technology and/or etching with hydrofluoric acid. The final prosthesis can be at a risk of chipping or cracking due to the relatively weak bond strength of the resin cementing the abutment. In addition, the luting procedure of the final prosthesis to the abutment is inconvenient and can be problematic when combined with the submucosal margin.

A lithium disilicate-reinforced liner has recently been developed to overcome the weak linkage of the final prosthesis of a glass–ceramic to a zirconia abutment designed to replace single anterior and posterior teeth. However, no scientific research has been conducted to elucidate the bond characteristics of the liner. The objective of this in vitro study was to investigate the effect of heat-bonding with the liner on the bond strength and fracture resistance of an implant-supported glass–ceramic crown bonded to a two-piece system zirconia abutment produced by a CAD/CAM procedure. The null hypothesis was set to test no significant difference between the liner-bonding and the cement-bonding of the implant-supported ceramic crown to the zirconia abutment on the initial chipping and fracture strengths of crowns, as well as the microtensile bond strength.

2. Materials and Methods

2.1. Bond Strength Test

2.1.1. Preparation of Specimen

Three CAD/CAM 3Y-TZP zirconia ceramic blocks (Zirtooth, O98FGJ1701, Hass, Gangneung, Korea) were machined to fabricate 6 zirconia ceramic specimens with dimensions of 10 × 10 × 5 mm, and they were sintered in an electric furnace (Programat EP3000/G2, Ivoclar Vivadent, Schaan, Liechtenstein) at 1450 °C. In addition, lithium disilicate-reinforced Amber Mill-Q glass ceramic blocks

(Hass, Gangneung, Korea) were machined to fabricate 6 glass ceramic specimens matching the zirconia ceramic specimens (10 mm × 10 mm × 5 mm).

For bonding with the liner, the liner (Hass, Gangneung, Korea), consisting of 5–15 wt.% Li_2O, 55–65 wt.% SiO_2, and 5–25 wt.% other trace elements of oxides and colorants was applied on the surfaces of 3 zirconia ceramic specimens. Then, heat-bonding with 3 lithium disilicate-reinforced glass ceramic specimens was conducted at 800 °C.

For bonding with a resin cement, 3 lithium disilicate-reinforced glass ceramic specimens were acid-etched with 9.5% hydrofluoric acid gel (Bisco, IL, USA) for 30 s, washed with distilled water, dried, and silane primer (Espe™ Sil, 3M/ESPE, Seefeld, Germany) coated. The surfaces of 3 zirconia ceramic specimens were roughened using an air-borne particle abrasion technology with 50 μm alumina particles (Hi-aluminas, Shofu, Japan) under the pressure of 3 atm at a distance of 10 mm [23]. After placing the specimens in an electric furnace (Programat EP3000/G2), the temperature was raised to 1000 °C at a rate of 50 °C/min, held for 10 min, and then cooled to 25 °C in the furnace to restore the phase transformations occurred during the blasting procedure. Then, a 10-methacryloyloxydecyl dihydrogen phosphate (MDP)-containing primer (Z-PRIME™ plus, BISCO, Schaumburg, IL, USA) was coated on the surface of zirconia ceramic specimens [24]. The cement bonding between the lithium disilicate-reinforced glass and zirconia ceramics was performed using a self-adhesive dual-cure resin cement (Rely-X™ U200, 3M/ESPE, Neuss, Germany) with an equal amount of base and catalyst pastes mixed for 20 s. The cement was applied to the prepared ceramic surface and a pressure of 49 N was applied to the ceramic specimens under a constant-load device. After the removal of excess cement, the resin cement was photopolymerized using an light emitting diode (LED) curing unit (G-Light, GC Corporation, Tokyo, Japan) for 20 s from each of the four directions for a total of 80 sec under a light intensity of 550 mW/cm^2 at a distance of 2 mm. The specimens were kept under pressure for additional 10 min, relieved from the constant-load device and immersed in distilled water for 24 h.

The bonded ceramic blocks were serially sectioned perpendicular to the bonded surface. The first cut was made through each specimen using a high-speed diamond cutting machine (Accutom-50, Struers Inc, Cleveland, OH, USA) to produce a 1 mm thick plate, and the sliced plate was cut using a low-speed diamond cutting machine (Metsaw-LS, Topmet, Daejeon, Korea) after rotating 90° for second set of cuts. Twelve specimens (1.0 mm × 1.0 mm × 10 mm) were prepared with 3 ceramic blocks per each group by selecting 4 specimens from the middle of a bonded ceramic block (10 mm × 10 mm × 10 mm). Thus, a total of 24 specimens (12 × 2 groups) were used for the bond strength test. Figure 1 schematically shows the preparation process of specimens for the microtensile bond test.

2.1.2. Microtensile Bond Strength Test

The prepared ceramic specimens were attached to the grip of metal holders, mounted, and subjected to tensile force in a universal testing machine (Instron, Model 5569, Instron Co., Norwood, MA, USA) at a crosshead speed of 0.5 mm/min [25,26]. The microtensile bond strength was calculated in MPa with the failure load (N) divided by the cross-sectional area (mm^2) of each test specimen.

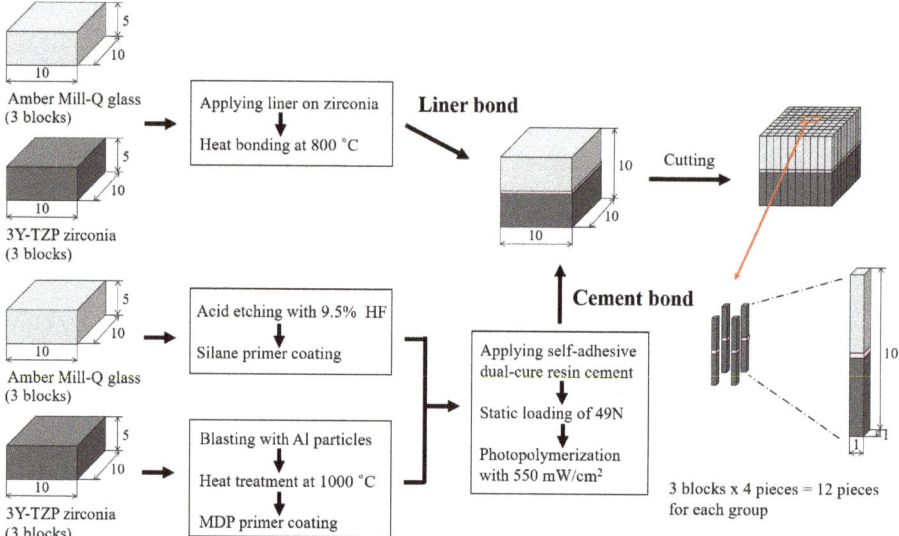

Figure 1. Schematic summary for the preparation process of specimens for the microtensile bond test.

2.2. Fracture Test for Implant-Supported Crowns

2.2.1. Preparation of Implant-Supported Crowns

The mandibular right first molar was scanned using a D900L scanner (3Shape, Copenhagen, Denmark) to create an implant-supported crown. Twenty ceramic crowns were produced by milling a lithium disilicate-reinforced glass ceramic block using CAD/CAM milling machine (CEREC MC, Sirona, Salzburg, Austria) to make 10 samples per group for the liner-bonded and cement-bonded groups. The internal aspect of each crown was acid-etched for bonding to a zirconia abutment presenting with a diameter of 4 mm and length of 8 mm. The bonding was performed using the same method as described in the section of the bond strength test for each liner-bonded and cement-bonded group. The zirconia abutments were connected to the titanium link abutments (SRN-SURO-H, GeoMedi Co, Ltd., Uiwang, Gyeonggi, Korea) with a dual-cure resin cement after applying an MDP containing primer.

The prepared crowns were coated with a thin layer of e.max Ceram glaze paste (Ivoclar/Vivadent, Schaan, Liechtenstein) and glazed in an electric furnace (Programat EP3000/G2). The temperature was raised to 730 °C at a rate of 35 °C/min and held for 1 min. For the liner-bonded group (MLG: Milled, liner-bonded and glazed), the ceramic crowns were glazed following the bonding with the zirconia abutments. However, for the cement-bonded group (MGC: Milled, glazed, and cement-bonded), the ceramic crowns were glazed prior to bonding to the abutments.

2.2.2. Fracture Test of Implant-Supported Crown

The prepared ceramic crown/abutment specimens were connected to titanium implant analogs (Geo 3D Analog, GeoMedi Co, Ltd., Uiwang, Gyeonggi, Korea) with titanium abutment screws and tightened to 30 N·cm twice at 30 seconds interval using a torque gauge (9810P, Aikoh Engineering Co, Higashiosaka, Osaka, Japan) [27]. The screw access holes were filled with a light-activated composite resin (Filteck™ Z350 XT, 3M ESPE, MN, USA), and the prepared specimens were immersed in distilled water at 37 °C for 24 h.

The crowns were supported and secured to a base retainer and loaded axially to fracture in a universal testing machine (Universal testing machine 4201, Instron, Canton, MA, USA). The compression load was transferred through a 5 mm diameter steel ball positioned in the central fossa of the crown.

The load was applied at a crosshead speed of 0.5 mm/min and continued to the point of fracture, and the fracture load of each ceramic crown was recorded [28,29]. The initial chipping strength was determined at the point when the load value showed slightly transient drop.

2.3. Surface Analysis

High-resolution field emission scanning electron microscopy (HR FE-SEM in KBSI Jeonju, SU8230, Hitachi, Japan) was used to investigate the topography of the liner-bonded surface, cross-sectional bonded interfaces, and fractured surfaces after etching with a 9.5% hydrofluoric acid gel for 30 s. The liner layer was removed with a 5% hydrofluoric acid solution for 30 min to identify the effects of liner treatment on zirconia ceramic. The distribution of the chemical elements at the bonded interface was analyzed using an energy dispersive X-ray spectrometer (EDS, Bruker, Germany), and the crystal structure on the liner-bonded surface was investigated by X-ray diffraction (XRD, Dmax III-A type, Rigaku, Japan).

2.4. Statistical Analysis

The statistical analysis was conducted by SPSS software (version 12.0, SPSS, Chicago, IL, USA). A Student's t-test was conducted to investigate significant differences between the experimental results for the 2 test groups ($p < 0.05$). A Weibull analysis was performed with the experimentally measured load-at-failure values for each group of bonded ceramic specimens.

3. Results

The null hypothesis on the microtensile bond strength was rejected. Table 1 summarizes the distribution characteristics of the Weibull analysis for the microtensile bond strength test. The Weibull modulus (m) was higher for the liner-bonded group (5.8) than for the cement-bonded group (4.1). The mean values of microtensile bond strength for the liner-bonded group (47.7 MPa) were significantly higher than for the cement-bonded group (19.6 MPa) ($p < 0.05$). In addition, the Weibull distribution showed a matched tendency with a single mode ($r^2 > 0.958$) (Figure 2).

Table 1. Weibull analysis data of a microtensile bond strength test for lithium disilicate-reinforced glass ceramic and zirconia of the liner-bonded group and the cement-bonded group.

Group Parameter	Liner-Bonded	Resin Cement-Bonded
m	5.836	4.133
σ_o	50.8	2.36
r^2	0.961	0.958
$\sigma_{f(mean)} \pm SD$	47.7 ± 8.7	19.6 ± 4.7
$\sigma_{f(min/med/max)}$	32.8/45.5/58.7	10.9/20.6/26.0
N	12	12

where m = Weibull modulus; σ_o = characteristic strength in MPa; r^2 = Weibull distribution regression coefficient squared; $\sigma_{f(mean)}$ = mean fracture strength in MPa; $\sigma_{f(min/med/max)}$ = minimum, median and maximum fracture strength in MPa; and N=number of samples.

The heat-bonding with the liner was found to induce a chemical interaction with the milled zirconia ceramic abutment. A large number of pores was found in the reaction layer of liner bonding, altering the surface topography of the zirconia ceramic (Figure 3a,b). The zone of the chemical reaction was measured approximately 3 μm across the interface of the bonding of the zirconia ceramic (Figure 3d,e).

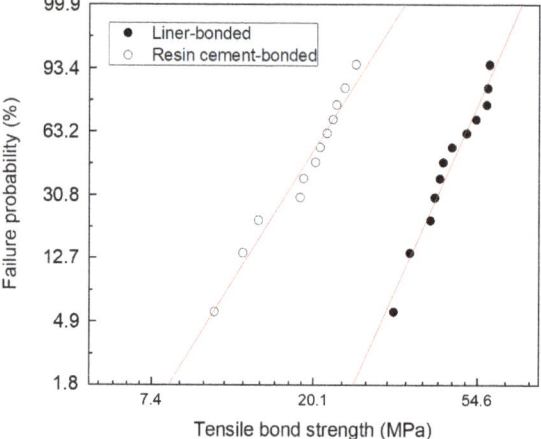

Figure 2. Weibull plots of a microtensile bond strength test for lithium disilicate-reinforced glass ceramic and zirconia of the liner-bonded group and the cement-bonded group.

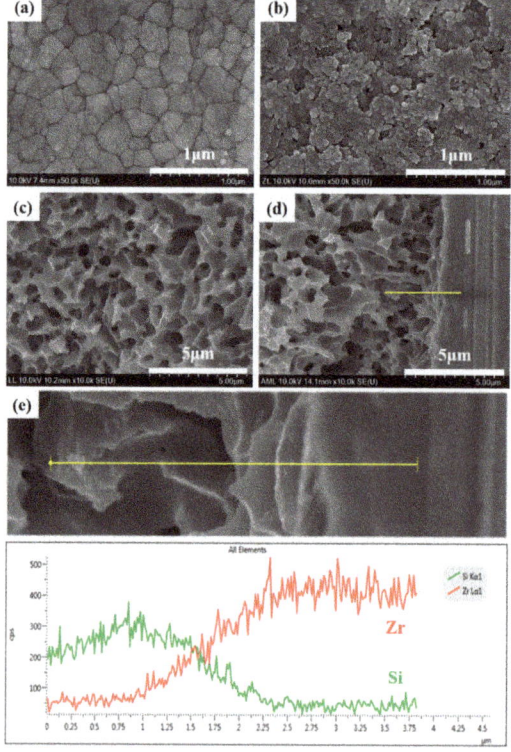

Figure 3. High-resolution field emission scanning electron microscopy (HR FE-SEM) images of the zirconia ceramic specimen demonstrating (**a**) crystal structure after sintering at 1450 °C, (**b**) its alteration of surface morphology with liner treatment (liner removed with acid etching with a 5% hydrofluoric acid solution for 30 min after a fracture test), (**c**) liner-bonded zirconia surface, and (**d**) cross-sectional view of liner-bonded interfacial interaction zone, and (**e**) energy dispersive X-ray spectrometer (EDS) line analysis data of Si and Zr after acid etching with 9.5% hydrofluoric acid gel for 30 s.

The chemical reaction of the liner against the zirconia abutment was confirmed by XRD diffraction analysis, where the peaks were noted as corresponding to lithium metasilicate (Li_2SiO_3) and zirconium silicate ($ZrSiO_4$), as well as zirconia (ZrO_2) and silica (SiO_2) (Figure 4).

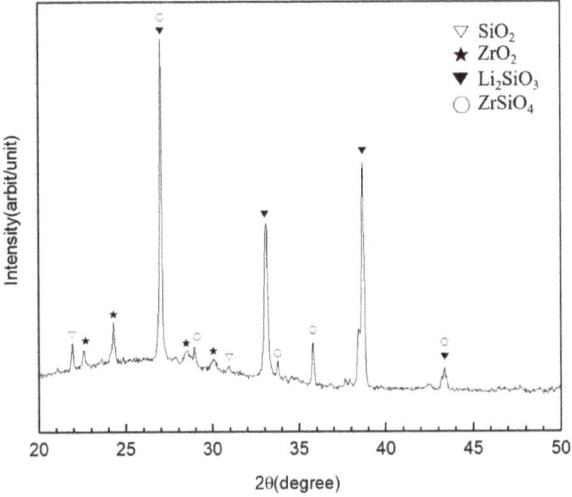

Figure 4. XRD diffraction analysis of the lithium disilicate-reinforced liner-bonded interface of the zirconia ceramic.

The interfacial layer created by the liner-bonding was found to be more consistent than that created by the cement-bonding. The liner-bonded interfacial layer was thicker with a higher mean value and standard deviation (33.6 ± 5.2 µm) than the cement-bonded interfacial layer (13.3 ± 1.6 µm) (Figure 5). Neither a pore nor a gap was noted in the liner-bonded interfacial zone, whereas numerous micro-pores and micor-gaps were found in the cement-bonded interfacial layer.

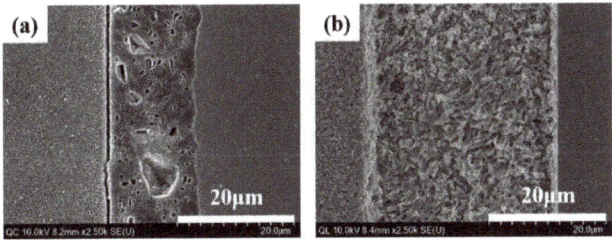

Figure 5. HR FE-SEM images of bonded interface between lithium disilicate-reinforced glass ceramic and zirconia for (**a**) the cement-bonded group and (**b**) the liner-bonded group.

When the debonded surfaces of the liner-bonded and liner-bonded groups were visually evaluated with high magnification images of HR FE-SEM, the mode of failure of the liner-bonded group was mixed with cohesive fracture propagated through the liner (Figure 6). In contrast, the mode of failure of the cement-bonded group was adhesive where the fracture occurred at the interface between the zirconia ceramic and the cement layer (Figure 7).

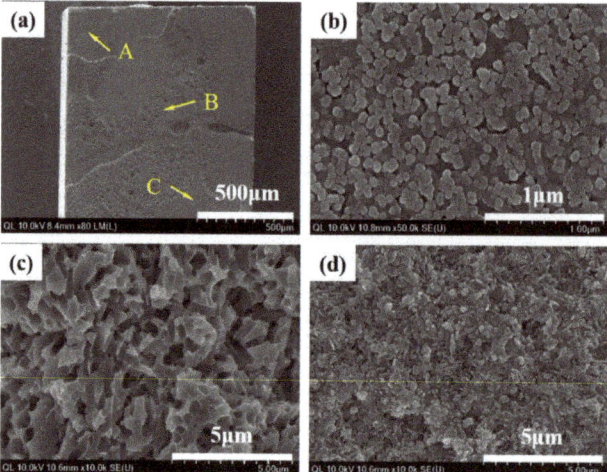

Figure 6. HR FE-SEM images of the liner-bonded group demonstrating mode of failure with the microtensile bond strength test. (**a**) Fracture surface of specimen demonstrating inhomogeneous pattern of fracture, (**b**) magnification of point A demonstrating that the zirconia surface layer reacted with the liner, (**c**) the magnification of point B demonstrating the microstructure of the liner, and (**d**) the magnification of point C demonstrating the presence of needle-shaped lithium disilicate crystals in the lithium disilicate-reinforced glass ceramic.

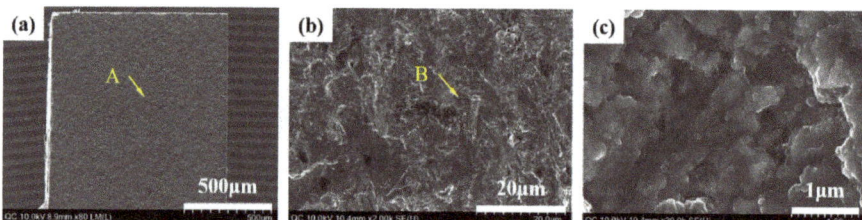

Figure 7. HR FE-SEM images of the cement-bonded group demonstrating mode of failure with the microtensile bond strength test. (**a**) Fractured surface of specimen demonstrating the homogeneous pattern of fracture, (**b**) the magnification of point A demonstrating the irregular structure of the zirconia ceramic, and (**c**) the magnification of point B demonstrating the crystal structure of zirconia.

The null hypothesis was rejected on the initial chipping and fracture strengths but accepted on the fracture strength. The mean values and standard deviations of the initial chipping strength and fracture strength of implant-supported glass ceramic crowns bonded to zirconia ceramic abutments were 843.8 ± 317.5 N and 1929.6 ± 191.1 N for the liner-bonded (MLG) groups and 341.0 ± 90.2 N and 1711.1 ± 275.4 N for the resin cement-bonded (MGC) group (Figure 8). The initial chipping strength of the MLG group was significantly higher than that of the MGC group ($p < 0.05$), although no significant difference was found in the fracture strength.

When the fractured surfaces of the lithium disilicate-reinforced glass ceramic crown and the zirconia ceramic abutment were visually evaluated with a high magnification images of HR FE-SEM, the mode of failure of the MLG group was mixed with adhesive and cohesive fractures propagated through the liner (Figures 9 and 10b–d). However, the mode of failure was adhesive in the MGC group, and the fractures consistently occurred at the interface between the cement layer and the zirconia ceramic abutment (Figures 9 and 10e,f).

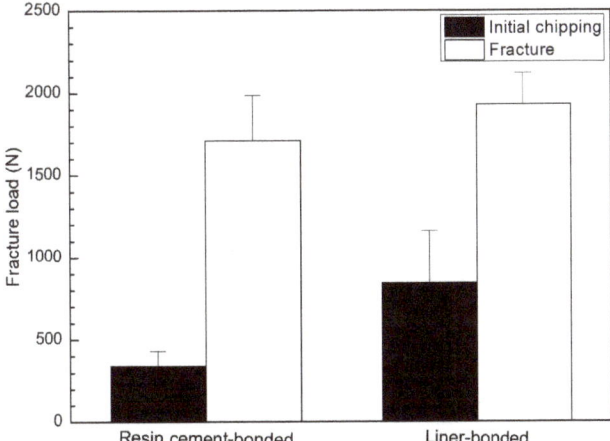

Figure 8. Graphical illustration for initial chipping and fracture strengths of crowns for the resin cement-bonded and liner-bonded groups.

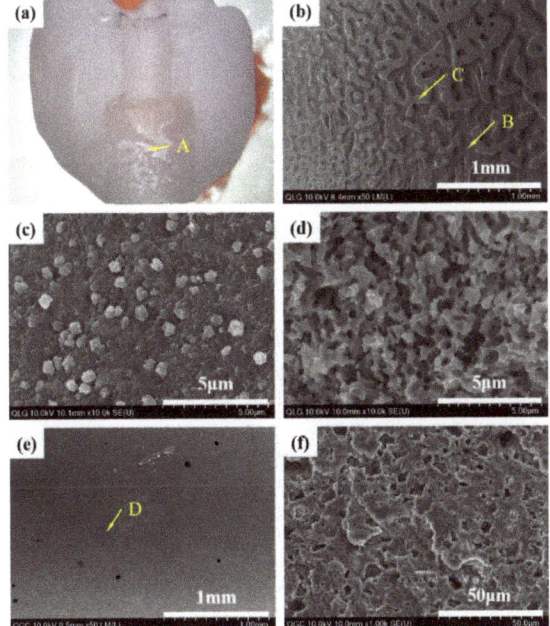

Figure 9. Photographic (**a**) and HR FE-SEM images of the lithium disilicate-reinforced glass ceramic crown that demonstrate the mode of failure; (**b**) the magnification of point A for the (milled, liner-bonded and glazed) MLG group: Inhomogeneous pattern of fracture surface; (**c**) the magnification of point B: Cohesive fracture occurred through the liner; (**d**) the magnification of point C: Adhesive fracture occurred at the interface with liner, and (**e**) magnification of point A for the milled, glazed, and cement-bonded (MGC) group: Homogeneous pattern of fracture surface; (**f**) the magnification of point D: Cement layer indicating adhesive fracture occurred at the interface with the zirconia ceramic abutment.

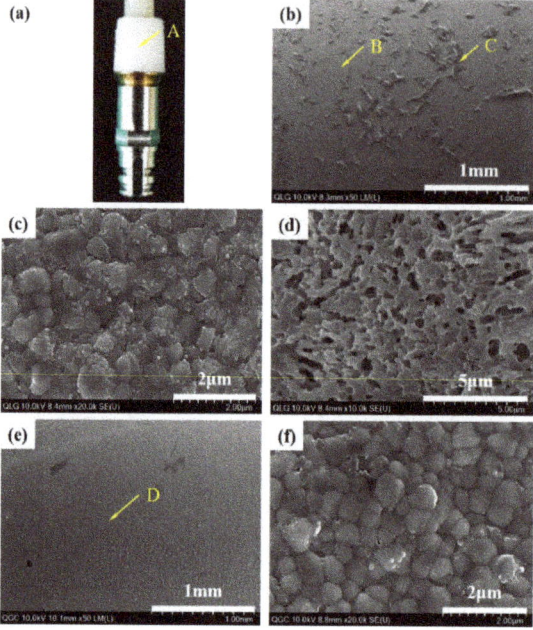

Figure 10. Photographic (**a**) and HR FE-SEM images (**b–d**) of the zirconia ceramic abutment; (**b**) the magnification of point A for the MLG group: Inhomogeneous pattern of fracture surface; (**c**) the magnification of point B: Adhesive fracture occurred at the interface with liner; (**d**) the magnification of point C: Cohesive fracture occurred through the liner; (**e**) the magnification of point A for the MGC group: Homogeneous pattern of fracture surface; (**f**) the magnification of point D: Zirconia crystals indicating that an adhesive fracture occurred at the interface with resin cement.

4. Discussion

The liner-bonded ceramic crowns were superior to the cement-bonded ceramic crowns in developing a higher resistance to fracture. The bond strength of the lithium disilicate-reinforced glass ceramic to the zirconia ceramic was higher when the bonding was established with the lithium-disilicate liner than with an adhesive cement. The mean values of microtensile bond strength for liner bond and cement bond were 19.6 and 47.7 MPa, respectively, in this study. As an ideal biomaterial would have bonding forces included in the interval of 5–50 MPa, even if these values are mostly theoretical, both the liner and cement bond tested in this study exhibited bond strength within these limits [30]. The liner-bonded ceramic crowns resisted a higher loading without demonstrating a chipping than the cement-bonded crowns, although no significant difference was found in the fracture strength between the liner-bonded group and the cement-bonded-group. As confirmed in Figures 3 and 5, the strong chemical bonding between the lithium disilicate liner and the zirconia ceramic was induced in the liner-bonded group, but there were lots of pores in the cement and mircro-gaps at the interface of the lithium disilicate-reinforced glass ceramic and cement in the cement-bonded group. Thus, it can be explained that the lower chipping strengths of the cement-bonded crowns than those of the liner-bonded crowns resulted from the stress concentration at defects of the interface when compression load was applied to the crown [31].

The bonding of ceramic restorations commonly involves acid etching and/or sandblasting procedures [32,33]. In the previous studies, it was reported that the shear bond strength of pre-sintered zirconia and the veneering porcelain did not change by conditioner treatment, but the failure mode was improved after thermal cycling [32]. It was also identified that the sandblasted zirconia surfaces had significantly higher shear bond strengths than non-treated and chemically etched surfaces without thermocycling, irrespective of conditioner treatment, and the highest shear bond strength and improved failure mode was confirmed by the application of both conditioner treatment and sandblasting [33]. Zirconia ceramics do not demonstrate the typical undercut features created by acid etching since they have chemical stability and almost no glass matrix. The silane treatment is not as effective as in silicate ceramics. The zirconia ceramics are usually roughened using an air-borne particle abrasion technology [34,35]. This procedure exposes sharp asperities of surface structures and increases surface area for bonding [36,37]. The resin cement should wet the surface and adhere to the irregular surface of ceramic restoration. This type of bonding relies on the surface characteristics of ceramic restorations, as well as the wettability and chemical compositions of resin cements, including the mechanisms of polymerization. According to Dérand et al. and Lüthy et al., the implant-supported zirconia restorations can fail at the interface when bonding is established with resin cements containing only 2,2-bis[p-(2′-hydroxy-3′-methacryloxypropoxy)phenylene]propane (bis-GMA) or 1,6-bis(methacryloxy-2-ethoxycarbonylamino)-2,4,4-trimethylhexane (UDMA) monomer [38,39]. In another study, it was identified that the shear bond strength of resin cements containing MDP of a zirconia ceramic was significantly improved by the application of a zirconia primer on both polished and blasted zirconia surfaces when comparing the results of the non-treatment of zirconia primer [40].

The one-piece system of zirconia abutments was found to present with some degree of misfit at the implant–abutment interface because of the challenge of the manufacturing process of the ceramic abutment [41,42]. Under loading, these abutments can induce an abrasion with micromotion at the implant–abutment interface [43]. The abutment screw can be loosened and lead to a bacterial colonization at the interface [6,44], as well a fracture failure of the implant-supported restoration.

The two-piece system of zirconia abutments involves using a titanium base to establish an interfacial connection with similar materials to the implant to avoid possible complications resulting from the use of dissimilar materials [18,22]. These abutments were found to demonstrate a higher bond strength when the abutment was roughened with 110 μm alumina particles at 2.5 atm, coated with a ceramic primer, and bonded with a dual-cure self-adhesive resin cement [45]. However, the increased bond strength was possible when the titanium base was included when roughening the zirconia core by sandblasting. The bonded interface was intimate without demonstrating a gap between the abutment and cement layer. According to Ebert et al. [46], the retention of cement-bonded zirconia abutments was significantly higher when the bonded interfacial gap was less than 30 μm.

The fracture resistance was higher when the zirconia abutment was designed to receive the support of the titanium base [17]. Gehrke et al. [7] used CAD/CAM technology to design zirconia abutments connected to an internal-hex titanium implant. The test specimens were thermocycled and cyclic loaded to fracture. They found that the two-piece system sustained a significantly higher fracture load than the one-piece system. Though the axial force generated during chewing typically does not exceed 220 N [47], the threshold for failure is estimated to be approximately 400 N according to the research conducted by Andersson et al. [48] and Att et al. [44].

Sandblasting is a common method of surface treatment used to bond zirconia ceramic restorations. The effect of this mechanical method, however, is limited to creating desired undercut features and increasing bond strength because of the dense polycrystalline structure of the zirconia ceramic [49,50]. However, the heat-bonding with the lithium disilicate liner was found to induce a chemical bonding to the zirconia ceramic, as indicated by the alteration of the crystal structure of zirconia at the interface of bonding. The zone of the chemical interaction with bonding revealed components of Si and Zr oxides which agreed with the previous research conducted by Jang et al. [31] and Aboushelib et al. [51].

The mode of failure was mixed with cohesive fracture propagated through the liner in the MLG group, whereas the failure was adhesive at the bonded interface of the ceramic and resin cement in the MGC group. The cement layer was thin at the interface and contained numerous pores and gaps. Meanwhile, no gap or pore was noted at the interface of the heat-bonding with the liner. These characteristics of bonding and mode of failure might have played a critical role in developing a higher fracture resistance of the implant-supported glass ceramic crown combined with a two-piece system of a zirconia abutment produced by a CAD/CAM procedure. However, a clinical study is yet to be conducted to determine the clinical success of the glass ceramic crown bonded to the zirconia abutment by means of heat-bonding with the liner.

5. Conclusions

The bond strength and fracture resistance of a milled lithium disilicate-reinforced glass ceramic bonded to a milled zirconia ceramic are affected by the modification of bonding procedures. The bond strength and fracture resistance were significantly higher when the bonding was established by means of heat-bonding with the lithium-disilicate liner than by a resin cement. The pattern of interfacial failure for the liner-bonded group was mixed with the fracture propagated through the liner. However, after visual inspection, the cement-bonded group demonstrated an adhesive failure at the interface of bonding. The results of this study suggest that heat-bonding with the liner can be an alternative to bond CAD/CAM-produced glass ceramic crowns to zirconia ceramic abutments in order to reduce the risk of crown dislodgement and fracture.

Author Contributions: Conceptualization, T.-S.B.; data curation, Y.-S.J. and J.-J.L.; formal analysis, Y.-S.J. and S.-H.O.; investigation, Y.-S.J., S.-H.O. and T.-S.B.; supervision, M.-H.L. and T.-S.B.; validation, Y.-S.J., M.-H.L. and T.-S.B.; writing–original draft, Y.-S.J. and S.-H.O.; writing–review and editing, Y.-S.J. and W.-S.O.

Funding: This research was supported by the Technological Innovation R&D Program (S2410617) funded by the Small and Medium Business Administration (SMBA, Korea).

Acknowledgments: This study has reconstructed the data for dissertation of doctor of Sang-Hun Oh. Yong-Seok Jang and Sang-Hoon Oh contributed equally to this work and are considered as joint first author.

Conflicts of Interest: The authors declare no conflict of interest.

References

1. Sailer, I.; Sailer, T.; Stawarczyk, B.; Jung, R.E.; Hämmerle, C.H.F. In vitro study of the influence of the type of connection on the fracture load of zirconia abutments with internal and external implant-abutment connections. *Int. J. Oral Maxillofac. Implant.* **2009**, *24*, 850–858.
2. Wadhwani, C.; Piñeyro, A.; Avots, J. An Esthetic Solution to the Screw-Retained Implant Restoration: Introduction to the Implant Crown Adhesive Plug: Clinical Report. *J. Esthet. Restor. Dent.* **2011**, *23*, 138–143. [CrossRef] [PubMed]
3. Wilson, T.G., Jr. The positive relationship between excess cement and peri-implant disease: A prospective clinical endoscopic study. *J. Periodontol.* **2009**, *80*, 1388–1392. [CrossRef] [PubMed]
4. Sailer, I.; Zembic, A.; Jung, R.E.; Hämmerle, C.H.F.; Mattiola, A. Single-tooth implant reconstructions: Esthetic factors influencing the decision between titanium and zirconia abutments in anterior regions. *Eur. J. Esthet. Dent.* **2007**, *2*, 3.
5. Zembic, A.; Sailer, I.; Jung, R.E.; Hämmerle, C.H.F. Randomized-controlled clinical trial of customized zirconia and titanium implant abutments for single-tooth implants in canine and posterior regions: 3-year results. *Clin. Oral Implant. Res.* **2009**, *20*, 802–808. [CrossRef] [PubMed]
6. Nakamura, K.; Kanno, T.; Milleding, P.; Ortengren, U. Zirconia as a dental implant abutment material: A systematic review. *Int. J. Prosthodont.* **2010**, *23*, 4.
7. Gehrke, P.; Johannson, D.; Fischer, C.; Stawarczyk, B.; Beuer, F. In vitro fatigue and fracture resistance of one- and two-piece CAD/CAM zirconia implant abutments. *Int. J. Oral Maxillofac. Implant.* **2015**, *30*, 546–554. [CrossRef]
8. Kosmač, T.; Oblak, Č.; Marion, L. The effects of dental grinding and sandblasting on ageing and fatigue behavior of dental zirconia (Y-TZP) ceramics. *J. Eur. Ceram. Soc.* **2008**, *28*, 1085–1090. [CrossRef]
9. Denry, I.; Kelly, J.R. State of the art of zirconia for dental applications. *Dent. Mater.* **2008**, *24*, 299–307. [CrossRef]
10. Park, J.I.; Lee, Y.; Lee, J.H.; Kim, Y.L.; Bae, J.M.; Cho, H.W. Comparison of fracture resistance and fit accuracy of customized zirconia abutments with prefabricated zirconia abutments in internal hexagonal implants. *Clin. Implant. Dent. Relat. Res.* **2013**, *15*, 769–778. [CrossRef]

11. Hamilton, A.; Judge, R.B.; E Palamara, J.; Evans, C. Evaluation of the fit of CAD/CAM abutments. *Int. J. Prosthodont.* **2013**, *26*, 370–380. [CrossRef] [PubMed]
12. Martínez-Rus, F.; Ferreiroa, A.; Özcan, M.; Bartolomé, J.F.; Pradíes, G. Fracture resistance of crowns cemented on titanium and zirconia implant abutments: A comparison of monolithic versus manually veneered all-ceramic systems. *Int. J. Oral Maxillofac. Implant.* **2012**, *27*, 6.
13. Klotz, M.W.; Taylor, T.D.; Goldberg, A.J. Wear at the titanium-zirconia implant-abutment interface: A pilot study. *Int. J. Oral Maxillofac. Implant.* **2011**, *26*, 970–975.
14. Stimmelmayr, M.; Edelhoff, D.; Güth, J.-F.; Erdelt, K.; Happe, A.; Beuer, F. Wear at the titanium–titanium and the titanium–zirconia implant–abutment interface: A comparative in vitro study. *Dent. Mater.* **2012**, *28*, 1215–1220. [CrossRef] [PubMed]
15. Foong, J.K.; Judge, R.B.; Palamara, J.E.; Swain, M.V. Fracture resistance of titanium and zirconia abutments: An in vitro study. *J. Prosthet. Dent.* **2013**, *109*, 304–312. [CrossRef]
16. Elsayed, A.; Wille, S.; Al-Akhali, M.; Kern, M. Comparison of fracture strength and failure mode of different ceramic implant abutments. *J. Prosthet. Dent.* **2017**, *117*, 499–506. [CrossRef]
17. Elsayed, A.; Wille, S.; Al-Akhali, M.; Kern, M. Effect of fatigue loading on the fracture strength and failure mode of lithium disilicate and zirconia implant abutments. *Clin. Oral Implant. Res.* **2018**, *29*, 20–27. [CrossRef]
18. Gehrke, P.; Alius, J.; Fischer, C.; Erdelt, K.J.; Beuer, F. Retentive strength of two-piece CAD/CAM zirconia implant abutments. *Clin. Implant. Dent. Relat. Res.* **2014**, *16*, 920–925. [CrossRef]
19. Truninger, T.C.; Stawarczyk, B.; Leutert, C.R.; Sailer, T.R.; Hämmerle, C.H.; Sailer, I. Bending moments of zirconia and titanium abutments with internal and external implant–abutment connections after aging and chewing simulation. *Clin. Oral Implant. Res.* **2012**, *23*, 12–18. [CrossRef]
20. Hjerppe, J.; Lassila, L.V.J.; Rakkolainen, T.; Narhi, T.; Vallittu, P.K. Load-bearing capacity of custom-made versus prefabricated commercially available zirconia abutments. *Int. J. Oral Maxillofac. Implant.* **2011**, *26*, 1.
21. Canullo, L. Clinical outcome study of customized zirconia abutments for single-implant restorations. *Int. J. Prosthodont.* **2007**, *20*, 5.
22. Canullo, L.; Coelho, P.G.; Bonfante, E.A. Mechanical testing of thin-walled zirconia abutments. *J. Appl. Oral Sci.* **2013**, *21*, 20–24. [CrossRef]
23. Cheng, C.-W.; Yang, C.-C.; Yan, M. Bond strength of heat-pressed veneer ceramics to zirconia with various blasting conditions. *J. Dent. Sci.* **2018**, *13*, 301–310. [CrossRef]
24. Yue, X.; Hou, X.; Gao, J.; Bao, P.; Shen, J. Effects of MDP-based primers on shear bond strength between resin cement and zirconia. *Exp. Ther. Med.* **2019**, *17*, 3564–3572. [CrossRef] [PubMed]
25. Sato, T.; Anami, L.; Melo, R.; Valandro, L.; Bottino, M. Effects of Surface Treatments on the Bond Strength Between Resin Cement and a New Zirconia-reinforced Lithium Silicate Ceramic. *Oper. Dent.* **2016**, *41*, 284–292. [CrossRef]
26. Mahmoodi, N.; Hooshmand, T.; Heidari, S.; Khoshro, K. Effect of sandblasting, silica coating, and laser treatment on the microtensile bond strength of a dental zirconia ceramic to resin cements. *Lasers Med. Sci.* **2016**, *31*, 205–211. [CrossRef]
27. Yilmaz, B.; Gilbert, A.; Seidt, J.; McGlumphy, E.; Clelland, N. Displacement of Implant Abutments Following Initial and Repeated Torqueing. *Int. J. Oral Maxillofac. Implant.* **2015**, *30*, 1011–1018. [CrossRef] [PubMed]
28. Schriwer, C.; Skjold, A.; Gjerdet, N.R.; Øilo, M. Monolithic zirconia dental crowns. Internal fit, margin quality, fracture mode and load at fracture. *Dent. Mater.* **2017**, *33*, 1012–1020. [CrossRef] [PubMed]
29. Mores, R.T.; Borba, M.; Corazza, P.H.; Della Bona, Á.; Benetti, P. Influence of surface finishing on fracture load and failure mode of glass ceramic crowns. *J. Prosthet. Dent.* **2017**, *118*, 511–516. [CrossRef] [PubMed]
30. Scribante, A.; Contreras-Bulnes, R.; Montasser, M.A.; Vallittu, P.K. Orthodontics: Bracket Materials, Adhesives Systems, and Their Bond Strength. *BioMed Res. Int.* **2016**, *2016*, 1–3. [CrossRef]
31. Jang, Y.-S.; Noh, H.-R.; Lee, M.-H.; Lim, M.-J.; Bae, T.-S. Effect of Lithium Disilicate Reinforced Liner Treatment on Bond and Fracture Strengths of Bilayered Zirconia All-Ceramic Crown. *Materials* **2018**, *11*, 77. [CrossRef] [PubMed]
32. Sawada, T.; Spintzyk, S.; Schille, C.; Zöldföldi, J.; Paterakis, A.; Schweizer, E.; Stephan, I.; Rupp, F.; Geis-Gerstorfer, J. Influence of Pre-Sintered Zirconia Surface Conditioning on Shear Bond Strength to Resin Cement. *Materials* **2016**, *9*, 518. [CrossRef] [PubMed]

33. Spintzyk, S.; Yamaguchi, K.; Sawada, T.; Schille, C.; Schweizer, E.; Ozeki, M.; Geis-Gerstorfer, J. Influence of the Conditioning Method for Pre-Sintered Zirconia on the Shear Bond Strength of Bilayered Porcelain/Zirconia. *Materials* **2016**, *9*, 765. [CrossRef]
34. Kern, M.; Wegner, S.M. Bonding to zirconia ceramic: Adhesion methods and their durability. *Dent. Mater.* **1998**, *14*, 64–71. [CrossRef]
35. Blatz, M.B.; Chiche, G.; Holst, S.; Sadan, A. Influence of surface treatment and simulated aging on bond strengths of luting agents to zirconia. *Quintessence Int.* **2007**, *38*, 9.
36. Ramos-Tonello, C.M.; Trevizo, B.F.; Rodrigues, R.F.; Magalhães, A.P.R.; Furuse, A.Y.; Lisboa-Filho, P.N.; Borges, A.F.S.; Tabata, A.S. Pre-sintered Y-TZP sandblasting: Effect on surface roughness, phase transformation, and Y-TZP/veneer bond strength. *J. Appl. Oral Sci.* **2017**, *25*, 666–673. [CrossRef] [PubMed]
37. He, M.; Zhang, Z.; Zheng, D.; Ding, N.; Liu, Y. Effect of sandblasting on surface roughness of zirconia-based ceramics and shear bond strength of veneering porcelain. *Dent. Mater. J.* **2014**, *33*, 778–785. [CrossRef] [PubMed]
38. Dérand, P.; Dérand, T. Bond strength of luting cements to zirconium oxide ceramics. *Int. J. Prosthodont.* **2000**, *13*, 131–135.
39. Lüthy, H.; Loeffel, O.; Hammerle, C. Effect of thermocycling on bond strength of luting cements to zirconia ceramic. *Dent. Mater.* **2006**, *22*, 195–200. [CrossRef]
40. Shin, Y.-J.; Shin, Y.; Yi, Y.-A.; Kim, J.; Lee, I.-B.; Cho, B.-H.; Son, H.-H.; Seo, D.-G. Evaluation of the shear bond strength of resin cement to Y-TZP ceramic after different surface treatments. *Scanning* **2014**, *36*, 479–486. [CrossRef]
41. Baldassarri, M.; Hjerppe, J.; Romeo, D.; Fickl, S.; Thompson, V.P.; Stappert, C.F.J. Marginal accuracy of three implant-ceramic abutment configurations. *Int. J. Oral Maxillofac. Implant.* **2012**, *27*, 3.
42. Smith, N.A.; Turkyilmaz, I. Evaluation of the sealing capability of implants to titanium and zirconia abutments against Porphyromonas gingivalis, Prevotella intermedia, and Fusobacterium nucleatum under different screw torque values. *J. Prosthet. Dent.* **2014**, *112*, 561–567. [CrossRef] [PubMed]
43. Han, J.; Zhao, J.; Shen, Z. Zirconia ceramics in metal-free implant dentistry. *Adv. Appl. Ceram.* **2017**, *116*, 138–150. [CrossRef]
44. Att, W.; Kurun, S.; Gerds, T.; Strub, J.R. Fracture resistance of single-tooth implant-supported all-ceramic restorations: An in vitro study. *J. Prosthet. Dent.* **2006**, *95*, 111–116. [CrossRef] [PubMed]
45. Von Maltzahn, N.F.; Holstermann, J.; Kohorst, P. Retention Forces between Titanium and Zirconia Components of Two-Part Implant Abutments with Different Techniques of Surface Modification. *Clin. Implant. Dent. Relat. Res.* **2016**, *18*, 735–744. [CrossRef] [PubMed]
46. Ebert, A.; Hedderich, J.; Kern, M. Retention of zirconia ceramic copings bonded to titanium abutments. *Int. J. Oral Maxillofac. Implant.* **2007**, *22*, 921–927.
47. Proeschel, P.; Morneburg, T. Task-dependence of Activity/Bite-force Relations and its Impact on Estimation of Chewing Force from EMG. *J. Dent. Res.* **2002**, *81*, 464–468. [CrossRef] [PubMed]
48. Andersson, B.; Taylor, Å.; Lang, B.R.; Scheller, H.; Schärer, P.; Sorensen, J.A.; Tarnow, D. Alumina ceramic implant abutments used for single-tooth replacement: A prospective 1-to 3-year multicenter study. *Int. J. Prosthodont.* **2001**, *14*, 432–438. [PubMed]
49. Liu, D.; Matinlinna, J.P.; Pow, E.H. Insights into porcelain to zirconia bonding. *J. Adhes. Sci. Technol.* **2012**, *26*, 1249–1265.
50. Fischer, J.; Grohmann, P.; Stawarczyk, B. Effect of zirconia surface treatments on the shear strength of zirconia/veneering ceramic composites. *Dent. Mater. J.* **2008**, *27*, 448–454. [CrossRef] [PubMed]
51. Aboushelib, M.N.; Kleverlaan, C.J.; Feilzer, A.J. Selective infiltration-etching technique for a strong and durable bond of resin cements to zirconia-based materials. *J. Prosthet. Dent.* **2007**, *98*, 379–388. [CrossRef]

 © 2019 by the authors. Licensee MDPI, Basel, Switzerland. This article is an open access article distributed under the terms and conditions of the Creative Commons Attribution (CC BY) license (http://creativecommons.org/licenses/by/4.0/).

Article

Survival Probability, Weibull Characteristics, Stress Distribution, and Fractographic Analysis of Polymer-Infiltrated Ceramic Network Restorations Cemented on a Chairside Titanium Base: An In Vitro and In Silico Study

João P. M. Tribst [1,*], Amanda M. O. Dal Piva [1], Alexandre L. S. Borges [1], Lilian C. Anami [2], Cornelis J. Kleverlaan [3] and Marco A. Bottino [1]

[1] Department of Dental Materials and Prosthodontics, São Paulo State University (Unesp/SJC), Institute of Science and Technology, São José dos Campos 12245-000, Brazil; amodalpiva@gmail.com (A.M.O.D.P.); alexandre.borges@unesp.br (A.L.S.B.); marco.bottino@unesp.br (M.A.B.)
[2] Department of Dentistry, Santo Amaro University (UNISA), São Paulo 04743-030, Brazil; lianami@gmail.com
[3] Department of Dental Materials Science, Academic Centre for Dentistry Amsterdam (ACTA), Universiteit van Amsterdam and Vrije Universiteit, 1081 LA Amsterdam, The Netherlands; c.kleverlaan@acta.nl
* Correspondence: joao.tribst@gmail.com; Tel.: +55-12-981222061

Received: 6 March 2020; Accepted: 10 April 2020; Published: 16 April 2020

Abstract: Different techniques are available to manufacture polymer-infiltrated ceramic restorations cemented on a chairside titanium base. To compare the influence of these techniques in the mechanical response, 75 implant-supported crowns were divided in three groups: CME (crown cemented on a mesostructure), a two-piece prosthetic solution consisting of a crown and hybrid abutment; MC (monolithic crown), a one-piece prosthetic solution consisting of a crown; and MP (monolithic crown with perforation), a one-piece prosthetic solution consisting of a crown with a screw access hole. All specimens were stepwise fatigued (50 N in each 20,000 cycles until 1200 N and 350,000 cycles). The failed crowns were inspected under scanning electron microscopy. The finite element method was applied to analyze mechanical behavior under 300 N axial load. Log-Rank ($p = 0.17$) and Wilcoxon ($p = 0.11$) tests revealed similar survival probability at 300 and 900 N. Higher stress concentration was observed in the crowns' emergence profiles. The MP and CME techniques showed similar survival and can be applied to manufacture an implant-supported crown. In all groups, the stress concentration associated with fractographic analysis suggests that the region of the emergence profile should always be evaluated due to the high prevalence of failures in this area.

Keywords: dental implant–abutment design; dental implants; dental materials; finite element analysis; material testing; ceramics

1. Introduction

Restorations performed using the computer-aided design and manufacturing facility (CAD/CAM) became increasingly popular in dental applications [1]. Due to the wide variety, CAD/CAM materials could be generally divided into two main categories: ceramics and composites [2]. Comparing both categories, indirect composite restorations are more resilient, easier to finish and polish, less abrasive to the antagonist, and allow an easy occlusal adjustment [3]. In contrast, the ceramic restorations present better biocompatibility, superior aesthetics, greater wear resistance and greater color stability [4].

Combining the positive properties of CAD/CAM ceramics and composite materials, a hybrid material was developed, which is also known as a polymer-infiltrated ceramic network (PIC) [5]. This material has a relatively low elastic modulus compared to conventional ceramics [6], besides presenting better marginal integrity and machinability [7]. Among the available CAD/CAM blocks from this new class of materials, the Vita Enamic (Vita Zahnfabrick, Bad Säckingen, Germany) stands out for its better long-term color stability [1] and the ability to deform during a load application prior to fracture [8], which ensures proper aesthetics and strength for the rehabilitation. These characteristics are a consequence of the feldspar ceramic involved in a resin matrix [9] based in urethane dimethacrylate (UDMA) and triethylene glycol dimethacrylate (TEGDMA) [5]. The ceramic portion consists of 58–63% of SiO_2, 20–23% of Al_2O_3, 9–11% of Na_2O, 0.5–2% de B_2O_3, and less than 1% of Zr_2O and CaO [10]. This combination provides adequate wear resistance, flexural strength, and elastic modulus close to dentine tissue [2,11,12]. PICs also have a hardness value between dentin and enamel [11], a maximum fracture load near 2000 N [13], and longitudinal clinical reports with high success rates [14,15]. In observing implant-supported prosthesis, the manufacture of metal–ceramic restorations is defined as the gold standard for prosthetic rehabilitation [16]. However, the monolithic crowns in PIC appear to be a reliable option [16].

Despite the reliability of implant-supported PIC restorations for cemented full-crown design, implant dentistry could present different limitations and aesthetics requirements [17]. Sometimes, the use of a conventional titanium abutment to link the crown and the implant could not be the ideal digital workflow [18]. Since the titanium is a metallic substrate, this abutment can generate a gray zone effect on the peri-implant marginal mucosa [19]. To reduce this effect, the hybrid abutment emerged as an alternative to be used to link the ceramic crown and the implant [20]. The hybrid abutment consists of two parts: a titanium base (Tibase) and ceramic mesostructure. The first one is responsible for keeping the connection between the implant and abutment in metal, and the second one is responsible for improving the peri-implant mucosa aesthetics [20–26]. For this technique, the crown is indicated to be cemented on the mesostructure (two-piece design).

To manufacture the mesostructure, the ceramic blocks containing a central hole (implant-solution CAD/CAM blocks) are required to allow the connection with the Tibase and the screw access to the implant. The literature reports the possibility of performing mesostructures in zirconia, lithium disilicate [20,25,27] and PIC [22–24,28]. The implant solution CAD/CAM blocks also can be machined as a crown without the mesostructure. This approach named one-piece design [29] simplifies the chairside process, requires the use of only one CAD/CAM block, and allows the manufacture of screw-retained crowns [20,28,29]. The comparison of the load-to-fracture of these two designs was performed for zirconia and lithium disilicate restorations [20,26,27]. However, it is well known that dental ceramics fail under fatigue, in consequence of the slow crack growth in stressed areas [30,31]. For this reason, this study investigated the survival probability, Weibull characteristics, and stress distribution of PIC crowns cemented on a chairside titanium base manufactured using different techniques. The null hypothesis was that there would be no difference between the designs for the analyzed parameters.

2. Materials and Methods

2.1. Specimens Preparations

Seventy-five (75) morse-taper implants (Conexão Sistemas de Prótese, Arujá, SP, Brazil) were installed in polyurethane resin, which is a validated material to simulate the bone tissue in in vitro studies due to its elastic behavior and stiffness [32]. For that, the polyurethane resin was manipulated using an identical volume of base and catalyst homogenized in a rubber bowl. The mixture was poured into (25 × 20 mm) polyvinyl chloride cylindrical support. After the complete resin cure, the polyurethane surface was polished with silicon carbide papers (P600 and P1200) under water cooling in an orbital gridding machine (Buehler, Ecomet 250, Lake Bluff, IL, USA). A sequence of surgical drills was used, according to the manufacturer's indication (Conexão Sistemas de Prótese,

Arujá, SP, Brazil), to make a synthetic surgical alveolus perpendicular to the surface and centered in the polyurethane cylinder. In each cylinder, with the aid of a manual torque wrench, the implants (4.1 × 10 mm) were installed (40 N·cm) keeping 3 mm of the implant not embedded in the resin, following the ISO 14801:2016 for mechanical implant testing.

All Tibases (Conexão Sistemas de Prótese, Arujá, Brazil) were sandblasted with 50 µm aluminum oxide (Al_2O_3) particles at a pressure of 1.5 bar, using an implant analog to assist the laboratorial handling. After, they were cleaned in an ultrasonic bath (5 min with isopropyl alcohol) and received a layer of Alloy Primer (Kuraray Noritake Dental Inc., Okayama, Japan) for 60 s. All titanium bases surfaces were gently blow dried, and the screw access holes were protected with a Teflon tape. A thin layer of titanium dioxide-based powder (Ivoclar Vivadent, Schaan, Liechtenstein) was sprayed onto each of the titanium bases for scanning (inEos Blue, inLab SW4.2, Sirona, Benshein, Germany) and subsequent restorations manufacture. The sets were randomly divided into three experimental groups ($n = 25$), according to the prosthesis manufacturing technique.

2.1.1. Crown Cemented on a Mesostructure (CME) Prosthesis Design: Two-Piece Prosthetic Solution Composed by a Crown Cemented on the Hybrid Abutment

In this technique, the first structure to be manufactured is the mesostructure. For that, the inLab software (Sirona Dental Systems, Bensheim, Alemanha) was used to design the mesostructure and its insertion axis. The data were sent to the equipment Cerec inLab (5884742 D329, Sirona for Dental Systems, Benshelm, Germany), and 25 structures were milled using Vita Enamic IS-14 blocks (Vita Zahnfabrik, Bad Säckingen, Germany) that contain a central access hole. The mesostructures were separated from the remaining blocks with the aid of a diamond blade and fine-grained diamond bur under abundant irrigation. Next, the mesostructures were polished with pink (10,000 rpm) and gray rubbers (8000 rpm) from a Vita Enamic Polishing Set (Vita Zahnfabrick) and cleaned in an ultrasonic bath (5 min in distilled water) to remove debris from the polishing rubbers or surface contaminant that may interfere on the adhesive procedure. To complete the hybrid abutment (mesostructure + Tibase), the mesostructure intaglio surface received a silane agent (Clearfill Ceramic Primer, Kuraray Noritake Dental Inc., Okayama, Japan) for 60 s. The self-etching resin cement (Panavia F 2.0, Kuraray Noritake Dental Inc., Okayama, Japan) was manipulated and applied on the Tibase and the mesostructure, which were held in position with 750 g. The excess cement was removed with a microbrush and light cured (Valo, Ultradent Products, South Jordan, UT, USA) following the manufacturer instructions. The hybrid abutments were installed on the implants with 30 N·cm torque. Next, the mesostructures were scanned for conventional full crown preparations. Then, twenty-five (25) crowns were designed, milled in PIC blocks, polished, and cemented on the hybrid abutment (Figure 1). Both cementation lines were polished with the Vita ENAMIC polishing kit at 5000 rpm.

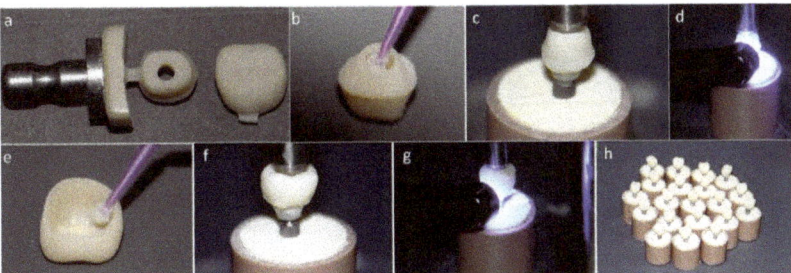

Figure 1. (a) Mesostructure machined in implant solution block and monolithic crown machined in conventional block. (b) Silanization of the internal surface of the mesostructure. (c) Luting of the mesostructure on the titanium base (Tibase) using self-etching resin cement. (d) Photoactivation of resin cement. (e) Crown silanization. (f) Luting of the crown on the mesostructure. (g) Light curing. (h) Crown cemented on a mesostructure (CME) group crowns completed.

2.1.2. Monolithic Crown (MC) Prosthesis Design: One-Piece Prosthetic Solution Composed by a Crown Direct Cemented on a Titanium Base

In this technique, first, the Tibases were installed on the implants with 30 N·cm torque. Then, the crowns were designed, polished, cleaned, and cemented directly on the Tibases as a conventional abutment. The cementation line was polished with the Vita Enamic polishing kit at 5000 rpm (Figure 2).

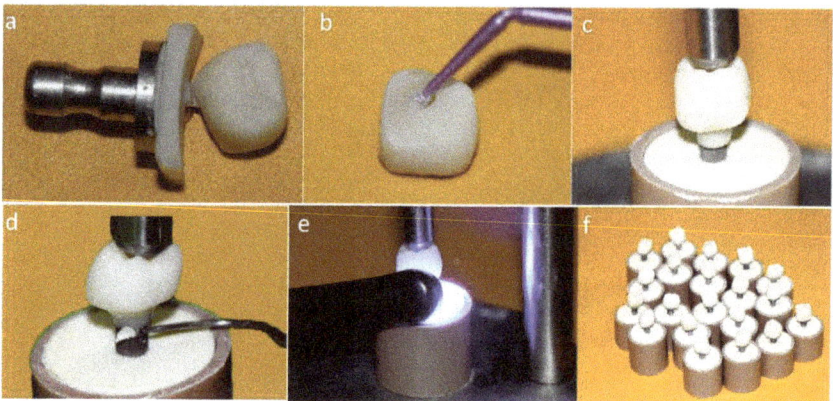

Figure 2. (**a**) Monolithic crown machined in conventional block; (**b**) Silanization of the internal surface of the crown; (**c**) Luting of the crown on the titanium base (Tibase) using self-etching resin cement; (**d**) Removal of excess cement; (**e**) Light curing; (**f**) MC group crowns completed.

2.1.3. MP Prosthesis Design: One-Piece Prosthetic Solution Composed by a Crown Cemented on a Tibase with Screw Access Hole

For the monolithic crown with perforation (MP) group, the crowns were manufactured in one piece using Vita ENAMIC IS-16 blocks (Vita Zahnfabrik, Bad Säckingen, Germany). The crowns with the occlusal perforation were polished, cleaned, and cemented on the Tibases. Through the screw access hole, the sets (crown and Tibase) were positioned on the implants and with a manual torque wrench fixed with 30 N·cm of torque. For each specimen, the screw access hole was conditioned with 5% hydrofluoric acid for 60 s, washed with a water jet for 20 s, and dried with air jets. Then, the bonding agent was applied, and the access was sealed with composite resin. The cementation line and the composite resin interface were polished with a polishing kit (Vita Zahnfabrik, Bad Säckingen, Germany) at 5000 rpm (Figure 3).

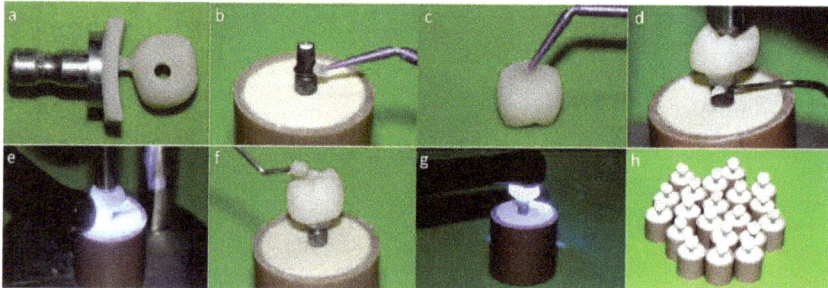

Figure 3. (**a**) Monolithic crown machined in an implant solution block; (**b**) Alloy primer application in the titanium base (Tibase); (**c**) Silanization of the internal surface of the crown; (**d**) Removal of excess cement; (**e**) Light curing; (**f**) Sealing of the screw access hole with composite resin; (**g**) Monolithic crown with perforation; (**h**) (MP) group crowns completed.

For all groups, the crowns (Vita ENAMIC, Vita Zahnfabrik, Bad Säckingen, Germany) were manufactured with identical anatomy, minimum occlusal thickness at 1 mm, 0.8 mm cervical wall around the base, 0.4 mm cervical terminus, and 80 µm space for the cement layer.

2.2. Fatigue Test

The specimens were stored in distilled water for a period of 24 h prior to the fatigue test. Five samples of each group were submitted to the single load to fracture test (SLF) in an universal testing machine (EMIC DL 1000, EMIC, São José dos Pinhais, PR; 1 mm/min speed, 1000 kgf load cell). From the mean load value (1200 N), the fatigue profile used in the stepwise test was determined. Twenty (20) specimens of each group were tested until failure in an adapted fatigue tester (Fatigue Tester, ACTA, The Netherlands). The fatigue load was delivered (6 mm diameter, stainless steel, water, 25 °C, 1.4 Hz) on the occlusal fossa [13,33]. The test started with 300 N during 5000 cycles (40% of the SLF). After each step of 20,000 cycles, the load was increased [34] in 50 N until the maximum load of 1200 N and 350,000 cycles. The specimens were checked for cracks and/or fractures in each step (Figure 4).

Figure 4. Sample positioned to perform the survival test.

2.3. Fractographic Analysis

The failures were classified according to the patterns obtained after the fatigue test [33,34]. To determine the fracture features methodologically, the ceramic fragments were evaluated to identify the direction of crack propagation and location of the origin [33–35] with the aid of a stereomicroscope (Stereo Discovery.V20, Carl Zeiss, LLC, USA). For that, the regions of interest were divided into quadrants, and the representative specimens were subjected to photomicrographs of greater magnification in each of these quadrants by scanning electron microscopy (SEM). For this, the specimens were cleaned in an ultrasonic bath with isopropyl alcohol for 10 min, dried, gold sputtered, and analyzed under scanning electron microscopy (SEM; Evo LS15, Oberkochen, Carl Zeiss, Germany) to identify the size and origin of the critical defect. The micrographies were merged to enable the crown fractographic analysis overview.

2.4. Nonlinear Finite Element Analysis

The three-dimensional (3D) industrial designs of the implant, Tibase, and prosthetic screw were provided by the manufacturer as stereolitographic files (Conexão Sistemas de Prótese, Arujá, Brazil) and imported to the modeling software (Rhinoceros version 5.0 SR8, McNeel North America, Seattle, WA, USA). Then, through automatic reverse engineering, the polygonal models were converted into 3D models formed by NURBS (Non Uniform Rational Basis Spline). In sequence, a cylinder was created corresponding to the in vitro polyurethane cylinder (20 × 25 mm). Then, the implant was centered perpendicularly to the cylinder, containing 3 mm of exposed threads similar to the in vitro test. A Boolean difference was used to ensure the juxtaposition between these structures. The model finished as a volumetric solid containing an implant, Tibase, screw, and fixation cylinder was tripled to obtain three models with identical geometries.

One specimen from each in vitro group was scanned and exported in STL format. The crowns' 3D models were submitted to the BioCad protocol [36] to perform a volumetric model whose geometry corresponded exactly to the in vitro specimens. Then, this same procedure was repeated to create the mesostructure 3D model. All cement layers were standardized with 80 μm.

For a static structural analysis, the models were checked and imported as a STEP file to the analysis software (ANSYS 17.2, ANSYS Inc., Houston, TX, USA). The contacts were considered nonlinear, containing 0.30 μ friction between the structures in titanium [36]. The number of tangent faces between solids was equivalent to assist the analysis convergence. Through an automatic creation, an initial mesh with tetrahedral elements was created; the absence of mesh defined as obsolete by the software was verified prior to final mesh refinement (Figure 5). The 10% convergence test was used to determine the mesh control to ensure the least possible influence on the results of the mathematical calculation [37–39]. Each piece of material information was inserted for each solid component in isotropic and homogeneous behavior, requiring the modulus of elasticity and Poisson ratio (Table 1) [5,39–41]. During contour definitions, the loading was performed in the occlusal region of the crown. The applied load was 300 N (Figure 5a) on the Z-axis [23]. The fixation location was defined under the surface of the polyurethane cylinder, simulating the sample holder in one plane (Figure 5b). A pre-tension was also applied with 30 N simulating the torque (Figure 5d) during the prosthetic screw tightening [42].

Table 1. Mechanical properties of the materials used in this study.

Material	Elastic Modulus (GPa)	Poisson Ratio	Reference
Titanium	110	0.33	[39]
Polymer infiltrated ceramic	30	0.28	[5]
Polyurethane	3.6	0.3	[40]
Resin cement	18.3	0.3	[41]

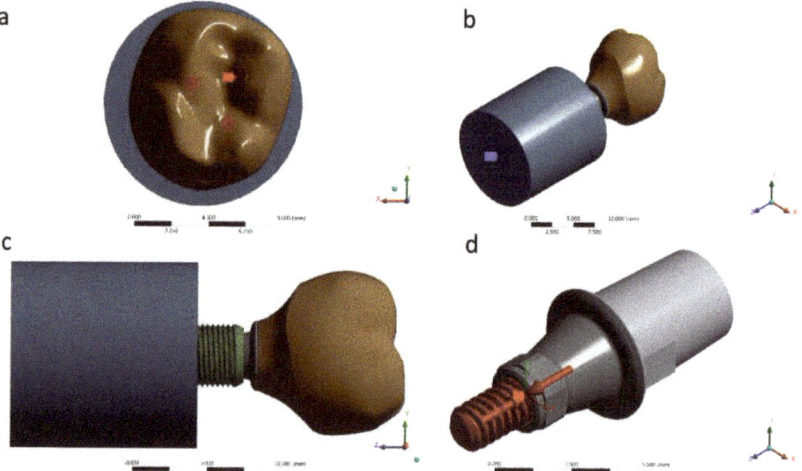

Figure 5. (a) Occlusal loading; (b) Fixing the system; (c) Mesh generated; (d) Pre-tension of 30 N·cm.

The composite resin shrinkage for sealing the screw access hole in the MC group was simulated simultaneously in the analysis using the thermal analogy [43]. The solutions were obtained in total deformation, von-Misses stress, maximum principal stress, and microstrain, for each group. The results were presented on an identical scale of values for visual comparison, as well as the absolute values were plotted on graphs for quantitative analysis of the peaks.

2.5. Data Analysis

Survival data were statistically analyzed by Kaplan–Meier and Mantel–Cox tests (Log-Rank and Wilcoxon tests) [44]. Data distribution and reliability analysis were assessed by Weibull analysis associated with two parameters: shape and scale showing the probability distribution of the material to fail in a certain fatigue time using the statistical software (Minitab 16.1.0, State College, PA, USA), with 95% confidence interval. The results obtained in the finite element analysis were exposed and descriptively evaluated through color graphics corresponding to the stress concentration.

3. Results

3.1. Fatigue Test

Weibull analysis showed difference between the mean values of characteristic strength according to the groups (Table 2). Weibull probability plots versus number of cycles reported at the sample failure during the fatigue test are present in Figures 6 and 7. Log-Rank ($p = 0.17$) and Wilcoxon ($p = 0.11$) revealed a similar survival probability between the manufacturing techniques at 300 N and 900 N, according to the confidence interval. However, at 600 N, MP group showed higher survival probability than the MC group, whereas the CME group showed an intermediate behavior (Table 2).

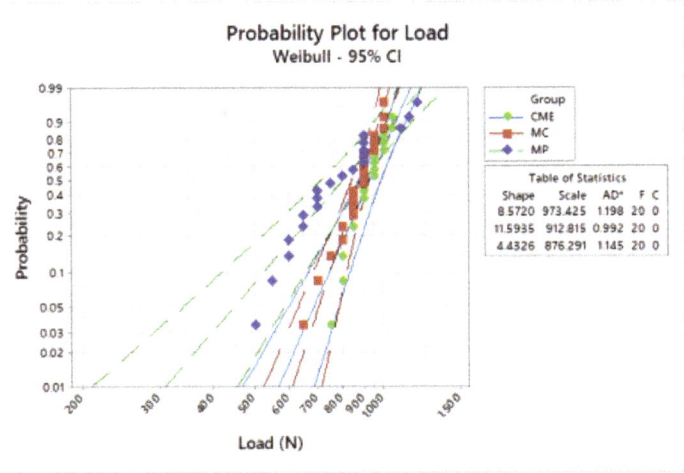

Figure 6. Weibull plot: survival probability versus load (N) reported on sample failure during the fatigue test.

Table 2. Crowns survival in fatigue at different load missions.

	Survival Probability (%)	MP	MC	CME
300 N	Upper bound	88	87	85
	Average	84 [A]	82 [A]	80 [A]
	Lower bound	79	77	74
600 N	Upper bound	50	36	40
	Average	44 [A]	30 [B]	33 [AB]
	Lower bound	37	24	27
900 N	Upper bound	7	6	9
	Average	5 [A]	1 [A]	6 [A]
	Lower bound	2	0.5	3

Similar capital letters correspond to no statistical significance between groups in the same row, according to the confidence interval.

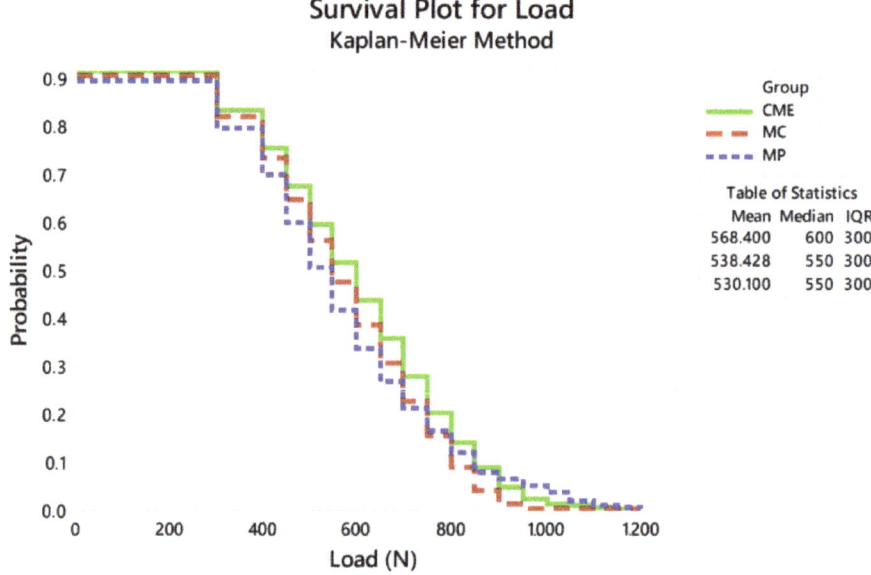

Figure 7. Survival plot using the Kaplan–Meier method.

Regardless of the similar survival and characteristic strength between CME and MP, MP showed lower data variation, being the most reliable technique (Table 3).

Table 3. Weibull modulus (*m*), characteristic strength (σ), and confidence intervals. The statistical differences were determined based on the confidence interval (CI).

Groups	*m* (CI)	σ (CI)
MP	8.5 (6.2–1.6)	973.4 (921.9–1027.7)
CME	4.4 (3.2–6.1)	912.8 (877.2–949.8)
MC	11.6 (8.1–16.4)	876.3 (789.1–973.1)

All groups showed cracks, wear facets, and bulk fractures. For the MP group, 15% of the samples presented a bulk fracture in two or more pieces, and 85% of the samples showed chipping failure in the crown emergence profile. For the MC group, 20% of the samples presented a bulk fracture in two or more pieces, and 80% of the samples showed chipping failure in the crown emergence profile. For CME, 20% of the samples showed factures only in the crown exposing the mesostructure, 10% of the samples failed as a bulk fracture with the crack involving the crown and mesostructure, and 70% of the samples failed in the emergence profile without involving the crown (Figure 8). For MC and MP, the fractographic analysis showed that the failure was originated at the cervical area, propagating to the top of the restoration, which was confirmed by the fracture features (Figures 9 and 10). For CME, the specimens in which the fracture was restricted only in the crown, the fractures features suggested the crack propagation direction from the marginal side with several secondary events in the occlusal surface (Figures 11–13).

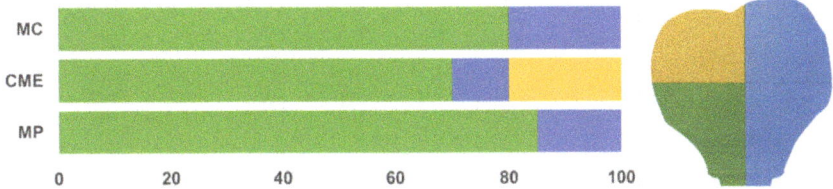

Figure 8. Quantitative analysis of the failures in % regarding the groups and the fracture location. In green, cracks found in the cervical region of the crown. In yellow, cracks found in the occlusal region. In blue, catastrophic failure.

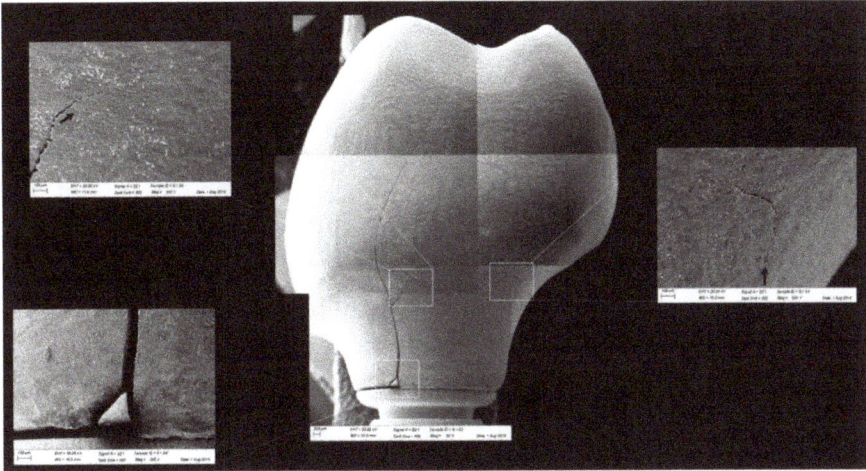

Figure 9. Fractographic analysis of a representative specimen from the MC group. The failure originated in the cervical region and propagated (black arrows) to the top of the restoration without separation of the fractured parts.

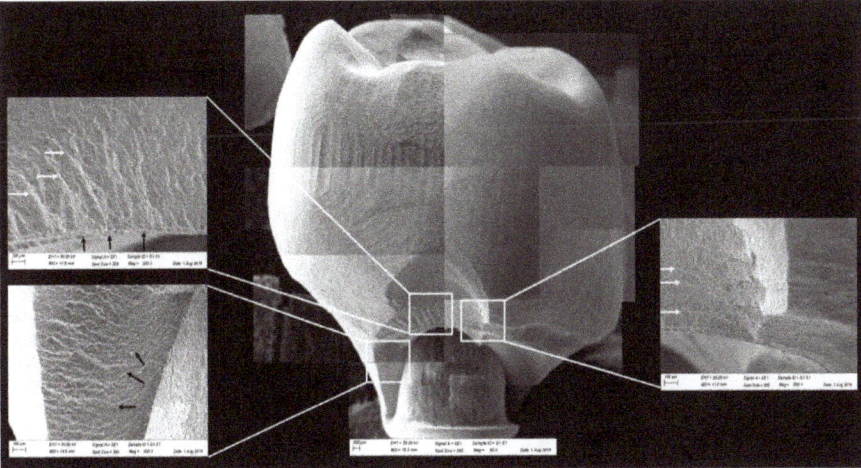

Figure 10. Fractographic analysis of a representative specimen from MP group. The failure originated in the cervical region and propagated to the top of the restoration. The white arrows indicate the hackle lines and the black arrows indicate the twist hackle marks.

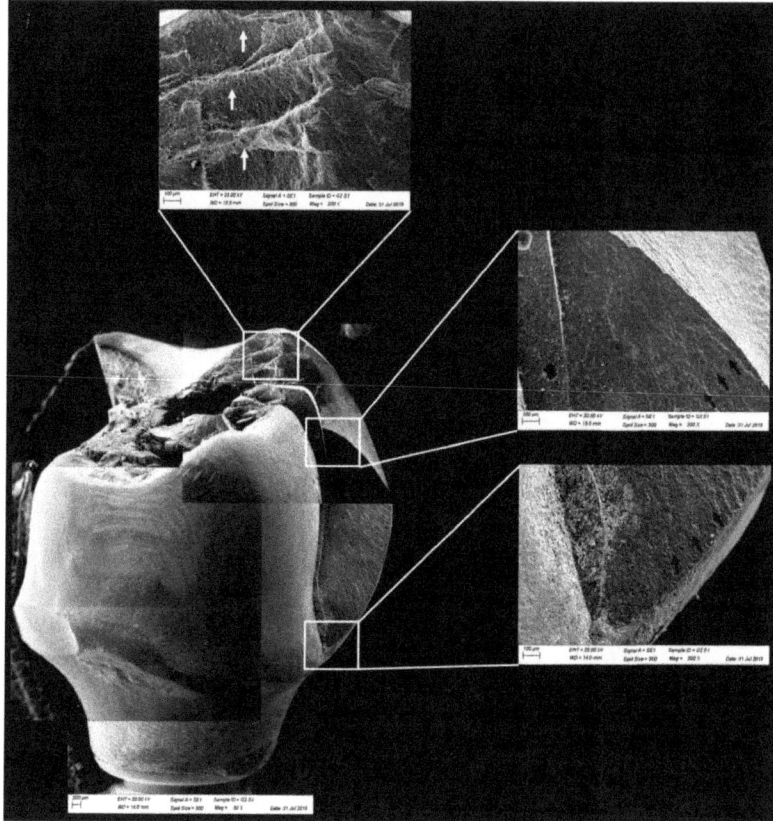

Figure 11. Fractographic analysis of a representative specimen from the CME group. The crown failed without the mesostructure involvement. The white arrows indicate the arrest lines and the black arrows indicate the hackle lines.

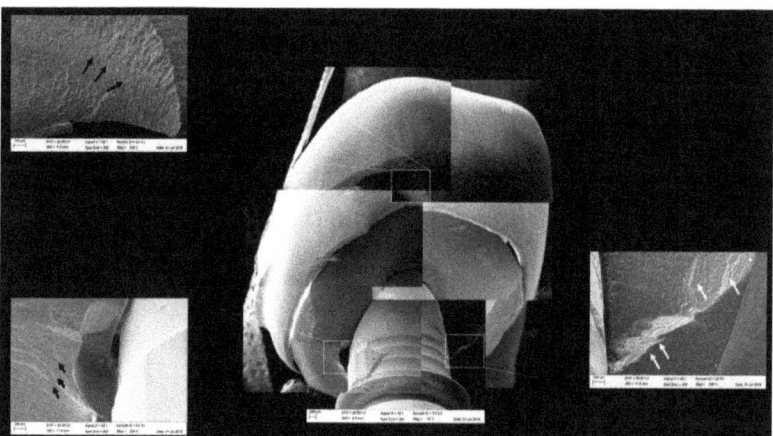

Figure 12. Fractographic analysis of a representative specimen from the CME group. The mesostructure failed without the involvement of the crown. The black arrows indicate the hackle lines, the white arrows indicate the arrest lines, and the red arrows indicate compression curls.

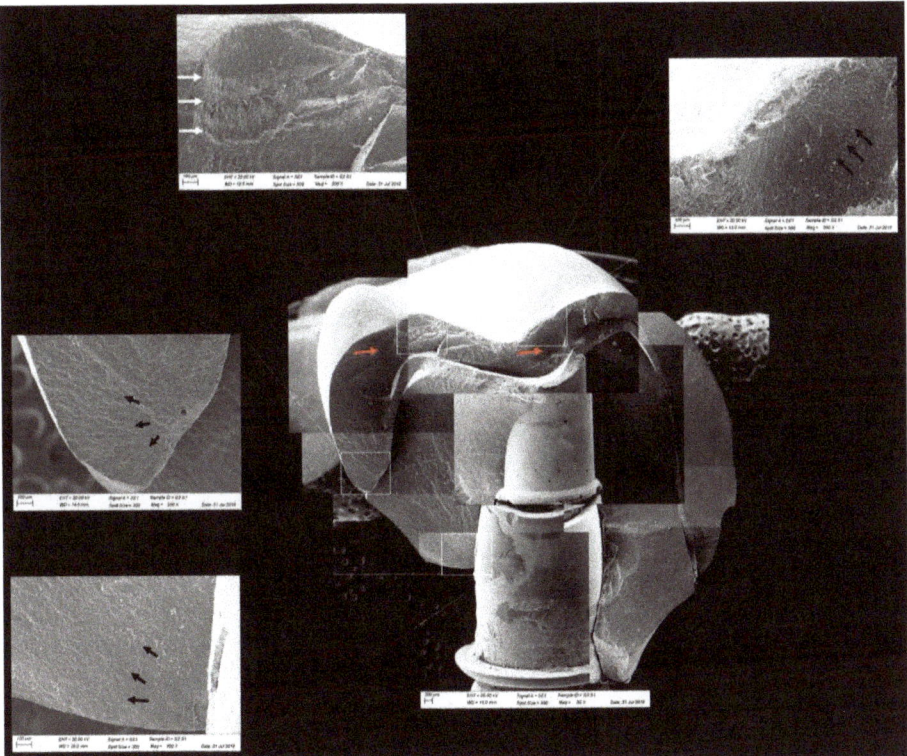

Figure 13. Fractographic analysis of a representative specimen from the CME group. The failure originated in the cervical region and propagated to the top of the restoration. The black arrows indicate the hackle lines, the white arrows indicate the arrest lines, and the red arrows indicate compression curls.

3.2. Nonlinear Finite Element Analysis

Observing the von-Mises failure criterion (Figure 14), which demonstrates the total resulting stress in the structures, it was possible to observe that the cervical region was the most involved regardless of the restoration technique, with the composite resin sealing of the group MP presenting a new stress area as well as the cement layer between the crown and mesostructure for the CME group.

In observing the tensile stress concentration in the crown, the numerical simulation showed a very similar mechanical behavior between the tested groups (Figure 15), with the highest stress concentration in the cervical region of the crown emergence profile. CME also presented stress concentration on the crown intaglio surface, which is compatible with the failure mode of 20% of the samples during the in vitro test (Figure 8). Furthermore, MP specimens showed high stress concentration in the composite resin used to seal the screw access hole. The stress peaks (Figure 16) corroborate with the colorimetric maps of the results (Figures 14 and 15), not allowing to assume a significant difference between the groups (10%).

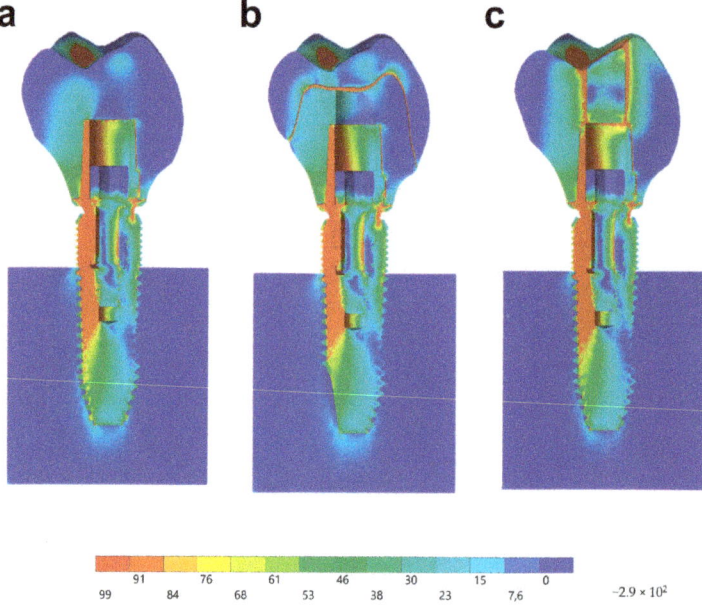

Figure 14. Stress distribution assessed by the von-Mises criterion according to the design of the restoration. (**a**) MC, (**b**) CME, and (**c**) MP.

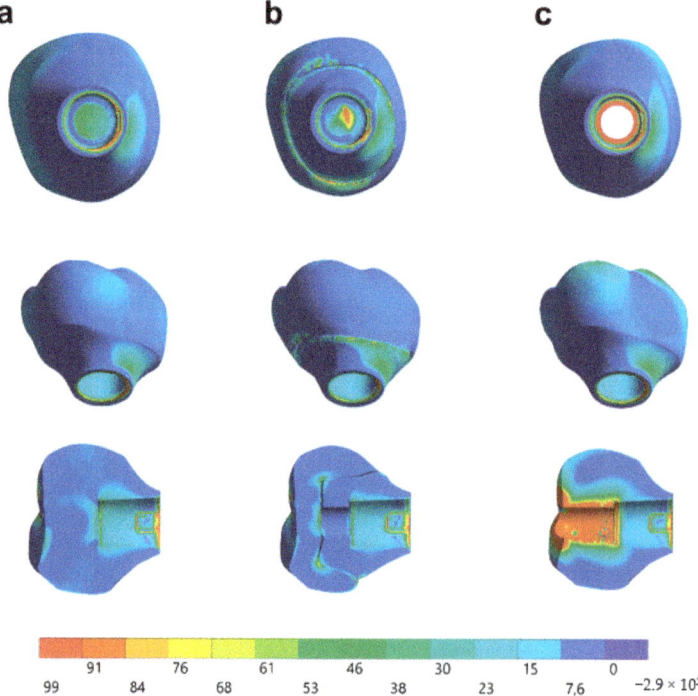

Figure 15. Stress distribution assessed by the maximum principal stress criterion according to the design of the restoration. (**a**) MC, (**b**) CME, and (**c**) MP.

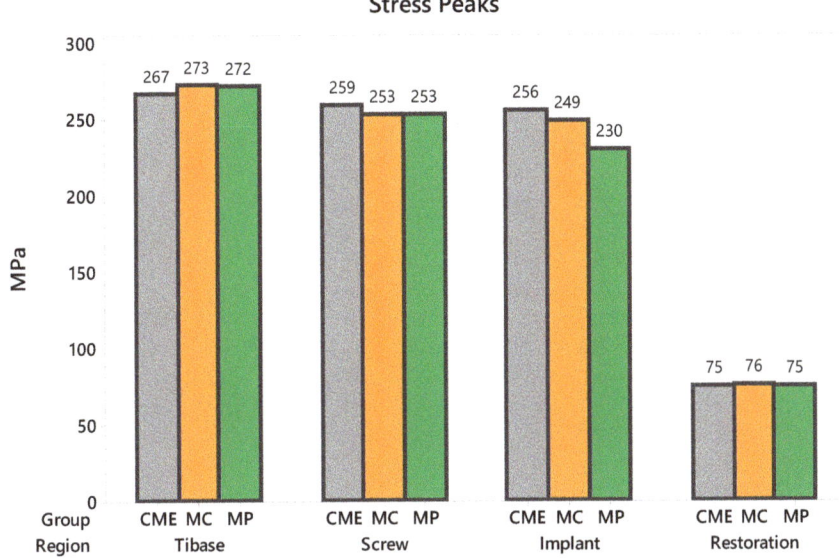

Figure 16. Stress peaks for each structure regarding the different groups. The restoration stress was calculated using the maximum principal stress criterion and the metallic structures using von-Mises criterion.

4. Discussion

This study evaluated the biomechanical behavior and survival probability of implant-supported polymer-infiltrated ceramic (PIC) restorations manufactured using different techniques. The results demonstrated that there is no difference between the groups for the stress distribution and that the reliability was similar between the groups at 300 N and 900 N. Therefore, we partially accept the null hypothesis.

The survival probability of restorations using Tibase as a connection between the implant and the crown is still scarce in the scientific literature [28,45,46]. The indication of this technique depends on the use of CAD/CAM blocks for implant solution that present a connective hole (screw access hole) created by the manufacturer [25,28]. According to the ceramic block size, it is possible to perform a two-piece restoration containing the mesostructure and the crown or a one-piece restoration [15,20,21,26–28,46]. These two restorative techniques were simulated in the present study, respectively, for the CME and MP groups.

The use of mesostructure to build the hybrid abutment to replace the conventional titanium abutments has already been reported [21,26,45]. Up to date, in vitro [26,47,48] and in vivo [49] studies and case reports [50,51] reported the use of lithium disilicate or polycrystalline ceramics to manufacture the mesostructure. Meanwhile, one case report [15] and in silico investigations [22–25] showed the possibility to manufacture the hybrid abutment using PIC material cemented on a Tibase, as performed in this study. Other authors [15] clinically evaluated the CME and MP groups using PIC. The authors suggested that the MP design has a disadvantage in aesthetics compared to the CME due to the presence of the screw access hole filled with composite resin. A literature review [28] suggested the possibility of using PIC to perform a mesostructure; however, the authors affirmed that the lack of longitudinal data does not allow its indication as the material first choice. Based on this, the present study results should assist the clinicians with understanding the biomechanical behavior of this treatment modality.

Comparing the restorative modalities evaluated in the present study, it was possible to observe that there is no difference between CME and MP, which are restorations created using implant solution CAD/CAM blocks and a Tibase. Both designs have already been compared for restorations made with different ceramic material [45]. The authors evaluated 10 anterior lithium disilicate restorations performed for each group and fatigued at 20 Hz until fracture [45], and they found that the MP design was statistically superior. In the present study, 20 restorations were manufactured for each group, the fatigue was adjusted to a maximum of 1.4 Hz, and the survival of MP and CME were not statistically different when PIC is the restorative material. In addition, this study used a posterior total crown with axial loads in order to restrict the failures only in the ceramic material, which was the object of the study.

The literature is not concise regarding the best protocol to use the Tibase. For example, another study [21] compared lithium disilicate anterior crowns using CME and MP techniques. However, the authors only performed the load to fracture using an universal testing machine with $n = 8$ specimens per group. The authors did not find any statistical difference between both designs, corroborating the findings for survival in the present study. Other study evaluated the load to failure of lithium disilicate and zirconia CME versus lithium disilicate MP crowns after aging [27]. Ten specimens per design were submitted to 2000 thermocycles (5–55 °C, water) and mechanical fatigue (150 N, 2 Hz, 100,000 cycles) aging. The authors observed that the lithium disilicate MP presented higher fracture load than a lithium silicate crown cemented on a mesostructure also in lithium silicate or in zirconia.

Still comparing different designs for using a Tibase, previous research [26] compared CME versus MP made in zirconia and lithium disilicate with $n = 8$/gr. The authors simulated premolars fatigued with 1.2 million thermomechanical cycles associated with 120 N load. After aging, the crowns were tested under an SLF test. Besides the cracks in the ceramic crown, the authors found plastic deformation on the Tibase, screw, and implant. This occurred as a consequence of the static load during compression test, which did not allow the slow crack growth in the ceramic material; this is the mechanism of failure for dental ceramics. For example, if the subject of the study is the ceramic design, it should be ideal that the test could provide the predominance of failures in this region, i.e., it will be possible to understand the restoration weakness and to promote new restorative designs. For this reason, the present study performed the stepwise test to determine at which load step the material has the highest probability of failure.

According to the literature [52], the screw access hole for retrieving ceramic implant-supported screw-retained crowns may decrease their fracture resistance. To prove this statement, the authors tested ceramic crowns in six different groups: monolithic zirconia, veneered zirconia, and lithium disilicate with and without screw access holes. The authors did not use a Tibase; instead, they used a custom abutment and the screw access hole was performed during the crown manufacturing, unlike implant-solution CAD/CAM blocks that already present the perforation. The results of the present study do not agree with these authors [52], because analyzing a ceramic restoration on fatigue, the screw access hole sealed with composite resin does not increase the generated stresses in the cervical region or decrease the restoration characteristic strength. Supporting the results found in the present study, previous authors [20,21,45] reported that groups with screw access holes showed similar load to failure to the groups without access holes. In addition, the authors suggested that perforated crowns may be even more resilient than groups without the access hole. The present study did not find any influence of the access hole on the strength characteristic; however, MP showed higher reliability than MC, corroborating with Roberts and Bailey (2018) [27].

According to a literature review [46] describing the possibilities and limitations of metal-free implant-supported single-tooth restorations, the authors described that one of the main advantages in using a Tibase in a digital workflow is the possibility of performing an adequate emergence profile in the crown. The emergence profile of an implant-supported prosthesis plays an important role in achieving esthetics, which are obtained when the clinician uses a properly concave abutment to mold the soft tissues for an adequate esthetic profile [53]. When improperly designed, the abutment emergence profile will compromise the cervical area blood supply, ultimately resulting in a loss of health and volume of the peri-implant tissues. For that reason, the emergence profile in an implant-supported restoration should be narrow to allow a higher volume of soft tissue [54]. Up to date, there has not been any in vitro study that created a restoration design that presents a concave emergence profile in the cervical area during the hybrid abutment testing [20,21,26,45]. During the present experiment, the 3D designs were created with this concern to simulate the most realistic shape of a restoration surrounded by a well-prepared soft tissue.

Observing the stresses analysis with the finite element method, it is possible to observe that the region of the emergence profile is the area of highest stress concentration in the restoration. This occurred probably due to the increase in the peri-implant soft tissue volume that requires the reduction on the restorative material in that region [15,53,54]. Therefore, the cervical area is the critical area of this restorative modality, since it is located near the fulcrum point of the crown and also because it has the smallest ceramic volume. The finite element method to observe the mechanical behavior of implant-supported single crowns associated with hybrid abutments has previously been reported for studies that compared different combinations of ceramic materials [22–24]. Therefore, it has been reported that to reduce the stress in the cervical region of the emergence profile, a flexible material should be used [22]. It is also recommended to use PIC to decrease the stress generated on the crown intaglio surface for CME design [24]. However, the present study demonstrated that MP restorations using only one CAD/CAM block for restoration are very similar to CME groups for the stress concentration in the cervical region. Moreover, the previous studies that simulated this restoration design [22–24] were not accompanied by in vitro experiments such as the present study, demonstrating that in fact, the region of stress concentration coincides with the possible failure origin observed in the fractographic analysis. At the occlusal region, the stress difference was caused by the composite polymerization shrinkage between the groups. Despite this, in the in vitro test, the survival was similar between the groups, and the failures were located at the cervical area.

The fractographic analysis followed the recommendations of the ADM (Academy of Dental Materials) guidance [55]. Thus, in a systematic protocol, all fractured restorations were observed under stereomicroscope to identify fracture features that could indicate the crack direction propagation. In this sense, the main failure mode observed was the crack/fracture of the cervical region in the emergence profile, while a smaller percentage of samples failed radially by splitting the crown into two or more pieces. However, for the CME group, different failure modes were also observed due to the presence of a second cement layer (between the mesostructure and the crown). In this group, in addition to failures in the cervical region, some specimens presented only crown fractures without mesostructure involvement. Some specimens presented mesostructure fracture without crown involvement, and two samples allowed the crack to propagate and separate the crown and mesostructure as a monolithic block. These failure modes in CME design have also been observed by previous studies that analyzed these restoration designs with other materials [26,27]. A fractographic analysis performed on failed zirconia mesostructure with Tibase during clinical use showed the same fracture pattern with the direction of crack propagation from the cervical area to occlusal [29], suggesting that the methods of the present study could provide failure modes similar to those found clinically.

In observing SEM micrographies, it was possible to note that all failures originated from the crown cervical region and spread toward the occlusal region. Secondary effects (damages) in the occlusal third near the compression region were also observed; however, they did not show fracture features that could suggest that the failure could have been originated in this region. It is important that the fractographic analysis is performed dividing each region of interest of the fractured restoration, analyzing each one in higher magnifications. This methodology is widely used to evaluate fractured dental crowns [56], disk-shaped specimens [55], and even for implant-supported crowns with Tibase [29] to assist in understanding how the failure occurred.

The higher variability of the failure modes observed for CME was reflected in the larger data variability in the Weibull test, increasing the slope of the data distribution line and the Weibull modulus.

Therefore, due to the possible different failure modes in this group, it has lower reliability, and therefore its clinical behavior is less predictable. However, its characteristic strength is not inferior to that of the MP and MC groups, which does not contraindicate this design as an option. The Weibull test approach has already been used for studies with dental ceramics [34] and dental implants [16], and it has been used as a statistic method to understand the relationship of data variation and final strength during fatigue [57].

Considering the reliability of the restoration itself, the lowest characteristic strength calculated was 789 N, which is about 34% less than the maximum fracture load initially calculated to determine the load variation in the fatigue test. This only reinforces that the slow crack growth during fatigue decreases the ultimate strength of ceramic restorations and should be taken into consideration during the experimental design of the test [16,31]. The Kaplan–Meier curve showed that the survival probability of the restorations decreases as the load increases, being less than 6% at 900 N, regardless of the design.

For this paper, the MC group simulated the use of a monolithic non-perforated crown as if the Tibase was a conventional abutment for cemented crowns. Although this design survived as well as the CME group and presented an adequate stress concentration, this design does not follow the manufacturer's recommendation [15]. The Tibase is not indicated as a conventional abutment because of the height of the platform, which cannot allow the complete removal of the resin cement from the cervical area in oral medium [15]. This group was created to elucidate whether the difference between the CME and MP groups would be due to the screw access hole of MP group or, due to the second cement layer on the CME design. No difference was calculated between these groups, suggesting that according to the manufacturer's recommendation, both techniques can create crowns of equal geometry, and the same clinical indication is correct. It is noteworthy that there was no loss of the composite resin used to seal the screw access of the MP group, which is not uncommon to be observed clinically for metal–ceramic crowns. Further studies evaluating the interface and bond strength between implant solutions CAD/CAM blocks and composite resin should be performed.

5. Conclusions

Using a digital workflow, the survival of an implant-supported restoration with PIC does not depend on the technique used to make it. The stress concentrations associated with fractographic analysis suggest that the emergence profile of the restoration should always be evaluated due to the high prevalence of failures in this area.

Author Contributions: All authors discussed and agreed upon the idea and made scientific contributions. Conceived and designed the study: J.P.M.T., A.M.O.D.P., L.C.A., C.J.K. and M.A.B. Data acquisition: J.P.M.T. and A.L.S.B. Experiment performing: J.P.M.T., A.M.O.D.P. and C.J.K. Writing the article: J.P.M.T., A.M.O.D.P. and L.C.A. Critical revision and final approval of the article: A.L.S.B., C.J.K. and M.A.B. All authors have agreed to the published version of the manuscript.

Funding: This research was funded by São Paulo Research Foundation (FAPESP) grants number 17/09104-4, 18/07404-3 and 17/23059-1.

Acknowledgments: The authors would like to thank São Paulo Research Foundation (FAPESP) with the grants n° 17/09104-4, n° 18/07404-3 and n° 17/23059-1.

Conflicts of Interest: The authors declare no conflict of interest. The funders had no role in the design of the study; in the collection, analyses, or interpretation of data; in the writing of the manuscript, or in the decision to publish the results.

References

1. Sagsoz, O.; Demirci, T.; Demirci, G.; Sagsoz, N.P.; Yildiz, M. The effects of different polishing techniques on the staining resistance of CAD/CAM resin-ceramics. *J. Adv. Prosthodont.* **2016**, *8*, 417–422. [CrossRef] [PubMed]
2. Coldea, A.; Swain, M.V.; Thiel, N. Mechanical properties of polymer-infiltrated-ceramic-network materials. *Dent. Mater.* **2013**, *29*, 419–426. [CrossRef] [PubMed]

3. Conrad, H.J.; Seong, W.-J.; Pesun, I.J. Current ceramic materials and systems with clinical recommendations: A systematic review. *J. Prosthet. Dent.* **2007**, *98*, 389–404. [CrossRef]
4. Denry, I.; Kelly, J. Emerging ceramic-based materials for dentistry. *J. Dent. Res.* **2014**, *93*, 1235–1242. [CrossRef]
5. Ramos, N.D.C.; Campos, T.M.B.; De La Paz, I.S.; Machado, J.P.B.; Bottino, M.A.; Cesar, P.F.; Melo, R. Microstructure characterization and SCG of newly engineered dental ceramics. *Dent. Mater.* **2016**, *32*, 870–878. [CrossRef]
6. Curran, P.; Cattani-Lorente, M.; Wiskott, H.W.A.; Durual, S.; Scherrer, S.S. Grinding damage assessment for CAD-CAM restorative materials. *Dent. Mater.* **2017**, *33*, 294–308. [CrossRef]
7. Goujat, A.; Abouelleil, H.; Colon, P.; Jeannin, C.; Pradelle, N.; Seux, D.; Grosgogeat, B. Mechanical properties and internal fit of 4 CAD-CAM block materials. *J. Prosthet. Dent.* **2017**, *119*, 384–389. [CrossRef]
8. Awada, A.; Nathanson, D. Mechanical properties of resin-ceramic CAD/CAM restorative materials. *J. Prosthet. Dent.* **2015**, *114*, 587–593. [CrossRef]
9. Della Bona, A.; Corazza, P.H.; Zhang, Y. Characterization of a polymer-infiltrated ceramic-network material. *Dent. Mater.* **2014**, *30*, 564–569. [CrossRef]
10. Gracis, S.; Thompson, V.P.; Ferencz, J.L.; Silva, N.R.; Bonfante, E.A. A new classification system for all-ceramic and ceramic-like restorative materials. *Int. J. Prosthodont.* **2015**, *28*, 227–235. [CrossRef]
11. Dirxen, C.; Blunck, U.; Preissner, R. Clinical Performance of a New Biomimetic Double Network Material. *Open Dent. J.* **2013**, *7*, 118–122. [CrossRef] [PubMed]
12. Homaei, E.; Farhangdoost, K.; Tsoi, J.K.-H.; Matinlinna, J.P.; Pow, E.H.N. Static and fatigue mechanical behavior of three dental CAD/CAM ceramics. *J. Mech. Behav. Biomed. Mater.* **2016**, *59*, 304–313. [CrossRef] [PubMed]
13. De Kok, P.; Kleverlaan, C.J.; De Jager, N.; Kuijs, R.; Feilzer, A.J. Mechanical performance of implant-supported posterior crowns. *J. Prosthet. Dent.* **2015**, *114*, 59–66. [CrossRef] [PubMed]
14. Peampring, C. Restorative management using hybrid ceramic of a patient with severe tooth erosion from swimming: A clinical report. *J. Adv. Prosthodont.* **2014**, *6*, 423–426. [CrossRef] [PubMed]
15. Kurbad, A. Final restoration of implants with a hybrid ceramic superstructure. *Int. J. Comput. Dent.* **2016**, *19*, 257–279.
16. Bonfante, E.A.; Suzuki, M.; Lorenzoni, F.C.; Sena, L.A.; Hirata, R.; Bonfante, G.; Coelho, P.G. Probability of survival of implant-supported metal ceramic and CAD/CAM resin nanoceramic crowns. *Dent. Mater.* **2015**, *31*, e168–e177. [CrossRef]
17. Glauser, R.; Sailer, I.; Wohlwend, A.; Studer, S.; Schibli, M.; Schärer, P. Experimental zirconia abutments for implant-supported single-tooth restorations in esthetically demanding regions: 4-year results of a prospective clinical study. *Int. J. Prosthodont.* **2004**, *17*, 285–290.
18. Aboushelib, M.N.; Salameh, Z. Zirconia implant abutment fracture: Clinical case reports and precautions for use. *Int. J. Prosthodont.* **2009**, *22*, 616–619.
19. Prestipino, V.; Ingber, A. All-Ceramic Implant Abutments: Esthetic Indications. *J. Esthet. Restor. Dent.* **1996**, *8*, 255–262. [CrossRef]
20. Elsayed, A.; Wille, S.; Al-Akhali, M.; Kern, M. Comparison of fracture strength and failure mode of different ceramic implant abutments. *J. Prosthet. Dent.* **2017**, *117*, 499–506. [CrossRef]
21. Elsayed, A.; Wille, S.; Al-Akhali, M.; Kern, M. Effect of fatigue loading on the fracture strength and failure mode of lithium disilicate and zirconia implant abutments. *Clin. Oral Implant. Res.* **2017**, *29*, 20–27. [CrossRef] [PubMed]
22. Tribst, J.P.M.; Piva, A.M.D.O.D.; Borges, A.L.S.; Bottino, M.A. Influence of crown and hybrid abutment ceramic materials on the stress distribution of implant-supported prosthesis. *Rev. de Odontol. da UNESP* **2018**, *47*, 149–154. [CrossRef]
23. Tribst, J.P.M.; Piva, A.M.D.O.D.; Borges, A.L.S.; Bottino, M.A. Different combinations of CAD/CAM materials on the biomechanical behavior of a two-piece prosthetic solution. *Int. J. Comput. Dent.* **2019**, *22*, 171–176. [PubMed]
24. Tribst, J.P.M.; Piva, A.M.O.D.; Özcan, M.; Borges, A.L.S.; Bottino, M.A. Influence of Ceramic Materials on Biomechanical Behavior of Implant Supported Fixed Prosthesis with Hybrid Abutment. *Eur. J. Prosthodont. Restor. Dent.* **2019**, *27*, 76–82. [PubMed]

25. Tribst, J.P.M.; Piva, A.M.D.O.D.; Anami, L.C.; Borges, A.L.S.; Bottino, M. A Influence of implant connection on the stress distribution in restorations performed with hybrid abutments. *J. Osseointegration* **2019**, *11*, 507–512.
26. Nouh, I.; Kern, M.; Sabet, A.E.; AboelFadl, A.K.; Hamdy, A.M.; Chaar, M.S. Mechanical behavior of posterior all-ceramic hybrid-abutment-crowns versus hybrid-abutments with separate crowns-A laboratory study. *Clin. Oral Implant. Res.* **2018**, *30*, 90–98. [CrossRef]
27. Roberts, E.E.; Bailey, C.W.; Ashcraft-Olmscheid, D.L.; Vandewalle, K.S. Fracture Resistance of Titanium-Based Lithium Disilicate and Zirconia Implant Restorations. *J. Prosthodont.* **2018**, *27*, 644–650. [CrossRef]
28. Edelhoff, D.; Schweiger, J.; Prandtner, O.; Stimmelmayr, M.; Güth, J.-F. Metal-free implant-supported single-tooth restorations. Part I: Abutments and cemented crowns. *Quintessence Int.* **2019**, *50*, 176–184.
29. Øilo, M.; Arola, D. Fractographic analyses of failed one-piece zirconia implant restorations. *Dent. Mater.* **2018**, *34*, 922–931. [CrossRef]
30. Scherrer, S.S.; Cattani-Lorente, M.; Vittecoq, E.; De Mestral, F.; Griggs, J.A.; Wiskott, H.A. Fatigue behavior in water of Y-TZP zirconia ceramics after abrasion with 30µm silica-coated alumina particles. *Dent. Mater.* **2010**, *27*, e28–e42. [CrossRef]
31. De Melo, R.M.; Pereira, C.; Ramos, N.D.C.; Feitosa, F.; Piva, A.M.D.O.D.; Tribst, J.P.M.; Ozcan, M.; Jorge, A.O.C.; Özcan, M. Effect of pH variation on the subcritical crack growth parameters of glassy matrix ceramics. *Int. J. Appl. Ceram. Technol.* **2019**, *16*, 2449–2456. [CrossRef]
32. Miyashiro, M.; Suedam, V.; Neto, R.T.M.; Ferreira, P.M.; Rubo, J.H. Validation of an experimental polyurethane model for biomechanical studies on implant supported prosthesis - tension tests. *J. Appl. Oral Sci.* **2011**, *19*, 244–248. [CrossRef] [PubMed]
33. Ramos, G.F.; Monteiro, E.; Bottino, M.; Zhang, Y.; De Melo, R.M.; Bottino, M.A. Failure probability of three designs of zirconia crowns. *Int. J. Periodontics Restor. Dent.* **2015**, *35*, 843–849. [CrossRef] [PubMed]
34. Anami, L.C.; Lima, J.; Valandro, L.F.; Kleverlaan, C.; Feilzer, A.; Bottino, M.A. Fatigue Resistance of Y-TZP/Porcelain Crowns is Not Influenced by the Conditioning of the Intaglio Surface. *Oper. Dent.* **2016**, *41*, E1–E12. [CrossRef]
35. Bottino, M.A.; Rocha, R.F.V.; Anami, L.C.; Özcan, M.; Melo, R.M. Fracture of Zirconia Abutment with Metallic Insertion on Anterior Single Titanium Implant with Internal Hexagon: Retrieval Analysis of a Failure. *Eur. J. Prosthodont. Restor. Dent.* **2016**, *24*, 164–168.
36. Idogava, H.T.; Noritomi, P.Y.; Daniel, G.B. Numerical model proposed for a temporomandibular joint prosthesis based on the recovery of the healthy movement. *Comput. Methods Biomech. Biomed. Eng.* **2018**, *21*, 1–9. [CrossRef]
37. Alkan, I.; Sertgöz, A.; Ekici, B. Influence of occlusal forces on stress distribution in preloaded dental implant screws. *J. Prosthet. Dent.* **2004**, *91*, 319–325. [CrossRef]
38. Tribst, J.P.M.; Piva, A.M.D.O.D.; De Melo, R.M.; Borges, A.L.S.; Bottino, M.A.; Ozcan, M. Short communication: Influence of restorative material and cement on the stress distribution of posterior resin-bonded fixed dental prostheses: 3D finite element analysis. *J. Mech. Behav. Biomed. Mater.* **2019**, *96*, 279–284. [CrossRef]
39. Benzing, U.R.; Gall, H.; Weber, H. Biomechanical aspects of two different implant-prosthetic concepts for edentulous maxillae. *Int. J. Oral Maxillofac. Implant.* **1995**, *10*, 188–198.
40. Souza, A.; Xavier, T.; Platt, J.; Borges, A.L.S. Effect of Base and Inlay Restorative Material on the Stress Distribution and Fracture Resistance of Weakened Premolars. *Oper. Dent.* **2015**, *40*, 158–166. [CrossRef]
41. Singh, S.V.; Gupta, S.; Sharma, D.; Pandit, N.; Nangom, A.; Satija, H. Stress distribution of posts on the endodontically treated teeth with and without bone height augmentation: A three-dimensional finite element analysis. *J. Conserv. Dent.* **2015**, *18*, 196–199. [CrossRef] [PubMed]
42. Tribst, J.P.M.; De Melo, R.M.; Borges, A.L.S.; Souza, R.O.D.A.E.; Bottino, M.A. Mechanical Behavior of Different Micro Conical Abutments in Fixed Prosthesis. *Int. J. Oral Maxillofac. Implant.* **2016**, *33*, 1199–1205. [CrossRef] [PubMed]
43. Correia, A.; Tribst, J.P.M.; Matos, F.D.S.; Platt, J.A.; Caneppele, T.M.F.; Borges, A.L.S. Polymerization shrinkage stresses in different restorative techniques for non-carious cervical lesions. *J. Dent.* **2018**, *76*, 68–74. [CrossRef] [PubMed]
44. Bewick, V.; Cheek, L.; Ball, J. Statistics review 12: Survival analysis. *Crit. Care* **2004**, *8*, 389–394. [CrossRef] [PubMed]

45. Kaweewongprasert, P.; Phasuk, K.; Levon, J.A.; Eckert, G.J.; Feitosa, S.; Valandro, L.F.; Bottino, M.C.; Morton, D. Fatigue Failure Load of Lithium Disilicate Restorations Cemented on a Chairside Titanium-Base. *J. Prosthodont.* **2018**, *28*, 973. [CrossRef] [PubMed]
46. Edelhoff, D.; Schweiger, J.; Prandtner, O.; Stimmelmayr, M.; Güth, J.-F. Metal-free implant-supported single-tooth restorations. Part II: Hybrid abutment crowns and material selection. *Quintessence Int.* **2019**, *50*, 260–269.
47. Bidra, A.S.; Rungruanganunt, P. Clinical Outcomes of Implant Abutments in the Anterior Region: A Systematic Review. *J. Esthet. Restor. Dent.* **2013**, *25*, 159–176. [CrossRef]
48. Sailer, I.; Asgeirsson, A.G.; Thoma, D.S.; Fehmer, V.; Aspelund, T.; Ozcan, M.; Pjetursson, B.E. Fracture strength of zirconia implant abutments on narrow diameter implants with internal and external implant abutment connections: A study on the titanium resin base concept. *Clin. Oral Implant. Res.* **2018**, *29*, 411–423. [CrossRef]
49. Mehl, C.; Gaβling, V.; Schultz-Langerhans, S.; Açil, Y.; Bähr, T.; Wiltfang, J.; Kern, M. Influence of Four Different Abutment Materials and the Adhesive Joint of Two-Piece Abutments on Cervical Implant Bone and Soft Tissue. *Int. J. Oral Maxillofac. Implant.* **2016**, *31*, 1264–1272. [CrossRef]
50. Pitta, J.; Fehmer, V.; Sailer, I.; Hicklin, S.P. Monolithic zirconia multiple-unit implant reconstructions on titanium bonding bases. *Int. J. Comput. Dent.* **2018**, *21*, 163–171.
51. Adolfi, D.; Tribst, J.P.M.; Adolfi, M.; Piva, A.M.D.O.D.; Saavedra, G.D.S.F.A.; Bottino, M.A. Lithium Disilicate Crown, Zirconia Hybrid Abutment and Platform Switching to Improve the Esthetics in Anterior Region: A Case Report. *Clin. Cosmet. Investig. Dent.* **2020**, *12*, 31–40. [CrossRef] [PubMed]
52. Hussien, A.N.M.; Rayyan, M.M.; Sayed, N.M.; Segaan, L.G.; Goodacre, C.J.; Kattadiyil, M.T. Effect of screw-access channels on the fracture resistance of 3 types of ceramic implant-supported crowns. *J. Prosthet. Dent.* **2016**, *116*, 214–220. [CrossRef] [PubMed]
53. Steigmann, M.; Monje, A.; Chan, H.; Wang, H.-L. Emergence profile design based on implant position in the esthetic zone. *Int. J. Periodontics Restor. Dent.* **2014**, *34*, 559–563. [CrossRef] [PubMed]
54. Schoenbaum, T.R.; Swift, E.J. Abutment Emergence Contours for Single-Unit Implants. *J. Esthet. Restor. Dent.* **2015**, *27*, 1–3. [CrossRef] [PubMed]
55. Scherrer, S.S.; Lohbauer, U.; Della Bona, A.; Vichi, A.; Tholey, M.; Kelly, J.R.; Van Noort, R.; Cesar, P.F. ADM guidance—Ceramics: Guidance to the use of fractography in failure analysis of brittle materials. *Dent. Mater.* **2017**, *33*, 599–620. [CrossRef] [PubMed]
56. Lohbauer, U.; Belli, R.; Cune, M.S.; Schepke, U. Fractography of clinically fractured, implant-supported dental computer-aided design and computer-aided manufacturing crowns. *SAGE Open Med. Case Rep.* **2017**, *5*, 1–9. [CrossRef]
57. Quinn, J.B.; Quinn, G.D. A practical and systematic review of Weibull statistics for reporting strengths of dental materials. *Dent. Mater.* **2009**, *26*, 135–147. [CrossRef]

© 2020 by the authors. Licensee MDPI, Basel, Switzerland. This article is an open access article distributed under the terms and conditions of the Creative Commons Attribution (CC BY) license (http://creativecommons.org/licenses/by/4.0/).

MDPI
St. Alban-Anlage 66
4052 Basel
Switzerland
Tel. +41 61 683 77 34
Fax +41 61 302 89 18
www.mdpi.com

Materials Editorial Office
E-mail: materials@mdpi.com
www.mdpi.com/journal/materials

www.ingramcontent.com/pod-product-compliance
Lightning Source LLC
LaVergne TN
LVHW070716100526
838202LV00013B/1109